Removal of EDCs and Pharmaceuticals in Drinking and Reuse Treatment Processes

Awwa
Research
Foundation

About the Awwa Research Foundation

The Awwa Research Foundation (AwwaRF) is a member-supported, international, nonprofit organization that sponsors research to enable water utilities, public health agencies, and other professionals to provide safe and affordable drinking water to consumers.

The Foundation's mission is to advance the science of water to improve the quality of life. To achieve this mission, the Foundation sponsors studies on all aspects of drinking water, including supply and resources, treatment, monitoring and analysis, distribution, management, and health effects. Funding for research is provided primarily by subscription payments from approximately 1,000 utilities, consulting firms, and manufacturers in North America and abroad. Additional funding comes from collaborative partnerships with other national and international organizations, allowing for resources to be leveraged, expertise to be shared, and broad-based knowledge to be developed and disseminated. Government funding serves as a third source of research dollars.

From its headquarters in Denver, Colorado, the Foundation's staff directs and supports the efforts of more than 800 volunteers who serve on the board of trustees and various committees. These volunteers represent many facets of the water industry, and contribute their expertise to select and monitor research studies that benefit the entire drinking water community.

The results of research are disseminated through a number of channels, including reports, the Web site, conferences, and periodicals.

For subscribers, the Foundation serves as a cooperative program in which water suppliers unite to pool their resources. By applying Foundation research findings, these water suppliers can save substantial costs and stay on the leading edge of drinking water science and technology. Since its inception, AwwaRF has supplied the water community with more than $300 million in applied research.

More information about the Foundation and how to become a subscriber is available on the Web at **www.awwarf.org.**

Removal of EDCs and Pharmaceuticals in Drinking and Reuse Treatment Processes

Prepared by:
Shane A. Snyder, **Eric C. Wert**, and **Hongxia (Dawn) Lei**
Water Quality Research and Development Division
Southern Nevada Water Authority, Henderson, NV 89015
and

Paul Westerhoff and **Yeomin Yoon**
Department of Civil and Environmental Engineering
Arizona State University, Tempe, AZ 85287

Sponsored by:
Awwa Research Foundation
6666 West Quincy Avenue, Denver, CO 80235-3098

Published by:

Awwa Research Foundation

American Water Works Association

IWA Publishing

DISCLAIMER

This study was funded by the Awwa Research Foundation (AwwaRF). AwwaRF assumes no responsibility for the content of the research study reported in this publication or for the opinions or statements of fact expressed in the report. The mention of trade names for commercial products does not represent or imply the approval or endorsement of AwwaRF. This report is presented solely for informational purposes.

CONTENTS

TABLES

FIGURES

FOREWORD

The Awwa Research Foundation is a nonprofit corporation that is dedicated to the implementation of a research effort to help utilities respond to regulatory requirements and traditional high-priority concerns of the industry. The research agenda is developed through a process of consultation with subscribers and drinking water professionals. Under the umbrella of a Strategic Research Plan, the Research Advisory Council prioritizes the suggested projects based upon current and future needs, applicability, and past work; the recommendations are forwarded to the Board of Trustees for final selection. The foundation also sponsors research projects through the unsolicited proposal process; the Collaborative Research, Research Applications, and Tailored Collaboration programs; and various joint research efforts with organizations such as the U.S. Environmental Protection Agency, the U.S. Bureau of Reclamation, and the Association of California Water Agencies.

This publication is a result of one of these sponsored studies, and it is hoped that its findings will be applied in communities throughout the world. The following report serves not only as a means of communicating the results of the water industry's centralized research program but also as a tool to enlist the further support of the nonmember utilities and individuals.

Projects are managed closely from their inception to the final report by the foundation's staff and large cadre of volunteers who willingly contribute their time and expertise. The foundation serves a planning and management function and awards contracts to other institutions such as water utilities, universities, and engineering films. The funding for this research effort comes primarily from the Subscription Program, through which water utilities subscribe to the research program and make an annual payment proportionate to the volume of water they deliver and consultants and manufacturers subscribe based on their annual billings. The program offers a cost-effective and fair method for funding research in the public interest.

A broad spectrum of water supply issues is addressed by the foundation's research agenda: resources, treatment and operations, distribution and storage, water quality and analysis, toxicology, economics, and management. The ultimate purpose of the coordinated effort is to assist water suppliers to provide the highest possible quality of water economically and reliably. The true benefits are realized when the results are implemented at the utility level. The foundation's trustees are pleased to offer this publication as a contribution toward that end.

David E. Rager
Chair, Board of Trustees
Awwa Research Foundation

Robert C. Renner, P.E.
Executive Director
Awwa Research Foundation

ACKNOWLEDGMENTS

This research project would not have been possible without the contributions and support from a large number of people from various areas of the water industry. We would like to thank every agency, firm, and individual who has supported this project; however, it is not possible to list every individual here. The team would like to acknowledge the primary contributors, while noting that many others have also contributed. We remain indebted to our colleagues, without whom this project would not have been possible.

Awwa Research Foundation (AwwaRF)
- Kim Linton
- Rick Karlin

AwwaRF Project Advisory Committee
- Greg Leslie – University of New South Wales, Australia
- Lucy McGovern - City of Camarillo Water Department, California
- Gordon Wheale – UK Water Industry Research (UKWIR)

AwwaRF Technical Review Committee
- Thomas Speth – US Environmental Protection Agency
- Michele Prevost – University of Montreal
- James Crook – Environmental Engineering Consultant

Southern Nevada Water Authority, Henderson, Nevada
- Brett Vanderford
- Rebecca Trenholm
- Janie Holady
- Fernando Rosario-Ortiz
- David Rexing
- Linda Parker
- Joseph Leising
- Ron Zegers
- Frankie Lewis
- Oscar Quiñones

Arizona State University
- Heath Mash (currently with US EPA)

Black & Veatch
- Robert Hulsey
- Bruce Long
- Vasu Veerapaneni
- Rick Bond
- Jessica Edwards-Brandt
- Willard Pack

Cincinnati Water, Ohio
- Jeff Vogt

City of Henderson, Nevada
- Mike Neher
- Dennis Porter

Keni Whalen

City of Las Vegas, Nevada

Dan Fischer

David Mendenhall

Bruce Dacko

City of Laredo, Texas - Utilities Water Treatment Department

Tony Moreno

City of North Tonawanda, New York

Paul Drof

Clayton County Water Authority, Jonesboro, Georgia

Alice Cook

Guy Pihera

Clark County Water Reclamation District, Las Vegas, Nevada

Doug Drury

Bill Shepherd

Devin Morgan

Kansas City, Missouri

Mary Lappin

John Reddy

Louisville Water Company, Kentucky

Mark Campbell

Rengao Song

Metropolitan Water District of Southern California

Alex Mofidi

Connie Lee

MWH

Samer Adham

Jay DeCarolis

Joan Oppenheimer

Northern Kentucky Water, Ft. Thomas, Kentucky

Mary Carol Wagner

Bari Joslyn

Orange County Water District – California

Mike Wehner

Joanne Daugherty

Passaic Valley Water Commission, Clifton, New Jersey

Linda Pasquariello

Linda Tatro

Joe Bella

Penn State University, University Park, Pennsylvania

Fred Cannon

Adam Redding

Philadelphia Water, Pennsylvania

Leah Gaffney

Jason Hunt

Chris Crockett

Trojan UV Technologies
Mihaela Stefan
Adam Festger
Christian Williamson
University of California, Berkeley
David Sedlak
United Water
Gregg Oelker
Ron Artley
West Basin Water District, El Segundo, California
Uzi Daniel
Rich Nagel

EXECUTIVE SUMMARY

INTRODUCTION

Endocrine disrupting chemicals (EDCs) and pharmaceuticals and personal care products (PPCPs) are groups of emerging contaminants that have been detected at trace concentrations (i.e., <0.1 μg/L) in waters around the world. EDCs are comprised of a vast group of chemicals that impact estrogen, androgen, and/or thyroid hormone function in animals. The EDCs that have received the greatest attention are those that can mimic or block the effect of endogenous estrogens. Natural and synthetic estrogens have been reported in US wastewaters since the 1960s; however, these hormones did not become widely recognized contaminants until their occurrence in wastewater effluents was linked to reproductive impacts in fish. Likewise, pharmaceuticals were first reported in surface waters during investigations of US waterways in the 1970s. Personal care products represent yet another group of contaminants that have found their way into source waters. These products include detergents, antimicrobials, over-the-counter medicines, and various household chemicals. Although various EDCs and pharmaceuticals and personal care products (PPCPs) were detected in surface waters for nearly five decades, concern was minimal because concentrations were minute, and environmental protection efforts were focused mainly on legacy contaminants (e.g., DDT, PCBs, etc.). Since the early reports of EDCs and PPCPs in US waters, analytical technology has advanced such that many previously undetectable environmental contaminants are now identifiable and quantifiable at trace levels. In the past decade, several reports have been published showing that EDCs and PPCPs can and will occur in surface and ground waters globally. The vast majority of available data relates to occurrence of these emerging contaminants in wastewater effluents. Only sparse data were available on the occurrence and removal in drinking water treatment plants.

RESEARCH OBJECTIVES

This study was initiated to determine the removal of various EDCs and PPCPs during a variety of conventional and advanced water treatment processes. In order to complete this evaluation, robust analytical methods were developed for representative target analytes. This report presents the results of this research endeavor. This project is unique in the diversity of contaminants and treatment processes evaluated. Specific objectives of the research included the following.

1. Review of available data on the occurrence and treatment of EDCs and PPCPs.
2. Selection of target compounds with a broad range of physico-chemical parameters with likely occurrence in source waters.
3. Development of robust analytical methodology with sufficient sensitivity to evaluate occurrence and removal in drinking water treatment plants.
4. Bench- and pilot- scale evaluations of conventional and advanced treatment processes using natural waters spiked with target compounds.
5. Evaluation of the occurrence and removal of target compounds in full-scale drinking water and water reuse facilities with subsequent comparisons to bench- and pilot-scale data.

6. Use of quantitative structure-activity relationship (QSAR) models to predict target compound properties and removal via sorption, volatilization, and biodegradation.
7. Recommendations for utilities.

APPROACH

A suite of target analytes was selected based on a literature review of occurrence and to ensure inclusion of chemicals with a wide variety of physicochemical properties (e.g., molecular size; polarity; aromaticity; acidic, basic, and/or neutral functional groups; and volatility). The choice of chemicals also was based upon inclusion of representative chemicals from several key classes of hormones, pharmaceuticals, and personal care products. In order to further condense the universe of possible contaminants for evaluation, only compounds for which analytical standards were commercially available were considered. The final list of target compounds was refined by excluding those chemicals that could not be analyzed using a single solid-phase extraction (SPE) with subsequent analysis by gas chromatography and/or liquid chromatography with tandem mass spectrometric detection (GC-MS/MS and LC-MS/MS, respectively).

The target compounds were spiked at ng/L concentrations into various natural waters, and their removal by physical, chemical, and biological water treatment processes was evaluated in batch mode (bench-scale) or dynamically in a flow-through mode (pilot-scale). Full-scale drinking water and water reuse treatment facilities were assessed by analyzing samples of raw water, water representing unit processes, and finished water. Observations of removal from full-scale facilities were compared to those made at bench- and pilot-scale. QSAR models were employed to make predictions of sorption, biodegradation, and volatilization potentials.

REPORT

The report chapters are organized according to specific treatment processes or experimental methods. Chapters 1-3 discuss the project background, research approach, and experimental methods. Chapters 4-11 discuss each treatment process evaluated, presenting bench-, pilot-, and full-scale results. Chapter 12 presents results from predictive computer modeling. Chapter 13 provides occurrence summary data from drinking water facilities and conclusions regarding the research project along with recommendations for utilities. A detailed summary of each chapter is provided in the subsequent sections.

Chapter 1 provides background information used to develop the research approach. The chapter includes a literature review providing a historical perspective of trace organic contaminants in addition to occurrence data and information on various approaches for analysis and treatment. Occurrence literature showed utilities with wastewater-impacted source waters have the highest probability of contamination. While EDCs and PPCPs have been detected in US waters for over 30 years, it is only in the past decade that information linking these chemicals to impacts in fish has brought the issue to the forefront. Current research is focusing on human toxicological relevance, which will provide additional insight to regulatory agencies and drinking water utilities regarding this issue.

Chapter 2 describes the research approach for this project. The project consisted of two phases. The first phase included the criteria for the selection of 36 target EDCs and PPCPs representative of the types of chemicals known or likely to occur in source waters with a broad range of physical-chemical structure/behavior. The second phase of the project involved testing of physical, chemical, and biological treatment processes to determine removal potential at environmentally relevant concentrations of the characteristic EDCs and PPCPs. Detailed information regarding target compound selection and data analysis methods is provided in this chapter.

Chapter 3 provides the experimental and analytical methods used throughout the project. A summary of the treatment processes evaluated is presented along with detailed methodologies for the bench scale and pilot scale treatment evaluations. A list of full-scale treatment facilities evaluated is provided, which describes the unit processes employed. The basis for chemical doses and contact times are provided in this chapter with process schematics included in Appendix A. This chapter describes the analytical methods developed and implemented for this project. Solid-phase extraction followed by gas chromatography with tandem mass spectrometry (GC-MS/MS) and liquid chromatography with tandem mass spectrometry (LC-MS/MS) was used. Method reporting limits (MRLs) for the target compounds were generally 1 – 10 ng/L.

Chapter 4 discusses our investigation of coagulation and softening for EDC/PPCP removal. Coagulation at bench-scale was conducted using both alum and ferric chloride. Pilot- and full-scale investigations are compared to bench experiments. Few EDCs and PPCPs could be removed by coagulation or softening alone. In bench-scale experiments, hydrophobic compounds were able to bind to particles and were subsequently removed during settling. In summary, coagulation was found to be largely ineffective for EDC/PPCP removal at bench-, pilot-, and full-scale.

Chapter 5 discusses the investigation of activated carbon in both granular and powdered forms for EDC/PPCP removal. Powdered activated carbon (PAC) was found to be highly effective for removing target compounds; however, PAC dose and contact time were critical in order to achieve efficient contaminant removal. Water quality also was found to impact contaminant removal by activated carbon, while initial spiking concentration did not seem to impact percent removal. Rapid small-scale column tests (RSSCTs) were used to evaluate GAC at bench-scale. All target compounds could be removed by GAC, but water-soluble contaminants could break through the GAC column relatively quickly. GAC with modified surface characteristics had significantly higher capacity for EDC/PPCP removal as compared to conventional GAC. At full-scale, water utilities that routinely regenerate GAC experienced excellent removal of target compounds, while those with GAC that had been in service for longer periods of time saw little or no removal of contaminants. The x-ray contrast media iopromide and the pharmaceuticals ibuprofen, meprobamate, sulfamethoxazole, and diclofenac were some of the compounds found to be most recalcitrant for activated carbon removal. Steroid hormones and other hydrophobic contaminants were effectively removed by activated carbon with minimal GAC breakthrough after >50,000 bed volumes. In summary, activated carbon can be effective for removing organic contaminants. PAC contact time and dose must be optimized and GAC must be regenerated or replaced on a regular basis to ensure efficient EDC/PPCP removal.

Chapter 6 discusses our investigation of free chlorine and chloramine for the oxidation of EDCs and PPCPs. Free chlorine was found to oxidize approximately half of the EDCs and PPCPs investigated. Hormones with a phenol functional group (i.e., 17β-estradiol, estrone, and ethynyl estradiol) were rapidly oxidized by free chlorine, while hormones with ketone functional groups were only partial oxidized by free chlorine. Chloramine was much less efficient at removing EDCs/PPCPs when compared to free chlorine. Sulfamethoxazole, trimethoprim, and erythromycin antibiotics are a few compounds that showed high removal by free chlorine but low removal by chloramine. Water quality did not greatly impact compound removal by chlorine or chloramines as long as the chlorine/chloramine dose met the initial demand of the water. Full-scale sampling confirmed observations from spiked bench- and pilot-scale experiments. Utilities using free chlorine disinfection achieved good removal of many EDCs and PPCPs, depending on individual contaminant structure. Utilities using chloramines for primary disinfection experienced significantly less removal of organic contaminants when compared to utilities using free chlorine.

Chapter 7 provides results from ozone and ozone advanced oxidation process (AOP) testing. Ozone and ozone AOP were conducted at bench-, pilot-, and full-scale using both drinking water sources and wastewater effluents. Ozone was found to be highly effective for the removal of most target compounds. A flame-retardant compound, tris-chloroethylphosphate (TCEP), was found to be the most challenging compound to oxidize. This is expected considering the purpose of this chemical is to resist oxidation. The herbicide atrazine, the x-ray contrast media iopromide, and the antianxiety pharmaceutical meprobamate were also resistant to oxidation, with less than 50% removal at typical drinking water ozone doses. The addition of hydrogen peroxide for advanced oxidation provided only a minor increase in removal of select compounds and is not likely to be cost effective, considering the minimal improvement gained. While ozone demand and decay varied widely, percent removal was in good agreement in all cases when ozone dose was adjusted to compensate for demand. The implementation of ozone for water treatment is likely the most cost effective measure for removing the majority of EDCs and PPCPs.

Chapter 8 provides data on UV and UV AOP for contaminant removal. UV processes were evaluated at bench-, pilot-, and full-scale. UV at a typical disinfection dose of 40 mJ/cm2 was ineffective for the removal of most target compounds. The antibiotic sulfamethoxazole, the antimicrobial triclosan, and the pharmaceutical diclofenac were removed by greater than 50% using a UV dose of approximately 40 mJ/cm2, while the remaining 33 target compounds exhibited poor removal. UV at higher doses (> 400 mJ/cm2) was able to remove several target compounds by direct photolysis. The combination of UV (> 400 mJ/cm2) combined with hydrogen peroxide (> 3 mg/L) provided excellent removal of most target compounds. Removal observed for target compounds using UV AOP closely resembled that of ozone. In order to oxidize the majority of EDC/PPCPs, a UV dose significantly greater than commonly employed for disinfection will be required. UV AOP will provide excellent removal for nearly all target compounds. Hydrogen peroxide residuals after AOP applications will quench chlorine/chloramine secondary disinfectants; therefore utilities must consider the impact of hydrogen peroxide on secondary disinfectant demand.

Chapter 9 presents the results from membrane testing at bench-, pilot-, and full-scale. Ultrafiltration (UF) and microfiltration (MF) were of little value for contaminant removal in general. Some data suggest that UF/MF membranes can adsorb target compounds either directly to the membrane surface or to organic particles on the membrane. Utilities should not count on UF/MF systems for EDC/PPCP removal. While a high degree of dissolved contaminant removal is not expected with UF/MF, these pretreatment systems can aid in lowering oxidant demands and UV transmittance by superior filtration as compared to conventional filtration. Nanofiltration (NF) provided good removal of many target compounds at all scales. Reverse osmosis (RO) was found to be highly effective for removing all target compounds. Trace quantities of some target compounds were able to breach RO membranes. In a double-pass RO system, all target compounds were removed to less than detection. NF and RO systems will provide good to excellent EDC/PPCP removal. Brines generated from NF/RO systems will contain higher levels of these contaminants, so brine disposal should be taken into consideration with NF/RO applications for contaminant removal.

Chapter 10 presents bench- and pilot-scale evaluates of magnetic ion-exchange media (MIEX). MIEX was found to be ineffective for the removal of most EDC/PPCP compounds. Diclofenac and triclosan were well removed (>80%), while naproxen had moderate removal (>50%) using MIEX. All other target compounds were removed by less than 50% with most less than 20%. MIEX did not appear to be a viable process for removal of the vast majority of EDCs and PPCPs, but is effective for contaminants that are negatively charged at ambient pH (i.e., triclosan and diclofenac).

Chapter 11 of this report presents results from water treatment processes which use biodegradation. A series of tests were performed to evaluate which of the target compounds were readily biodegradable. In these investigations, it is difficult to separate observed removal from biodegradation from that of absorption. The processes evaluated removed several of the target compounds. Batch biologically-active sand tests showed that acetaminophen, caffeine, gemfibrozil, ibuprofen, and estrogen hormones were rapidly biodegraded, while atrazine, carbamazepine, iopromide, and others were slowly degraded. A simulated aquifer storage and recovery experiment using soil columns revealed similar trends, although most compound removal was superior in the soil column as compared to batch tests. This is likely due to adsorption to soil particles as opposed to a higher degree of biological activity. Removal of target compounds during soil permeation of reuse water used for irrigation was evaluated. Few of the target compounds were detected in the source water, but those detectable were rapidly attenuated except the antibiotic sulfamethoxazole. Biofiltration on anthracite and activated carbon filters was evaluated at pilot-scale. Removal of target compounds by biologically-active anthracite was generally low. Some hydrophobic compounds were well removed, most likely by adsorption to the biofilm. Biologically-active carbon was quite effective for removal of most target compounds, presumably by adsorbing to the activated carbon as opposed to biodegradation. A riverbank filtration (RBF) pilot facility was tested for removal of target compounds. RBF was found to be effective for the majority of target compounds. Atrazine, carbamazepine, dilantin, and TCEP were poorly removed by RBF. A computer model for biodegradation was also applied and agreed reasonably well with removal trends observed in bench-, pilot-, and full-scale.

Chapter 12 presents information on the predictive capability of computer models relative to EDCs and PPCPs. The goal of this chapter is to develop potential tools to prescreen these compounds and help focus research on more persistent contaminants during seven typical water treatment processes. Computer software was used to calculate molecular geometry, and to predict the compound's reactivity and physicochemical properties, such as hydrogen bonding, water solubility and partitioning coefficient between two different media. A variety of models were evaluated and the one best suitable for EDCs/PPCPs was selected each for adsorption, air stripping, hydrolysis and biodegradation. The predictive values for these processes were provided and, if possible, validated by experimental data. Monte Carlo (MC) statistical simulations were utilized to develop QSAR models for predicting compound removal by ozonation, chlorination and UF processes via the incorporation of bench-scale and pilot-scale data with removal mechanism. Future research needs were discussed in order to further calibrate and validate these models.

Chapter 13 provides full-scale occurrence data and a summary of the project results. Twenty full-scale drinking water facilities were evaluated for EDC/PPCP occurrence and removal. DEET was the common contaminant of both raw and finished drinking waters, with 100% occurrence in raw and 90% occurrence in finished waters. Carbamazepine, dilantin, sulfamethoxazole, and meprobamate were the pharmaceuticals which most frequently occurred in raw drinking water (>80% of samples), while meprobamate, dilantin, ibuprofen, and iopromide were the pharmaceuticals of greatest occurrence in finished drinking water (>65% of samples). Atrazine occurred in the largest concentration of any contaminants in both raw and finished drinking water, but was well below the maximum contaminant limit of 3000 ng/L. Hormones occurred infrequently or not at all, with no detections of 17β-estradiol or ethynyl estradiol at a reporting limit of 1.0 ng/L. Pharmaceuticals rarely occurred in concentrations greater than 10 ng/L in raw and finished drinking waters. This chapter summarizes removal by compound and unit process. Based upon occurrence information and treatment process performance, specific emerging contaminant indicators are suggested for measuring individual treatment process efficacy for contaminant removal.

CONCLUSIONS

1. A large number of EDCs and PPCPs can be measured with a high degree of accuracy using chromatography coupled with tandem mass spectrometry (GC/MS/MS and LC/MS/MS).
2. Careful measures must be taken to ensure proper sample preservation and oxidant quenching to avoid erroneous analytical and treatment results.
3. Bench- and pilot-scale treatment studies using spiked target compounds accurately predicted observations in full-scale treatment plants.
4. Several target analytes were detected in raw and finished drinking waters across the US.
5. Conventional coagulation, flocculation, and sedimentation are ineffective for removing the majority of target EDCs and PPCPs.
6. Free chlorine disinfection can remove many target compounds depending on the structure of the contaminant.
7. Chloramines are less effective than free chlorine at EDC/PPCP removal.
8. Ozone is much more effective than chlorine and is able to significantly remove the majority of target analytes.
9. Ultraviolet (UV) irradiation at disinfection doses is ineffective for removing most EDCs and PPCPs; however, high energy UV at oxidative doses can be highly effective.
10. Advanced oxidation processes (AOPs) i.e., ozone/peroxide and UV/peroxide, are highly effective at removing the majority of contaminants evaluated.
11. Activated carbon is highly effective for removal of target analytes; however, exhausted activated carbon is ineffective.
12. Magnetic ion-exchange is ineffective for the removal of most EDCs and PPCPs.
13. Reverse osmosis and nanofiltration are highly effective for removing EDCs and PPCPs, while ultrafiltration and microfiltration are largely ineffective.
14. Biological removal and sorption processes can reduce the concentrations of many target analytes during riverbank filtration, biological filtration, and soil aquifer treatment.
15. Treatment trains combining various advanced processes are the most effective for removing trace concentrations of EDCs and PPCPs (e.g., reverse osmosis, ozone, AOPs, activated carbon).

16. Sensitive analytical methods are available to detect ultra-trace levels of contaminants in drinking water, such that non-detects for all contaminants will likely never be achievable.

RECOMMENDATIONS

Since EDCs and PPCPs occur in US drinking water only at minute concentrations, it is highly unlikely that most of these chemicals will pose any credible threat to human health via drinking water exposure. Utilities that wish to reduce the concentrations of the vast majority of microcontaminants should consider advanced processes such as ozonation, AOPs, GAC, reverse osmosis or tight nanofiltration membranes, or a combination of processes. However, conventional drinking water plants using free chlorine disinfection will be able to destroy most natural and synthetic estrogenic steroid hormones, which are the compounds of greatest concern due to their biological potency. The overall removal of contaminants through a drinking water plant will be the sum of the multi-barriers included in the treatment train. This report provides information on trace contaminants that are applicable for monitoring as unit process performance indicators based upon demonstrated ability for removal and significant probability of occurrence at detectable levels. Water utilities should urge regulatory agencies to consider human health protection, and not simply trace occurrence, in establishment of subsequent monitoring requirements and regulatory limits.

FUTURE RESEARCH

The results of this study show that some EDCs and PPCPs are detectable in raw and finished drinking waters of the US. While some treatment processes are clearly superior to others for reducing the concentrations of these trace contaminants, modern analytical tools are able to detect minute levels of EDCs and PPCPs. Future research should focus on determining the toxicological significance of trace occurrence of various contaminants in order to establish sensible analytical detection limits and treatment goals.

CHAPTER 1
INTRODUCTION

BACKGROUND

In the early 1900s, scientists became aware that certain chemicals could mimic the natural hormones of animals. These early reports focused mainly on compounds isolated from plants that lead to reduced fertility in some species of animals (Bennetts, Underwood, and Shier 1946). Through the 1940s, several publications demonstrated that natural and synthetic chemicals interacted with the primary male (androgenic) and female (estrogenic) hormonal systems. Decades later, scientists realized that environmentally relevant concentrations of certain pesticides were able to impact the reproductive system of animals (Welch, Levin, and Conney 1969). The term endocrine disrupting chemicals (EDCs) was later coined to include all chemicals capable of mimicking (agonists) or blocking (antagonists) the endogenous hormonal system of animals. Although many compounds are now commonly referred to as EDCs, there is no definitive list of EDCs, and assays for determination of whether a particular chemical, or mixture of chemicals, results in adverse impacts to the endocrine system are still under development.

In 1965, scientists in the United States investigated the fate of steroid hormones in wastewater treatment plants (Stumm-Zollinger and Fair 1965). The first known reports of pharmaceuticals in natural waters also originated from studies of US waters in 1970s (Tabak and Bunch 1970; Garrison, Pope, and Allen 1975; Hignite and Azarnoff 1977). These pharmaceuticals included heart medications, pain relievers, and birth control medications. Despite these initial reports, pharmaceuticals in the environment received little attention until reports from the United Kingdom and the United States linked the occurrence of trace steroids to biological activity in fish and cellular bioassays. Modern analytical techniques have increased sensitivity and accuracy, allowing ultra-trace levels of a wide variety of contaminants to be identified and quantified. It is now evident that pharmaceuticals and steroids are ubiquitous trace contaminants of wastewater effluents globally. These contaminants now have been shown to reach source waters through upstream municipal effluent discharges, confined animal feeding operations, and groundwater contamination. Today, international research efforts are underway to determine the human and ecological health impacts of these environmental contaminants.

When this project began in early 2002, several reports had shown that various EDCs and pharmaceuticals can and will occur in natural waters (Buser, Muller, and Theobald 1998; Buser, Poiger, and Muller 1998; Ahrer, Scherwenk, and Buchberger 2001; Kolpin et al. 2002) and that wastewater discharge is a significant source for such contamination (Ternes 1998; Snyder et al. 1999; Snyder et al. 2001c). Far fewer reports investigated the efficacy of conventional and advanced water treatment processes to remove these contaminants. Available data on the occurrence of EDCs and PPCPs in drinking water were essentially nonexistent. The primary objective of this research project was to determine the removal potential of various water treatment processes for a structurally diverse group of EDC and PPCPs. Additional objectives included the acquisition of occurrence data for these contaminants in US drinking waters and the application of structural-activity models for predicting contaminant properties that influence their fate during water treatment.

Since EDCs and PPCPs represent thousands of possible contaminants, a reasonable group of representative compounds for thorough evaluation was needed. Comprehensive literature reviews for treatment data, analytical methods, and environmental occurrence for EDCs and PPCPs were conducted. The literature reviews revealed that most occurrence reports were based on classes of compounds (e.g., acidic pharmaceuticals) that could be analyzed by a single analytical method. Similarly, the majority of treatment data available were limited to a few classes of compounds, and often not specific to EDCs and PPCPs. This research project sought to create a list of target compounds that would include not only frequently occurring EDCs and PPCPs, but also compounds that would provide a great amount of structural diversity (e.g., molecular weight, water solubility, acidic/basic/neutral, aromatic/aliphatic). From the occurrence data evaluated, past experience of the research team, and structural properties, the research team chose a series of target analytes to investigate.

Before treatment studies could begin, analytical techniques previously developed by this team required modification to include all the target analytes. In the case of certain hormones and pharmaceuticals, government licenses had to be procured prior to purchasing purified standard materials. Our team chose to utilize both gas chromatography and liquid chromatography, each coupled to tandem mass spectrometers. An extraction method was developed to utilize a single solid-phase extraction (SPE), where the resulting extract was split between gas chromatography and liquid chromatography. This method avoided the use of tedious, time consuming, and often variable derivatization techniques to structurally alter polar compounds in order to render them suitable for gas chromatographic separation. The analytical method required sensitivity that would allow contaminant identification and quantitation at environmentally relevant levels (ng/L). The research team desired sub-ng/L method detection limits with corresponding reporting limits ranging from 1 – 10 ng/L. A great deal of knowledge was gained through the analytical method development phase of this project. For instance, some compounds biodegraded rapidly, even in glass bottles stored at near freezing temperatures. Also, it was determined that certain oxidant quenching agents alone were capable of destroying target compounds. Through an iterative series of spiking and holding studies, a robust and highly sensitive method for a wide variety of trace organic contaminants was developed.

The initial phase of the treatment study utilized batch experiments conducted as bench-scale experiments. For these experiments, three natural waters and one model water were spiked with target analytes. Nominal spiked concentrations were chosen to be two orders of magnitude greater than the method reporting limit in order to allow 99% (2-log) removal observations. Experiments were designed to evaluate coagulation (alum and ferric), chemical softening (lime and soda ash), carbon adsorption (granular and powdered activated carbon), oxidation (chlorine, chloramine, ozone, and ultraviolet irradiation), advanced oxidation (ozone/peroxide and ultraviolet/peroxide), membrane separation (ultrafiltration and nanofiltration), and magnetic ion exchange resin. These experiments were conducted using a limited set of conditions representative of full-scale treatment plants.

The second phase of the treatment investigation was conducted as dynamic flow-through pilot-scale experiments. Spiking of pilot-scale experiments was accomplished either by infusion of the spike solution into the influent water line followed by static mixing, or by spiking head tanks followed by mixing within the tank. In some cases, pilot testing was conducted on secondary or tertiary wastewater effluents, in which case spiking was not necessary since the "raw" water contained many of the target analytes. Pilot-scale investigations were conducted on a variety of source waters at different US locations and at one site in Canada. Processes evaluated at pilot-scale include: coagulation, powdered activated carbon, chlorination, chloramination, ozonation, ultraviolet irradiation, ozone/peroxide, ultraviolet/peroxide, magnetic ion-exchange, and several membrane processes. Pilot-scale experiments were designed to be comparable to bench-scale experiments by using similar operational conditions (i.e., oxidant dose).

To determine removal during full-scale water treatment, samples were collected from 20 drinking water utilities and six water reuse plants from across the US. In most cases, samples were collected between each unit process and from the raw and finished water. Although the drinking water utilities surveyed had relatively low concentrations of only a few target compounds, those compounds detected exhibited removals that compared favorably to observations made during bench-scale and pilot-scale experiments. The evaluation of unit processes was sometimes complicated by simultaneous removal processes, such as chlorination during coagulation and preoxidation before UV disinfection. Despite these occasional difficulties, most observed removal was similar to removal predicted during bench-scale and pilot- scale testing. While several target contaminants were detectable in nearly all raw and finished waters tested, the concentrations were extremely low. The most potent estrogenic EDC steroids (17β-estradiol and ethynyl estradiol) were not detected in any of the drinking waters evaluated. It is clear that certain EDCs and pharmaceuticals can be identified and quantified in drinking water; however, the detection of these trace organics is a function of extremely sensitive analytical techniques.

Quantitative structural-property relationship (QSPR) and quantitative structural-activity relationship (QSAR) models were evaluated for predicting physical-chemical properties of target analytes (e.g., water solubility, volatility, surface area) and removal by certain processes. In this research effort, literature methods for predicting four fundamental physical properties important to the fate and transport of microcontaminants in water were critically evaluated. These four properties were: (1) adsorption capacity onto activated carbon; (2) biodegradation tendency; (3) air-water partitioning; and (4) hydrolysis rate constant at elevated pH. Empirical observations made during bench-scale experiments were compared to predictions from the models evaluated. The general trends predicted by the models employed agreed with observations from bench-scale and field experiments.

From this research, it is obvious that no treatment process alone is capable of removing all EDCs and PPCPs to less than one ng/L. This study clearly demonstrates that certain water treatment processes can greatly reduce the concentration of many classes of contaminants, while other processes have little impact on contaminant removal. Multi-barrier process trains provide the greatest flexibility and generally offer a superior level of contaminant removal. Utilities desiring to remove microcontaminants through treatment should consider physical processes such as activated carbon or membrane technology, or chemical oxidation using ozone or advanced oxidation. Source water protection options should also be considered in order to minimize both ecological and human exposures. The advent of modern analytical instrumentation will undoubtedly uncover more contaminants at even lower concentrations; therefore, a strong emphasis on toxicological relevance should be considered as a major research need.

LITERATURE REVIEW

Thorough literature reviews were conducted at the onset of this project in early 2002. Literature reviews of EDC and PPCP analysis, occurrence, and treatment were performed in order to determine the universe of possible compounds to investigate. Both the review of analytical methods for EDCs/PPCPs and the review of treatment were published as peer-reviewed journal articles in 2003 (Snyder et al. 2003a; Snyder et al. 2003b). The main points from these literature reviews are provided in this report.

Methods for EDC and PPCP Analysis

Because EDCs and PPCPs represent structurally diverse classes of compounds, a plethora of analytical methods have been, or could be, applied for the identification and quantification of these chemicals in water. This review focuses on organic chemicals, while noting that the several inorganic EDCs have been identified (e.g., perchlorate). The literature review focused on methodologies applicable for trace analysis (sub-μg/L concentrations). The measurement of organic chemicals in water most commonly consists of an extraction of the chemicals from water, concentration of the resulting extract, chromatographic separation, and detection. The following sections provide general information on the analytical process as well as specific examples from the primary classes of EDCs/PPCPs. A more extensive review of specific methodologies and compounds is provided in Table A.1 of Appendix A. Information gained from the initial review of available literature on EDC/PPCP analysis influenced the methods developed and utilized throughout this project (Vanderford et al. 2003; Trenholm et al. 2006).

Extraction/Concentration

Some EDCs and pharmaceuticals have been found to induce biological responses in aquatic organisms at sub-μg/L concentrations (Arcand-Hoy, Nimrod, and Benson 1998; Routledge et al. 1998; Miles-Richardson et al. 1999; Snyder et al. 2001d; Lattier et al. 2002; van Aerle et al. 2002; Snyder et al. 2004; Saal et al. 2005). As a result, most analytical techniques for EDCs/PPCPs in water have been designed to achieve detection limits in ng/L. The majority analytical instruments are not able to directly detect compounds at these levels. Therefore, an extraction step is required to concentrate the target compounds to a detectable level. Conventional extraction techniques such as liquid-liquid (Yook, Hong, and Kim 1994; Holm et al. 1995; Romero, Ventura, and Gomez 2002), Soxhlet (Bennie et al. 1997; Pryor, Hay, and Walker 2002; Fatoki and Awofolu 2003), and steam distillation (Kubeck and Naylor 1990; Fowler, Haviland, and Kennedy 1998; Snyder et al. 2001a), have been used to extract EDC/PPCP compounds from water; however, solid-phase extraction (SPE) is by far the most common technique employed for sample enrichment. SPE utilizes a stationary phase that binds target compounds as the liquid phase (in this case water) is passed through the SPE device. Generally, water is pumped or pulled through the stationary phase which is packed into a cartridge or column. After extract, the stationary phase often is dried by passing nitrogen or air through the SPE column/cartridge. This drying step prevents much of residual water from entering the resulting extract, which can be detrimental for certain types of instrumental analyses. Solvents with a greater affinity for the target compounds than for the stationary phase are used to elute the compounds from the SPE column/cartridge. Often a multisolvent system is required to remove all target compounds from the SPE. The resulting extract volume depends on how much solvent is used in the elution process. This extract is generally concentrated further by evaporation with a gentle stream of nitrogen to one mL or less. For certain analyses, a cleanup step may be required, such as gel-permeation chromatography (to remove large molecular weight compounds) or silica gel or alumina columns that can fractionate the extract based on polarity, hence separating compounds based on polarity (Khim et al. 1999a; Khim et al. 1999b; Khim et al. 1999c; Snyder et al. 1999; Snyder et al. 2001b; Snyder et al. 2001d). More modern variants of SPE include solid-phase microextraction (SPME) and various on-line and automated SPE techniques. SPME involves a fine silica fiber that is coated externally with a stationary phase. This small (1 – 2 cm) fiber is immersed into the aqueous sample (generally 20 mL or less) and remains in contact with a stirred solution until a predetermined equilibrium is established. The SPME fiber is introduced directly in the analytical instrumentation through a sample injection system and compounds desorbed from the fiber using heat in gas chromatography or a solvent in liquid chromatography. Passive samplers also have been used in the monitoring of water for various EDCs. Semipermeable membrane devices (SPMDs) are passive samplers used for an integrated assessment of water contaminants. These devices generally consist of a low-density polyethylene tube filled with triolene or other lipophillic material and may remain in contact with water for an extended period of time (days – months). The SPMD can mimic the bioaccumulation of chemicals into aquatic organisms and provide relative water concentration data (Bevans et al. 1996; Bennett and Metcalfe 2000; Echols et al. 2000; Metcalfe et al. 2000; Petty 2000; Lindstrom et al. 2002). Typical SPMDs using non-polar liquids would be unlikely to accumulate more polar compounds (including many EDCs and PPCPs). However, efforts are underway to refine passive samplers for efficient uptake of polar compounds.

Gas Chromatography

Gas Chromatography (GC) has been used for many years for chemical analysis. Traditionally, GC works by injecting a liquid sample into a heated injection chamber, where the compounds are quickly vaporized and swept by a carrier gas, usually helium, onto a silica chromatography capillary column (i.e. 30 m long by 0.25 μm diameter) internally coated with a stationary phase. The chromatography column is positioned in a temperature-controlled oven. The individual compounds will separate as they flow through the column based on their boiling point, affinity for the stationary phase, and temperature profile of the GC oven. A detector then monitors the compounds as they leave the column. There are many different kinds of detectors including flame-ionization (FID), electron capture (ECD), nitrogen-phosphorus (NPD), flame-photometric (FPD), and mass spectrometer (MS). Each detector has its own advantages and disadvantages with regard to sensitivity, selectivity, and cost.

Mass spectrometry is one of the most useful and common techniques because of its good sensitivity and superior selectivity. Since MS detects analytes based on mass to charge ratio (m/z), analytes must first be ionized. The interface where this takes place is known as the ionization source. There are two major techniques to induce ionization in the source: electron ionization (EI) and chemical ionization (CI). In EI, the GC column effluent is directed through a beam of electrons created by a filament. This results in the loss or gain of an electron from the compound, creating a positively or negatively charged molecule, respectively. This type of ionization results in molecular fragmentation, which is related to the structural properties of the compound. Thus, each compound has a unique MS "fingerprint" that allows for the identification of the compound based on its fragmentation pattern. In CI, a gas (typically methane) is first ionized creating a radical that then collides with the analyte, which results in the gain or loss of a hydrogen atom. CI is considered a "soft" ionization process because it generally results in less fragmentation than EI. Even though CI can increase analyte sensitivity, it often provides less structural information.

There are several types of mass analyzers including ion trap, magnetic/electric sector, single and triple quadrupole, time-of-flight (TOF), Fourier-transform, and hybrid mass spectrometers. Some mass analyzers (i.e. ion trap, triple quadrupole) can perform tandem mass spectrometry (MS/MS). In MS/MS, the analyte is ionized and then fragmented to yield product ions. MS/MS transitions are unique for a given compound and result in lower background noise and subsequently higher signal to noise ratios. Techniques using MS/MS are favored in the analysis of complex matrices, such as wastewater effluents.

GC-MS has been used for a variety of applications, including the analysis of EDCs. GC-MS is amenable for the analysis of volatile and semi-volatile EDCs and PPCPs. Many EDCs/PPCPs have polar functional groups that require structural modification to become amenable to GC-MS. To improve performance, some compounds can be chemically modified by derivatization. Derivatization alters the structure of the target molecule such that the molecule has lower polarity and/or greater volatility. One drawback is that derivatization procedures can be time-consuming, labor intensive, and increase the probability of technical error resulting in larger variability. Moreover, many of the chemical agents used for derivatization are challenging to handle, as they are toxic and/or explosive.

Techniques such as SPME and derivatization have been developed to reduce the amount of sample preparation required and increase the applicability of GC-MS, respectively. GC-MS still requires a considerable amount of sample clean up and preparation in order to prevent costly maintenance and diminished performance. These disadvantages have led to the increased popularity of other types of analytical methods, such as liquid chromatography/mass spectrometry (LC-MS).

Liquid Chromatography

Liquid chromatography (LC) has long been used for the analysis of organic compounds in water. Typical LC systems consist of a solvent pump, sample injector, analytical column, and a detector. An LC system allows for the separation of analytes in the liquid phase such that time-resolved peaks relate to a compound's relative affinity for the stationary and mobile phases. Detection is most commonly accomplished using a non-specific detector, such as ultraviolet or visible (UV-VIS) transmittance or absorbance. LC methodology provides for the analysis of nearly all EDC/PPCP compounds including those that have extreme polarities (e.g., acidic compounds), are thermally labile (e.g. steroids), and/or have relatively large molecular weights (e.g. macrolide antibiotics). These types of compounds have generally been beyond the capabilities of GC based systems or required derivatization. Conversely, conventional LC systems (i.e., UV-VIS) lack the sensitivity and selectivity of GC-MS and therefore cannot be used for the detection of most EDCs/PPCPs in complex matrices at environmentally relevant concentrations. When LC is coupled with mass spectrometry (LC-MS), it becomes a powerful tool for the analysis of trace contaminants in water. This combination allows for the analysis of compounds that are not amenable to GC separation, while providing the sensitivity and selectivity of a mass spectrometer.

Unlike GC-MS, the separation of analytes in LC-MS occurs in the liquid phase. The analytes reach the mass spectrometer as a liquid rather than as a gas. Since ionization in a GC-MS occurs via the use of a filament at well below atmospheric pressure, a similar type of ionization for LC-MS is not practical. There are three main types of ionization used in LC-MS: electrospray ionization (ESI), atmospheric pressure chemical ionization (APCI), and atmospheric pressure photoionization (APPI).

ESI is accomplished by guiding the eluent from an LC system into a capillary tube upon which a potential has been applied. The eluent from the capillary is composed of fine droplets which gain a charge from an electric potential applied to the capillary. These charged droplets then encounter a heat source that rapidly evaporates the mobile phase, which transfers the charge from the droplet to the analyte. In APCI, atmospheric gases are ionized by a corona discharge and react with solvent molecules to form reagent ions. These reagent ions then react with analyte molecules to form charged species. Similarly to APCI, an APPI source sprays the analyte into a heated region for volatilization then analytes are bombarded with photons of a particular energy emitted by an APPI lamp. This creates molecular ions of the analytes while leaving the solvent uncharged. These ionization processes are "soft" and result in little fragmentation of the original compound.

As with GC-MS, there are a number of different mass spectrometers that can be coupled to an LC. Ion traps (Huang and Sedlak 2000), single and triple quadrupoles (Snyder et al. 2001c; Ternes 2001), TOF (Hirsch et al. 2001), and sector mass spectrometers (Snyder et al. 2001c) have all been used for the analysis of water contaminants.

Although LC-MS has made the analysis of many compounds easier, it has several limitations. LC-MS systems are relatively expensive and analysts often need specialized training and experience to become proficient in their use. Perhaps the most troublesome limitation is due to a phenomenon known as matrix suppression. Due to the nature of ESI, compounds that elute together compete for charge in the source. Thus, background interferences can alter the response of target compounds. To compensate for this, researchers have used the method of standard additions or isotope dilution. The standard addition method is time consuming, while isotopically-labeled EDC/PPCP standards are not readily available and those that are available can be quite costly. APCI is generally not susceptible to matrix suppression, but many compounds that ionize by ESI do not efficiently ionize using APCI. APPI is a relatively new technique that has shown promise in ionizing a broad spectrum of compounds without matrix suppression and research efforts using this source are ongoing. The coupling of LC and tandem mass spectrometry (LC-MS/MS) currently provides a robust methodology for a broad range of EDCs and PPCPs. LC-MS/MS provides a high degree of selectivity by monitoring ion mass transitions as opposed to discrete masses. As a result, the signal to noise ratio for analytes in challenging matrices is much greater, thus providing not only superior selectivity, but also greater sensitivity.

Methods Applicable to Target Analytes

The description of selection criteria used to develop the target analyte list for this project is provided in Chapter 2. A detailed review of analytical methods available for a large number of potential EDCs and PPCPs is provided in Appendix A. Table 1.1 provides examples of previously published analytical methods and detection limit ranges for target analytes considered for evaluation in this study. This review is limited to GC-MS and LC-MS methodologies.

EDC and PPCP Occurrence

Several excellent overviews of EDC and PPCP occurrence have been published previously (Stahlschmidt-Allner et al. 1997; Ankley et al. 1998; Halling-Sorensen et al. 1998; Daughton and Ternes 1999; Heberer 2002; Kolpin et al. 2002; Koschorreck, Koch, and Ronnefahrt 2002; Lange and Dietrich 2002; Shaw and McCully 2002). Each month, more data on this widely studied topic surfaces. The occurrence data used to build this target compound list focused on literature available by the end of 2001 (Fisher, Keasling, and Schueler 1952; Stumm-Zollinger and Fair 1965; Welch, Levin, and Conney 1969; Tabak and Bunch 1970; Hignite and Azarnoff 1977; Mueller and Kim 1978; Tabak, Bloomhuff, and Bunch 1981; Eckel, Ross, and Isensee 1993; Shore, Gurevitz, and Shemesh 1993; Holm et al. 1995; Heberer and Stan 1996; Buser, Muller, and Theobald 1998; Desbrow et al. 1998; Halling-Sorensen et al. 1998; Ternes et al. 1998; Daughton and Ternes 1999; Snyder et al. 1999; Metcalfe et al. 2000; Lindsey, Meyer, and Thurman 2001; Snyder et al. 2001c; Snyder et al. 2001d; Ternes, Bonerz, and Schmidt 2001). While a tremendous wealth of occurrence data has been published since 2001, this study provides only examples of representative publications since the initial literature review was completed. An abbreviated review is provided in the following paragraphs, while a thorough review is provided in in Appendix A.

The EPA has defined environmental EDCs as exogenous agents which interfere with the "synthesis, secretion, transport, binding, action, or elimination of natural hormones in the body that are responsible for the maintenance of homeostasis, reproduction, development, and/or behavior" (1997). This definition is quite broad and later in the same document the EPA states, "…cells within the brain are a potential target for environmental chemicals that have no impact on steroid hormones directly, but yet will lead to a disruption of endocrine function" (1997). There are numerous other definitions and opinions on what defines an EDC. It is generally accepted that the three major classes of endocrine endpoints are estrogenic (compounds which mimic or block natural estrogen), androgenic (compounds which mimic or block natural testosterone), and thyroidal (compounds with direct or indirect impacts to the thyroid gland). It is also important to note that EDCs vary greatly in structure and for the purposes of this study we deal only with organic EDCs while acknowledging that inorganic EDCs (e.g. perchlorate) may have important toxicological relevance.

Table 1.1
Update on method development for selected EDCs and PPCPs investigated in this project

Target Compound	Extraction	Chromatography	Detection Limit Range (ng/L)	Citation
Steroids				
Androstenedione	SPE	Derivatize GC/MS, GC/MS/MS	0.1-1	(Kolodziej, Gray, and Sedlak 2003; Villegas-Navarro, Rosas-L, and Reyes 2003)
Ethynyl estradiol	SPE, SPME	Derivatize GC/MS or GC/MS/MS, LC/MS, LC/MS/MS, HPLC/UV or electrochemical detector	0.2-400	(Penalver et al. 2002; Braun et al. 2003; Brossa et al. 2004; Quintana et al. 2004b; Zuehlke et al. 2004)
Estradiol	SPE, SPME	Derivatize GC/MS or GC/MS/MS, LC/MS, LC/MS/MS, HPLC/UV or electrochemical detector	0.2-400	(Penalver et al. 2002; Brossa et al. 2004; Hernando et al. 2004a; Lagana et al. 2004; Liu, Zhou, and Wilding 2004; Quintana et al. 2004b; Zuehlke et al. 2004)
Estriol	SPE	Derivatize GC/MS or GC/MS/MS, LC/MS, LC/MS/MS	0.3-7	(Benijts, Lambert, and De Leenheer 2004; Lagana et al. 2004; Quintana et al. 2004a; Rodriguez-Mozaz, Lopez de Alda, and Barcelo 2004)
Estrone	SPE, SPME	Derivatize GC/MS or GC/MS/MS, LC/MS, LC/MS/MS, HPLC/UV or electrochemical detector	0.1-700	(Penalver et al. 2002; Brossa et al. 2004; Hernando et al. 2004a; Lagana et al. 2004; Quintana et al. 2004b; Zuehlke et al. 2004)
Progesterone	SPE	Derivatize GC/MS, HPLC/DAD	1-20	(de Alda and Barcelo 2001; Villegas-Navarro, Rosas-L, and Reyes 2003; Labadie and Budzinski 2005)
Testosterone	SPE	Derivatize GC/MS, GC/MS/MS	0.2-1	(Kolodziej, Gray, and Sedlak 2003; Villegas-Navarro, Rosas-L, and Reyes 2003)
Antimicrobials				
Erythromycin	SPE	LC/MS, LC/MS/MS	0.03-50	(Zuccato et al. 2000; Hilton and Thomas 2003; Vanderford et al. 2003; Stolker et al. 2004; Yang and Carlson 2004)
Sulfamethoxazole	SPE	LC/MS, LC/MS/MS	6-104	(Vanderford et al. 2003; Göbel et al. 2004; Pitarch et al. 2004; Yang, Cha, and Carlson 2004)
Triclosan	SPE	GC/MS	0.16	(Weigel, Kallenborn, and Huhnerfuss 2004)
Trimethoprim	SPE	LC/MS, LC/MS/MS	1-27	(Cahill et al. 2004; Göbel et al. 2004; Sørensen and Elbæk 2004)
Analgesics				
Acetaminophen (Tylenol)	SPE	LC/MS, LC/MS/MS	8.6	(Vanderford et al. 2003; Cahill et al. 2004)
Diclofenac	SPE, SPME, LPME	Derivatize GC/MS, LC/MS/MS, CE/MS, HPLC/fluorescence detector	<1-45	(Ahrer and Buchberger 2001; Gonzalez-Barreiro et al. 2003; Marchese et al. 2003b; Quintana and Reemtsma 2004; Quintana, Rodil, and Reemtsma 2004; Rodríguez et al. 2004; Thomas and Foster 2004)

(continued)

Table 1.1 (Continued)

Target Compound	Extraction	Chromatography	Detection Limit Range (ng/L)	Citation
Ibuprofen (Advil)	SPE, LLE, SPEM, LPME	Derivatize GC/MS, CE/MS, LC/MS, LC/MS/MS, HPLC/fluorescence detector	<1-20	(Zwiener, Glauner, and Frimmel 2000; Gonzalez-Barreiro et al. 2003; Marchese et al. 2003a; Marchese et al. 2003b; Quintana, Rodil, and Reemtsma 2004; Rodríguez et al. 2004; Soliman, Pedersen, and Suffet 2004)
Naproxen	SPE, SPME, LPME	Derivatize GC/MS, CE/MS, LC/MS/MS, HPLC/fluorescence detector	<1-20	(Gonzalez-Barreiro et al. 2003; Marchese et al. 2003b; Rodriguez et al. 2003; Quintana, Rodil, and Reemtsma 2004; Rodríguez et al. 2004; Lin, Chen, and Ding 2005)
Psychoactive Compounds				
Caffeine	SPE	Derivatize GC/MS	0.26-19	(Thomas and Foster 2004; Weigel, Kallenborn, and Huhnerfuss 2004)
Carbamazepine	SPE, LPME, LLE	Derivatize GC/MS, LC/MS/MS, HPLC/fluorescence detector	1-50	(Gonzalez-Barreiro et al. 2003; Müller et al. 2003; Vanderford et al. 2003; Pitarch et al. 2004; Soliman, Pedersen, and Suffet 2004; Lin, Chen, and Ding 2005)
Diazepam (Valium)	SPE, LLE	GC/MS, LC/MS/MS	0.02-100	(Zuccato et al. 2000; Soliman, Pedersen, and Suffet 2004; van der Ven et al. 2004)
Dilantin	SPE	LC/MS/MS	1	(Vanderford et al. 2003)
Fluoxetine (Prozac)	SPE, SPME	Derivatize GC/MS, LC/MS, LC/MS/MS	1-18	(Vanderford et al. 2003; Cahill et al. 2004; Pablo et al. 2004)
Meprobamate	SPE	LC/MS/MS	1	(Vanderford et al. 2003)
Pesticides				
Atrazine	SPE	GC/MS, LC/MS, LC/MS/MS	0.2-20	(Brossa et al. 2002; Brossa et al. 2003; Lopez-Roldan, Lopez de Alda, and Barcelo 2004)
DDT	SPE, SPME	GC/ECD, GC/MS	3-8	(Brossa et al. 2002; Brossa et al. 2003; Dong, Zeng, and Yang 2005)
DEET	SPE	GC/MS	0.026-0.082	(Weigel, Kuhlmann, and Huhnerfuss 2002; Loraine and Pettigrove 2006)
Lindane (γ-BHC)	SPE, SPME	GC/ECD, GC/MS	1.3-20	(Brossa et al. 2002; Brossa et al. 2003; Dong, Zeng, and Yang 2005)
Metolachlor	SPE, SPME	GC/MS, GC/MS/MS	0.1-5	(Sabik et al. 2003; Scheyer et al. 2006)
Fragrances				
Galaxolide	SPE, SPME, Closed loop stripping analysis	GC/MS	1-6	(Garcia-Jares et al. 2002; Bruchet et al. 2005; Mitjans and Ventura 2005)

(continued)

Table 1.1 (Continued)

Target Compound	Extraction	Chromatography	Detection Limit Range (ng/L)	Citation
Musk Ketone	Closed loop stripping analysis	GC/MS	10	(Mitjans and Ventura 2005)
Polyclic Aromatic Hydrocarbons				
Benzo(a)pyrene	SPE, SPME	GC/MS	2-120	(Brossa et al. 2002; Brossa et al. 2003; Psillakis et al. 2003)
Fluorene	LLE, SPME	GC/FID, GC/MS	na, 10	(Mottaleb et al. 2003; Psillakis et al. 2003)
Heart Medications				
Gemfibrozil	SPE, LLE	Derivatize GC/MS, LC/MS, LC/MS/MS	1-90	(Reddersen and Heberer 2003; Cahill et al. 2004; Hernando et al. 2004b; Soliman, Pedersen, and Suffet 2004) (Vanderford et al. 2003)
Pentoxifylline	SPE	LC/MS/MS	1	(Vanderford et al. 2003)
Flame Retardant				
TCEP	SPE, LLE	GC/MS	1,40	(Kolpin et al. 2002; Fries and Puttmann 2003)
Sun Screen				
Oxybenzone	SPE	GC/MS	0.12	(Loraine and Pettigrove 2006)
X-ray Contrast Media				
Iopromide	SPE	LC/MS/MS	1-10	(Hirsch et al. 2000; Vanderford et al. 2003)

The estrogenic activity of synthetic compounds seemed to be of little interest to the environmental community until the discovery that the organochlorine pesticide DDT and its metabolites had endocrine disruptive properties (Fisher, Keasling, and Schueler 1952; Bitman et al. 1968; Welch, Levin, and Conney 1969; Bitman and Cecil 1970; Wrenn et al. 1970). The connection between endocrine disruption and reproductive failures in wildlife was not made until the 1980s, when it was reported that gulls living in areas contaminated with DDT exhibited deformed sex organs and skewed sex-ratios (Fry and Toone 1981; Fry et al. 1987). When gull eggs from areas without DDT contamination were exposed to DDT, the same deformities were observed. Furthermore, when gull eggs were injected with the endogenous estrogen 17β-estradiol (E2), the same sexual deformities were induced (Fry et al. 1987). This was one of the first documented connections between an environmental contaminant and reproductive impact acting via a hormone-mediated mechanism. Further evidence of the endocrine-disrupting properties of DDT and its metabolites were provided by researchers in Florida who linked reproductive disorders in alligators with DDT metabolites that were spilled into Lake Apopka (Guillette Jr. et al. 1994; Guillette Jr. et al. 1996).

Further evidence of the importance of endocrine disruption in wildlife populations has been reported since the discoveries related to DDT. For example, marine gastropods exposed to tributyltins, which leach from certain antifouling paints and PVC pipes, experienced severe population declines and reproductive disorders including imposex (Gibbs, Bryan, and Pascoe 1991). In some amphibian populations, supernumerary limbs and missing limbs have been attributed to certain pesticides and other anthropogenic chemicals (Ouellet et al. 1997; Sparling 2000). In particular, trace concentrations of the widely used herbicide atrazine have been implicated in endocrine disruption of frogs from the Midwestern United States (Hayes et al. 2002). Degradation products from the widely used alkylphenol polyethoxylate (APE) surfactants, which are wide-spread contaminants of wastewater treatment plant effluents, have been shown to be estrogenic (Mueller and Kim 1978; White et al. 1994; Routledge and Sumpter 1997; Giesy et al. 2000; U.S. Environmental Protection Agency 2006), bioaccumulative (Ahel, McEvoy, and Giger 1993; Liber et al. 1999; Lye et al. 1999; Snyder et al. 2001b), and become sequestered in sediment (Bennie et al. 1997; Long et al. 1998; Khim et al. 1999c). There are many other examples of exogenous compounds acting as EDCs. Likewise, many more EDCs will likely be discovered as screening and testing methods become available.

Endocrine disruption also can be caused by naturally occurring compounds. Estrogens from plant sources, known as phytoestrogens, have been reported since the 1930s and in some cases have been linked to reproductive failures in animals (Walker and Janney 1930; Levin, Burns, and Collins 1951; Brookbanks, Welch, and Coup 1969; Metzler and Pfeiffer 1995; Safe and Gaido 1998). For example, sheep grazing on certain strains of clover in New Zealand exhibited severe reproductive impairment due to phytoestrogens (Millington, Francis, and McKeown 1964; Adams 1998). Likewise, the inability of captive Cheetahs to reproduce at the Cincinnati Zoo was linked to a diet high in phytoestrogens (Setchell et al. 1987). Humans are exposed to these compounds as well via a variety of sources. In 1951, a vegetable oil was found to contain various phytoestrogens and phytoandrogens (Levin, Burns, and Collins 1951). Interestingly, many over-the-counter medicinal supplements, such as those recommended for estrogen replacement therapy in post-menopausal women, contain high levels of phytoestrogens. A 2002 publication indicated that phytoestrogens in the human diet are a likely cause of sperm count declines reported in North America and in Europe (Shaw and McCully 2002). More research is necessary to clearly demonstrate the relevance of phytoestrogens in the human diet, both by direct ingestion of vegetables and vegetable products, and possibly from bioaccumulation of phytoestrogens in meat and dairy products. Likewise, ecosystem health could be impacted from the degradation of vegetable matter and paper products in wastewater treatment plants (WWTPs) similar to estrogenicity demonstrated in pulp and paper mill effluents (Koistinen et al. 1998; Tremblay and Van Der Kraak 1999).

The issue of endocrine disruption in the aquatic environment surfaced mainly in the 1990s when reports from the United Kingdom and the United States indicated that fish living below wastewater treatment plants had reproductive abnormalities (Purdom et al. 1994; Bevans et al. 1996; Folmar et al. 1996; Harries et al. 1996; Jobling et al. 1998). These reproductive abnormalities include changes in the levels of sex steroids (e.g., estrogen and testosterone), gonadal histology (i.e. hemaphrodism and intersex), and increased levels of the female egg yolk precursor, vitellogenin, in male fish. These studies initiated a resurgence in research on hormone agonistic and antagonistic compounds, now collectively referred to as EDCs.

Initial attempts to identify the cause of feminization of fish exposed to wastewater effluent focused on synthetic organic chemicals with known estrogenic effects, such as plasticizers and APE surfactant degradation products. However, a series of studies employing *in vitro* bioassay-directed chemical fractionation implicated the human hormone, E2, and the synthetic birth control pharmaceutical, 17α-ethynyl estradiol (EE2), as the most potent estrogens in these complex mixtures (Desbrow et al. 1998; Snyder et al. 1999; Snyder et al. 2001d). Related research involving exposure of fish to E2 and EE2 under laboratory conditions at concentrations as low as ng/L demonstrated measurable changes in fish endocrine biomarkers (Arcand-Hoy, Nimrod, and Benson 1998; Kramer et al. 1998; Panter, Thompson, and Sumpter 1998; Routledge et al. 1998). These reports demonstrated that a pharmaceutical (EE2) could act as aquatic contaminant, inducing measurable endocrine disruptive effects in aquatic organisms. This cause-effect relationship has stimulated a great deal of new research in the identification of trace pharmaceuticals in the environment.

The unexpected consequences of trace concentrations of EDCs on wildlife raised concerns about the potential effects of these chemicals on humans. The connection between endocrine disruption in wildlife and potentially adverse health effects in humans was theorized by Theo Colburn, of the World Wildlife Fund, in her book "Our Stolen Future" (Colborn, Dumanoski, and Myers 1997). Perhaps the most well documented evidence of endocrine disruption in humans involved *in utero* exposure to the potent synthetic estrogen diethylstilbestrol (DES), which resulted in adverse reproductive impacts in human offspring (Herbst, Ulfelder, and Poskanzer 1971; Gill et al. 1979). In this case, DES exposure was not as an environmental contaminant, but as a pharmaceutical that was administered deliberately to pregnant women and has been used in certain agricultural applications (Sower, Schreck, and Evenson 1983; Rumsey and Hammond 1990). The unanticipated sensitivity of the developing human reproductive system to DES clearly demonstrates that the human embryo/fetus is not immune to insult by exogenous chemicals that act as hormones. Some researchers have attributed decreases in human sperm quality and quantity over the past five decades to EDCs in the environment (Sharpe and Skakkebaek 1993; Stone 1994; Carlsen et al. 1995). It has also been suggested that increases in breast, testicular, and prostate cancers reported over the past forty years are related to EDCs in the environment (Krishnan and Safe 1993; Ahlborg et al. 1995; Carlsen et al. 1995; 1997; Ashby 1997; Gillesby and Zacharewski 1998). This topic is still quite controversial and some scientists have produced data refuting these arguments (Safe and Gaido 1998). The estrogenic hormones are less likely to cause adverse effects in humans than they are in fish due to the differences in exposure, as fish may constantly be exposed to EDCs present in the aquatic environment, while humans are assumed to consume an average of 2 L/day of water.

Some scientists have suggested that drinking water disinfection byproducts (DBPs) may act as EDCs. Several reports, which include epidemiological studies, have reported increases in spontaneous abortions and cancers in humans as linked to elevated concentrations of halogenated DBPs (King and Marrett 1996; King, Dodds, and Allen 2000; King, Marrett, and Woolcott 2000). Additionally, the Endocrine Disruptor Screening and Testing Advisory Committee (EDSTAC), a Federal Advisory Committee, suggested that DBPs be included as one of the mixtures that EPA evaluate for endocrine impact (U.S. Environmental Protection Agency 2006). Since DBPs are often orders of magnitude greater in concentration than EDCs/PPCPs and have been implicated as human reproductive toxicants, efforts to control EDCs and PPCPs via drinking water oxidation may actually increase the concentrations of more potent DBPs. While some of these studies and the conclusions drawn from them are widely disputed, reports such as these have increased public awareness and concern regarding endocrine disruptors.

The first reports showing pharmaceuticals in the aquatic environment came from the United States in the 1970s (Tabak and Bunch 1970; Garrison, Pope, and Allen 1975; Hignite and Azarnoff 1977). In 1970, Tabak and Bunch published a study that investigated the degradation of natural and synthetic estrogens and androgens by microorganisms in activated sludge, which noted that low quantities of these compounds could pose an ecological threat (Tabak and Bunch 1970). Later studies have shown that natural and synthetic estrogens are ubiquitous contaminants of wastewater at trace concentrations (Tabak, Bloomhuff, and Bunch 1981; Aherne, English, and Marks 1985; Desbrow et al. 1998; Lee and Peart 1998; Snyder et al. 1999; Huang and Sedlak 2001). Other pharmaceuticals were also reported to occur in wastewater effluents in the 1970s (Garrison, Pope, and Allen 1975; Hignite and Azarnoff 1977). When compared with the number of prescribed pharmaceuticals every year, more research is needed to understand the environmental fate and ecological impacts of pharmaceuticals. A wide variety of PPCPs have been identified as environmental contaminants including: antibiotics, surfactants, x-ray contrast media, analgesics, antiseptics, and many others in wastewater, surface water, ground water and drinking water (Halling-Sorensen et al. 1998; Daughton and Ternes 1999).

In 2002, the US Geological Survey indicated that pharmaceutical contamination of US surface waters is widespread (Kolpin et al. 2002). Some reports have shown trace levels of PPCPs and some EDCs in potable drinking water (Heberer and Dumnbier 2000; Hirsch et al. 2000; Stackelberg et al. 2004). Table 1.2 provides occurrence data for the compounds selected for evaluation in this project in surface waters. Chapter 2 shows the structure and select chemical properties for these compounds. A more comprehensive occurrence database is provided in Appendix A, Table A.2. Sparse data exist to explain the toxicological relevance of trace pharmaceutical compounds. Public perception regarding the presence of PPCPs in water supplies is increasing, and studies regarding treatment options and human health relevance are urgently needed.

EDC and PPCP Treatment

The initial literature review of EDC/PPCP treatment was published in 2003 (Snyder et al. 2003b). Despite the limited number of compounds studied previously, coagulation/flocculation has repeatedly been shown ineffective in removing most PPCPs and EDCs. Carbamazepine and primidone, known to be persistent in the environment, were not affected by coagulation, even at an influent concentrations as high as 1000 ng/L (Ternes et al. 2002b). Likewise, the removal of other PPCPs and EDCs by coagulation also was not significant. Less than 20% of diclofenac, ibuprofen, naproxen and ketoprofen was removed during coagulation by ferric at various pH conditions (Ternes et al. 2002b; Vieno, Tuhkanen, and Kronberg 2005). These findings have been confirmed by experiments performed in this project with a much larger pool of chemicals, such as steroid hormones, personal care products, and a flame retardant (Westerhoff et al. 2005). While the majority of research shows minimal removals through coagulation (Ternes et al. 2002b; Ballard and MacKay 2005; Westerhoff et al. 2005), much higher removals of steroid hormones during coagulation using a special aluminum coagulant (PAX-18) were recently reported (Bodzek and Dudziak 2006). Future studies are needed to evaluate the effect of non-conventional coagulants on contaminant removal.

GAC and PAC have been shown efficient in removing hydrophobic organic contaminants despite the competitive effects from natural organic matter (NOM), which tend to decrease the adsorption of organic compounds. As expected, most EDCs and PPCPs investigated so far have shown excellent removals by GAC or PAC. However, a number of factors can affect the removal efficiency, most common of which are hydrophobicity, charge, and molecular size. As a result, the adsorption of neutral compounds by GAC/PAC is found related to the octanol-water partition coefficient (K_{ow}), which reflects the hydrophobicity of a compound. This is best illustrated by the relationship between log K_{ow} and extent of activated carbon removal demonstrated for bisphenol A, estradiol, and ethynyl estradiol (Yoon et al. 2003). Consistent with this theory, amitrol with a log K_{ow} value of -0.86 was poorly adsorbed by different types of activated carbon (Choi et al. 2005). Complex behaviors were reported for charged species, as some demonstrated much lower removals as compared to neutral species (Adams et al. 2002; Ternes et al. 2002b; Yoon et al. 2003; Vieno, Tuhkanen, and Kronberg 2005; Westerhoff et al. 2005), while others such as acifluorofen, endothall, and two haloacetic acids exhibit moderate removal regardless of charge (Speth and Miltner 1998). Since protonation or deprotonation can occur as pH changes, which subsequently affect molecular charge, the removal of charged species can be pH dependent. Herbicide diquat dibromide is a divalent cation and its removal by GAC is dramatically impacted by aqueous pH (Speth and Miltner 1998). Above pH 6.1, it has Freundlich K values in the thousands, which generally suggest an effective removal by GAC. However, as pH decreases, the K values drop rapidly to as low as 103 at pH 3. This phenomenon was explained by the ion exchange mechanism. Very large molecules tend to have low removal by GAC/PAC due to steric hindrance.

Table 1.2
Occurrence of selected EDCs and PPCPs in various water matrices reported in recent years

Target Compound	Water Matrix	Concentration Range	Citation
Steroids			
Androstenedione	No data available		
Ethynyl estradiol	Surface	n.d. –5.1	(Belfroid et al. 1999; Baronti et al. 2000; Huang and Sedlak 2001; Kuch and Ballschmiter 2001; Cargouet et al. 2004)
Estradiol	Surface	0.15 - 3.6	(Belfroid et al. 1999; Baronti et al. 2000; Huang and Sedlak 2001; Kuch and Ballschmiter 2001; Cargouet et al. 2004)
Estriol	Surface	0.6->40	(Barel-Cohen et al. 2006)
Estrone	Surface	0.1-4.1	(Belfroid et al. 1999; Kuch and Ballschmiter 2001; Hollert et al. 2005)
Progesterone	Surface	n.d.- 199	(Kolpin et al. 2002; Labadie and Budzinski 2005)
Testosterone	Surface	0.4-5550	(Barel-Cohen et al. 2006; Yang, Luan, and Lan 2006)
Antimicrobials			
Erythromycin	Surface and ground	n.d. – 1000	(Hirsch et al. 1999; Sacher et al. 2001; Calamari et al. 2003; Hilton and Thomas 2003; Ashton, Hilton, and Thomas 2004; Kolpin et al. 2004; Stackelberg et al. 2004)
Sulfamethoxazole	Surface and ground	n.d. – 1020	(Holm et al. 1995; Hirsch et al. 1999; Lindsey, Meyer, and Thurman 2001; Sacher et al. 2001; Gray et al. 2002; Hilton and Thomas 2003; Ashton, Hilton, and Thomas 2004; Kolpin et al. 2004; Stackelberg et al. 2004)
Triclosan	Surface	n.d. - 2300	(Kolpin et al. 2002; Lindstrom et al. 2002; Singer et al. 2002; Tixier et al. 2002; Sabaliunas et al. 2003; Kolpin et al. 2004; Stackelberg et al. 2004; Glassmeyer et al. 2005; Halden and Paull 2005)
Trimethoprim	Surface and ground	n.d. - 249	(Hirsch et al. 1999; Hilton and Thomas 2003; Ashton, Hilton, and Thomas 2004)
Analgesics			
Acetaminophen (Tylenol)	Surface	n.d. – 10,000	(Ternes 1998; Kolpin et al. 2002; Kolpin et al. 2004; Stackelberg et al. 2004; Glassmeyer et al. 2005)
Diclofenac	Surface and ground	n.d. - 568	(Heberer et al. 1997; Buser, Poiger, and Muller 1998; Ahrer, Scherwenk, and Buchberger 2001; Farré et al. 2001; Huang and Sedlak 2001; Öllers et al. 2001; Sacher et al. 2001; Hilton and Thomas 2003; Tixier et al. 2003; Ashton, Hilton, and Thomas 2004; Lindqvist, Tuhkanen, and Kronberg 2005; Vieno, Tuhkanen, and Kronberg 2005; Wang et al. 2006)
Hydrocodone	No data available		

(continued)

Table 1.2 (Continued)

Target Compound	Water Matrix	Concentration Range	Citation
Ibuprofen (Advil)	Surface	n.d. - 5040	(Heberer et al. 1997; Ternes 1998; Buser, Poiger, and Muller 1999; Stumpf et al. 1999; Farré et al. 2001; Öllers et al. 2001; Calamari et al. 2003; Hilton and Thomas 2003; Tixier et al. 2003; Ashton, Hilton, and Thomas 2004; Kolpin et al. 2004; Lindqvist, Tuhkanen, and Kronberg 2005; Vieno, Tuhkanen, and Kronberg 2005)
Naproxen	Surface	n.d. - 400	(Öllers et al. 2001; Tixier et al. 2003; Lindqvist, Tuhkanen, and Kronberg 2005; Vieno, Tuhkanen, and Kronberg 2005)
Psychoactive Compounds			
Caffeine	Surface	40-250	(Ternes, Bonerz, and Schmidt 2001; Poiger et al. 2003; Stackelberg et al. 2004; Glassmeyer et al. 2005)
Carbamazepine	Surface and ground	n.d. - 1500	(Seiler et al. 1999; Öllers et al. 2001; Sacher et al. 2001; Ternes, Bonerz, and Schmidt 2001; Heberer 2002; Tixier et al. 2003; Kolpin et al. 2004; Stackelberg et al. 2004; Glassmeyer et al. 2005)
Diazepam (Valium)	Surface	n.d. – 2.13	(Ternes 2001; Calamari et al. 2003)
Dilantin	No data available		
Fluoxetine (Prozac)	Surface	n.d. - 12	(Kolpin et al. 2002)
Meprobamate	No data available		
Pesticides			
Atrazine	Surface and ground	1000-20,000	(Rebich, Coupe, and Thurman 2004; Shomar, Muller, and Yahya 2006)
DDT	Surface	19-663	(Zhang et al. 2004)
DEET	Surface	3-268	(Loraine and Pettigrove 2006)
Lindane (γ-BHC)	Surface	n.d. - 110	(Kolpin et al. 2002)
Metolachlor	Surface	n.d. - 2020	(McConnell, Harman-Fetcho, and Hagy 2004; Rebich, Coupe, and Thurman 2004)
Fragrances			
Galaxolide	Surface	n.d. - 100	(Mitjans and Ventura 2004; Stackelberg et al. 2004)
Musk Ketone	Surface	0.08	(Peck and Hornbuckle 2004)
Polyclic Aromatic Hydrocarbons			
Benzo(a)pyrene	Surface	n.d. - 240	(Kolpin et al. 2002; Zhang et al. 2004)
Fluorene	Surface	21-141	(Zhang et al. 2004)

(continued)

Table 1.2 (Continued)

Target Compound	Water Matrix	Concentration Range	Citation
Heart Medications			
Gemfibrozil	Surface	n.d. - 1550	(Ternes 1998; La Farre et al. 2001)
Pentoxifylline	No data available		
Flame Retardant			
TCEP	Surface	75-100	(Kolpin et al. 2004; Stackelberg et al. 2004; Glassmeyer et al. 2005)
Sun Screen			
Oxybenzone	No data available		
X-ray Contrast Media			
Iopromide	Surface and ground	n.d. - 3600	(Putschew, Schittko, and Jekel 2001; Liebig, Moltmann, and Knacker 2006)

In addition to different removal mechanisms, adsorption kinetics and adsorbate concentration must also be considered in evaluating the efficiency of an activated carbon process. Using GAC, it was determined that higher initial contaminant concentrations resulted in higher percent removal for estradiol (Fuerhacker, Dürauer, and Jungbauer 2001). Water qualities also are important due to the impact on equilibrium time. In deionized water, less than three hours was required for estradiol to reach adsorption equilibrium on GAC, while 17 hours was needed when industrial waste was investigated (Fuerhacker, Dürauer, and Jungbauer 2001; Zhang and Zhou 2005).

Ozone is an oxidant, which is gaining popularity in water treatment applications because it is a powerful disinfectant and is efficient in removing organic compounds. Ozone is particularly effective in oxidizing compounds with unsaturated double bonds or moieties with electron donating properties. Oxidation of EDCs/PPCPs during ozonation depends highly on the chemical structures of these compounds and ozone dose. Several publications have indicated that ozone can efficiently remove these emerging contaminants, such as diclofenac, carbamazepine, bezafibrate, and estrogens (Zwiener and Frimmel 2000; Ternes et al. 2002a; Ternes et al. 2003; Huber et al. 2005a; McDowell et al. 2005; Snyder et al. 2006b). The extent of removals for these contaminants was shown consistent with magnitude of reaction rate constants (Huber et al. 2003). For instance, the reaction rate constant of bezafibrate is more than 100 times smaller than that of carbamzepine and diclofenac (Huber et al. 2003), which could explain the relatively inefficient (50%) removal of bezafibrate in contrast to the nearly complete removal of the other two pharmaceuticals (Ternes et al. 2002a). The rate at which ozone reacts with a particular contaminant is moiety specific and pH dependent. Antibacterial substance, azithromycin, reacted rapidly with a 2nd order reaction rate constant in the order of 10^6 $M^{-1}s^{-1}$ due to the high reactivity of neutral tertiary amine moiety. However, at low pH the protonation of the amine moiety decreased the reaction rate by several orders of magnitude (Dodd, Buffle, and Von Gunten 2006). Studies performed with three antibiotics also reported the effect of pH by monitoring the change in chemical oxygen demand (COD) during ozonation. Higher removal of COD at neutral pH was observed in contrast to the low removal at acidic pH (Balcioglu and Otker 2003). In addition to pH, other water qualities such as organic carbon concentration and temperature can affect a particular compound's reaction with ozone. For instance, different removal efficiency was reported for the same compound in river water as compared to distilled water (Zwiener and Frimmel 2000).

Despite the generally efficient oxidation of EDCs and PPCPs, some compounds were not well removed by ozone. Studies performed independently by two research groups have shown less than 40% removal of clofibric acid under several testing conditions (Zwiener and Frimmel 2000; Ternes et al. 2002b). Iodinated X-ray contrast media (diatrizoate, iopamidol, iopromide and iomeprol) are also resistant to ozone oxidation, with the ionic species most resistant (Ternes et al. 2003). At a relatively high ozone dose of 15 mg/L and a contact time of 18 min, which is much stronger than what would be normally adopted by water treatment facilities, less than 14% of diatrizoate was oxidized. Advanced oxidation processes with the addition of hydrogen peroxide (H_2O_2) to promote hydroxyl radical reactions help to improve the elimination of many recalcitrant compounds during ozonation (Zwiener and Frimmel 2000); however, the process may require optimization before any benefit can be noticed (Ternes et al. 2003).

Free chlorine is not as strong an oxidant as ozone, yet still efficient in removing a significant number of PPCPs and EDCs via oxidation, substitution, and addition reactions. Electron rich sulfamethazine and naproxen are susceptible to oxidation by free chlorine, while ibuprofen and ketoprofen, both of which are substituted aromatics with carboxylic groups, show very weak reactivity with free chlorine (Adams et al. 2002; Boyd, Zhang, and Grimm 2005). Because pH shifts the equilibrium distribution of hypochlorous acid and hypochlorite ion, naproxen was found to react faster with free chlorine at pH 5 and 7 than pH 9 (Boyd, Zhang, and Grimm 2005). Other pharmaceuticals are expected to have similar pH dependence when they react with free chlorine. The reactivity of chlorine dioxide lies between that of free chlorine and ozone. Based on a study with nine pharmaceuticals, chlorine dioxide was found to react faster than free chlorine but slower than ozone. Yet, ClO_2 is still an efficient way to oxidize some emerging contaminants, such as sulfonamide antibiotics and estrogens (Huber et al. 2005b).

UV at a typical disinfection dose (i.e., 40 mJ/cm^2) is not expected to oxidize organic compounds. No significant removal by UV disinfection was observed for seven representative antibiotics. However UV irradiation at increased energies (i.e. >400 mJ/cm^2) is capable of oxidizing some organic contaminants. At a UV fluence of 3000 mJ/cm^2, the above mentioned seven antibiotics were removed by 50-80% (Adams et al. 2002). Two steroid hormones and the plasticizer bisphenol A also showed some removal by direct UV photolysis with the medium pressured lamp performing better than the low pressure lamp due to its higher energy output (Rosenfeldt and Linden 2004). Similar to ozonation, high-energy UV irradiation can be combined with H_2O_2 for advanced oxidation. An UV advanced oxidation process (AOP) has been shown more efficient than UV alone to degrade bisphenol A, ethynyl estradiol, estradiol and carbamazepine (Rosenfeldt and Linden 2004; Vogna et al. 2004).

Photocatalytic degradation of recalcitrant pharmaceuticals with the catalyst TiO_2 was investigated for wastewater treatment (Doll and Frimmel 2004). The technique was shown to be very promising for the degradation of carbamazepine, clofibric acid, and two X-ray contrast media (iomeprol and iopromide). However, photocatalytic processes are not yet widely used in the drinking water and water reuse industries.

Membranes offer unique benefits for micropollutant removal as high rejections can be obtained without formation of byproducts. Numerous studies have reported rejections of micropollutants by NF membranes (Kimura et al. 2003; Nghiem, Schaefer, and Elimelech 2005a; Nghiem, Schaefer, and Elimelech 2005b; Snyder et al. 2006a; Yoon et al. 2006). The rejections of contaminants vary widely depending on the target pollutant, the type of membrane, the feed water, and the operating conditions. These factors were also found to affect contaminant rejection by reverse osmosis (RO). In a comparison between polyamide NF and RO, both membranes showed high removal (>90%) for a group of negatively charged disinfection byproducts and pharmaceuticals. For neutral species, despite the decrease in rejections for both membranes, RO performance was significantly better than NF, which was repetitively demonstrated to hold true for several EDCs and PPCPs (Adams et al. 2002; Drewes, Heberer, and Reddersen 2002; Kimura et al. 2003; Xu et al. 2005). Despite reports on membrane treatment efficiency of EDCs and PPCPs, much information is still needed. Only a limited number and type of compounds have been investigated and data supporting specific removal mechanisms are sparse. Size/steric exclusion, electrostatic repulsion, adsorption, and diffusion mechanisms have been identified as the key factors for organic contaminant transport through membranes (Ozaki and Li 2002; Kimura et al. 2003; Schafer, Nghiem, and Waite 2003; Ng and Elimelech 2004; Nghiem et al. 2004). Such interactions between the organic contaminant and the membrane material are determined by the coupled influence of the physicochemical characteristics of the contaminant, the membrane properties, and the solution chemistry. In addition, the presence of NOM and hydrodynamic conditions in the membrane system, which depend on the operating conditions, also affect membrane performance (Agenson, Oh, and Urase 2003; Park et al. 2004; Weber et al. 2004).

One key difference between physical removal and chemical removal processes is the potential for byproduct formation. Coagulation, activated carbon, and membranes eliminate contaminants by physical adsorption or separation, while chemical removal relies on oxidation reactions to "remove" contaminants. Therefore, the latter produces oxidation products, which in some cases are concerns for human and environmental exposure. For instance, deethylatrazine is a degradation product of atrazine, which has been found to exhibit acute lethal effect to aquatic organisms due to synergistic interactions with organophosphate insecticides (Kotrikla, Lekkas, and Bletsa 1997; Trimble and Lydy 2006).

As PPCPs and EDCs are becoming an increasing concern for both the public and the water industry, efforts dedicated to this area have increased dramatically since the beginning of this project. Reports on occurrence indicate that this is a ubiquitous issue, yet can be attenuated by various available water process technologies. This project will enrich our knowledge base on the removal of PPCPs and EDCs by evaluating currently available water treatment processes with a diversity of representative compounds to provide a more comprehensive understanding of PPCP and EDC fate during water treatment.

REFERENCES

1997. Special report on environmental endocrine disruption: an effects assessment and analysis. *Office of Research and Development. Washington, D.C. February 1997. EPA/630/R-96/012*

Adams, C., Y. Wang, K. Loftin, and M. Meyer. 2002. Removal of antibiotics from surface and distilled water in conventional water treatment processes. *Journal of Environmental Engineering-Asce*, 128(3): 253-260.

Adams, N. R. 1998. Natural and anthropogenic environmental estrogens: the scientific basis for risk assessment. Clover phyto-estrogens in sheep in Western Australia. *Pure and Applied Chemistry* 70(9): 1855-1862.

Agenson, K. O., J.-I. Oh, and T. Urase. 2003. Retention of a wide variety of organic pollutants by different nanofiltration/reverse osmosis membranes: controlling parameters of process. *Journal of Membrane Science* 225(1-2): 91-103.

Ahel, M., J. McEvoy, and W. Giger. 1993. Bioaccumulation of the Lipophilic Metabolites of Nonionic Surfactants in Freshwater Organisms. *Environmental Pollution* 79: 243-248.

Aherne, G. W., J. English, and V. Marks. 1985. The role of immunoassay in the analysis of microcontaminants in water samples. *Ecotoxicology and Environmental Safety* 9: 79-83.

Ahlborg, U. G., L. Lipworth, L. Titus-Ernustoff, C.-C. Hsieh, A. Hanberg, J. Baron, D. Trichopoulos, and H. O. Adami. 1995. Organochlorine compounds in relation to breast cancer, endometrial cancer, and endometriosis: An assessment of the biological and epidemiological evidence. *Critical Reviews in Toxicology* 25: 463-531.

Ahrer, W., and W. Buchberger. 2001. Combination of aqueous and non-aqueous capillary electrophoresis with electrospray mass spectrometry for the determination of drug residues in water. *Monatsh. Chem.*, 132(3): 329-337.

Ahrer, W., E. Scherwenk, and W. Buchberger. 2001. Determination of drug residues in water by the combination of liquid chromatography or capillary electrophoresis with electrospray mass spectrometry. *Journal of Chromatography A* 910: 69-78.

Ankley, G., E. Mihaich, R. Stahl, D. Tillitt, T. Colborn, S. McMaster, R. Miller, J. Bantle, P. Campbell, N. Denslow, R. Dickerson, L. Folmar, M. Fry, J. Giesy, E. Gray, P. Guiney, T. Hutchinson, S. Kennedy, V. Kramer, G. LeBlanc, M. Mayes, A. Nimrod, R. Patino, R. Peterson, R. Purdy, R. Ringer, P. Thomas, L. Touart, G. Van Der Kraak, and T. Zacharewski. 1998. Overview of a Workshop on Screening Methods for Detecting Potential (Anti-) Estrogenic/Andogenic Chemicals in Wildlife. *Environmental Toxicology and Chemistry* 17(1): 68-87.

Arcand-Hoy, L. D., A. C. Nimrod, and W. H. Benson. 1998. Endocrine-modulating substances in the environment: estrogenic effects of pharmaceutical products. *International Journal of Toxicology* 17: 139-158.

Ashby, J. 1997. A Hierachical Approach to the Evaluation of Chemicals for Estrogenic and other Endocrine-Disrupting Properties. *Environmental Toxicology and Pharmacology 3* 87-90.

Ashton, D., M. Hilton, and K. V. Thomas. 2004. Investigating the environmental transport of human pharmaceuticals to streams in the United Kingdom. *Sci. Total Environ.*, 333(1-3): 167-184.

Balcioglu, I. A., and M. Otker. 2003. Treatment of pharmaceutical wastewater containing antibiotics by O-3 and O-3/H2O2 processes. *Chemosphere* 50(1): 85-95.

Ballard, B. D., and A. A. MacKay. 2005. Estimating the removal of anthropogenic organic chemicals from raw drinking water by coagulation flocculation. *Journal of Environmental Engineering-Asce*, 131(1): 108-118.

Barel-Cohen, K., L. S. Shore, M. Shemesh, A. Wenzel, J. Mueller, and N. Kronfeld-Schor. 2006. Monitoring of natural and synthetic hormones in a polluted river. *Journal of Environmental Management*, 78(1): 16-23.

Baronti, C., R. Curini, G. D'Ascenzo, A. Di Corcia, A. Gentili, and R. Samperi. 2000. Monitoring natural and synthetic estrogens at activated sludge sewage treatment plants and in a receiving river water. *Environmental Science & Technology* 34(24): 5059-5066.

Belfroid, A. C., A. Van der Horst, A. D. Vethaak, A. J. Schafer, G. B. J. Rijs, J. Wegener, and W. P. Cofino. 1999. Analysis and occurrence of estrogenic hormones and their glucuronides in surface water and waste water in The Netherlands. *The Science of the Total Environment* 225: 101-108.

Benijts, T., W. Lambert, and A. De Leenheer. 2004. Analysis of Multiple Endocrine Disruptors in Environmental Waters via Wide-Spectrum Solid-Phase Extraction and Dual-Polarity Ionization LC-Ion Trap-MS/MS. *Analytical Chemistry* 76(3): 704-711.

Bennett, E. R., and C. D. Metcalfe. 2000. Distribution of degradation products of alkylphenol ethoxylates near sewage treatment plants in the lower great lakes, North America. *Environmental Toxicology and Chemistry* 19(4): 784-792.

Bennetts, H. W., E. J. Underwood, and F. L. Shier. 1946. A specific breeding problem of sheep on subterranean clover pastures in western Australia. *Aust. Vet. J.*, 22: 2-12.

Bennie, D. T., C. A. Sullivan, H. B. Lee, T. E. Peart, and R. J. Maguire. 1997. Occurence of alkylphenols and alkylphenol mono- and diethoxylates in natural waters of the Laurentian Great Lakes basin and the upper St. Lawrence River. *The Science of the Total Environment* 193: 263-275.

Bevans, H. E., S. L. Goodbred, J. F. Miesner, S. A. Watkins, T. S. Gross, N. D. Denslow, and T. Schoeb. 1996. Synthetic Organic Compounds and Carp Endocrinology and Histology in Las Vegas Wash and Las Vegas and Callville Bays of Lake Mead, Nevada, 1992 and 1995. *Water-Resources Investigations Report 96-4266*

Bitman, J., and H. C. Cecil. 1970. Estrogenic activity of DDT analogs and polychlorinated biphenyls. *Journal of Agricultural and Food Chemistry* 18: 1108-1112.

Bitman, J., H. C. Cecil, S. J. Harris, and G. F. Fries. 1968. Estrogenic activity of o,p'-DDT in the mammalian uterus and avian oviduct. *Science* 162(3851): 371-372.

Bodzek, M., and M. Dudziak. 2006. Removal of natural estrogens and synthetic compounds considered to be endocrine disrupting substances (EDs) by coagulation and nanofiltration. *Polish Journal of Environmental Studies*, 15: 35-40.

Boyd, G. R., S. Zhang, and D. A. Grimm. 2005. Naproxen removal from water by chlorination and biofilm processes. *Water Research*, 39: 668-676.

Braun, P., M. Moeder, S. Schrader, P. Popp, P. Kuschk, and W. Engewald. 2003. Trace analysis of technical nonylphenol, bisphenol A and 17α-ehynyl estradiol in wastewater using solid-phase microextraction and gas chromatography-mass spectrometry. *J. Chromatogr. A*, 988: 41-51.

Brookbanks, E. O., R. A. H. Welch, and M. R. Coup. 1969. Estrogens in pasture and a possible relation with mastitis. *New Zealand Vetinary Journal* 17: 159-160.

Brossa, L., R. M. Marce, F. Borrull, and E. Pocurull. 2002. Application of on-line solid-phase extraction-gas chromatography-mass spectrometry to the determination of endocrine disruptors in water samples. *Journal of Chromatography A* 963(1-2): 287-294.

Brossa, L., R. M. Marcé, F. Borrull, and E. Pocurull. 2003. Determination of endocrine-disrupting compounds in water samples by on-line solid-phase extraction-programmed-temperature vaporisation-gas chromatography-mass spectrometry. *J. Chromatogr. A*, 998: 41-50.

Brossa, L., E. Pocurull, F. Borrull, and R. M. Marce. 2004. Solid-phase extraction/high-performance liquid chromatography-electrospray mass spectrometry to determine endocrine disruptors in water samples. *Chromatographia* 59(7-8): 419-423.

Bruchet, A., C. Hochereau, C. Picard, V. Decottignies, J. M. Rodrigues, and M. L. Janex-Habibi. 2005. Analysis of drugs and personal care products in French source and drinking waters: the analytical challenge and examples of application. *Water Science and Technology*, 52: 53-61.

Buser, H. R., M. D. Muller, and N. Theobald. 1998. Occurrence of the pharmaceutical drug clofibric acid and the herbicide mecoprop in various Swiss lakes and in the North Sea. *Environmental Science & Technology* 32(1): 188-192.

Buser, H. R., T. Poiger, and M. D. Muller. 1998. Occurrence and fate of the pharmaceutical drug diclofenac in surface waters: Rapid photodegradation in a lake. *Environmental Science & Technology* 32(22): 3449-3456.

Buser, H. R., T. Poiger, and M. D. Muller. 1999. Occurrence and environmental behavior of the chiral pharmaceutical drug ibuprofen in surface waters and in wastewater. *Environmental Science & Technology* 33(15): 2529-2535.

Cahill, J. D., E. T. Furlong, M. R. Burkhardt, D. Kolpin, and L. G. Anderson. 2004. Determination of pharmaceutical compounds in surface- and ground-water samples by solid-phase extraction and high-performance liquid chromatography-electrospray ionization mass spectrometry. *Journal of Chromatography A* 1041(1-2): 171-180.

Calamari, D., E. Zuccato, S. Castiglioni, R. Bagnati, and R. Fanelli. 2003. Strategic survey of therapeutic drugs in the rivers Po and Lambro in northern Italy. *Environ. Sci. Technol.*, 37(7): 1241-1248.

Cargouet, M., D. Perdiz, A. Mouatassim-Souali, S. Tamisier-Karolak, and Y. Levi. 2004. Assessment of river contamination by estrogenic compounds in Paris area (France). *Science Of The Total Environment* 324(1-3): 55-66.

Carlsen, E., A. Giwercman, N. Keiding, and N. E. Skakkebaek. 1995. Declining sperm quality and increasing incidence of testicular cancer: Is there a common cause? *Environmental Health Perspectives* 130: 137-139.

Choi, K. J., S. G. Kim, C. W. Kim, and S. H. Kim. 2005. Effects of activated carbon types and service life on removal of endocrine disrupting chemicals: amitrole, nonylphenol, and bisphenol-A. *Chemosphere* 58(11): 1535-1545.

Colborn, T., D. Dumanoski, and J. P. Myers. 1997. Our stolen future. Plume Publishing: New York, New York.

Daughton, C. G., and T. A. Ternes. 1999. Pharmaceuticals and Personal Care Products in the Environment: Agents of Subtle Change? *Environmental Health Perspectives* 107(6): 907-938.

de Alda, M. J. L., and D. Barcelo. 2001. Determination of steroid sex hormones and related synthetic compounds considered as endocrine disrupters in water by fully automated on-line solid-phase extraction-liquid chromatography-diode array detection. *J. Chromatogr. A*, 911(2): 203-210.

Desbrow, C., E. J. Routledge, G. C. Brighty, J. P. Sumpter, and M. Waldock. 1998. Identification of estrogenic chemicals in STW Effluent. 1. Chemical fractionation and in vitro biological screening. *Environmental Science & Technology* 32(11): 1549-1558.

Dodd, M. C., M. O. Buffle, and U. Von Gunten. 2006. Oxidation of antibacterial molecules by aqueous ozone: Moiety-specific reaction kinetics and application to ozone-based wastewater treatment. *Environ. Sci. Technol.*, 40(6): 1969-1977.

Doll, T. E., and F. H. Frimmel. 2004. Kinetic study of photocatalytic degradation of carbamazepine, clofibric acid, iomeprol and iopromide assisted by different TiO2 materials-determination of intermediates and reaction pathways. *Water Research* 38(4): 955-964.

Dong, C., Z. Zeng, and M. Yang. 2005. Determination of organochlorine pesticides and their derivations in water after HS-SPME using polymethylphenylvinylsiloxane-coated fiber by GC-ECD. *Water Research*, 39: 4204-4210.

Drewes, J. E., T. Heberer, and K. Reddersen. 2002. Fate of pharmaceuticals during indirect potable reuse. *Water Science and Technology* 46(3): 73-80.

Echols, K. R., R. W. Gale, T. R. Schwartz, J. N. Huckins, L. L. Williams, J. C. Meadows, D. Morse, J. D. Petty, C. E. Orazio, and D. E. Tillitt. 2000. Comparing Polychlorinated Biphenyl Concentrations and Patterns in the Saginaw River Using Sediment, Caged Fish, and Semipermeable Membrane Devices. *Environmental Science & Technology* 34(19): 4095-4102.

Eckel, W. P., B. Ross, and R. K. Isensee. 1993. Pentobarbital found in ground water. *Ground Water* 31(5): 801-804.

Farré, M., I. Ferrer, A. Ginebreda, M. Figueras, L. Olivella, L. Tirapu, M. Vilanova, and D. Barceló. 2001. Determination of drugs in surface water and wastewater samples by liquid chromatography-mass spectrometry: methods and preliminary resutls including toxicity studies with *Vibrio fischeri*. *J. Chromatogr. A*, 2001: 187-197.

Fatoki, O. S., and R. O. Awofolu. 2003. Methods for selective determination of persistent organochlorine pesticide residues in water and sediments by capillary gas chromatography and electron-capture detection. *Journal of Chromatography A* 983(1-2): 225-236.

Fisher, A. L., H. H. Keasling, and F. W. Schueler. 1952. Estrogenic action of some DDT analogs. *Proc. Soc. Exptl. Biol. Med.* , 81: 439-441.

Folmar, L. C., N. D. Denslow, V. Rao, M. Chow, D. A. Crain, J. Enblom, J. Marcino, and L. J. Guillette Jr. 1996. Vitellogenin Induction and Reduced Serun Testosterone Concentrations in Feral Male Carp (*Cyprinus carpio*) Captured Near a Major Metropolitan Sewage Treatment Plant. *Environmental Health Perspectives* 104(10): 1096-1101.

Fowler, B. R., L. Haviland, and S. Kennedy. 1998. Analysis of alkylphenols in Fraser river water and suspended sediments. *poster*

Fries, E., and W. Puttmann. 2003. Monitoring of the three organophosphate esters TBP, TCEP and TBEP in river water and ground water (Oder, Germany). *J. Environ. Monit.*, 5(2): 346-352.

Fry, D. M., and C. K. Toone. 1981. DDT-induced feminization of gull embryos. *Science* 231: 919-924.

Fry, D. M., C. K. Toone, S. M. Speich, and R. J. Peard. 1987. Sex ratio skew and breeding patterns of gulls: demographic and toxicological considerations. *Studies of Avian Biology?* , 10: 26-43.

Fuerhacker, M., A. Dürauer, and A. Jungbauer. 2001. Adsorption isotherms on 17ß-estradiol on granular activated carbon (GAC). *Chemosphere* 44: 1573-1579.

Garcia-Jares, C., M. Llompart, M. Polo, C. Salgado, S. Macias, and R. Cela. 2002. Optimisation of a solid-phase microextraction method for synthetic musk compounds in water. *J. Chromatogr. A*, 963(1-2): 277-285.

Garrison, A. W., J. D. Pope, and F. R. Allen. 1975. GC/MS analysis of organic compounds in domestic wastewaters. In *Chem. Congr. North Am. Cont.*: Ann Arbor Science.

Gibbs, P. E., G. W. Bryan, and P. L. Pascoe. 1991. TBT-induced imposex in the dogwhelk, *Nucella lapillus*: Geographical uniformity of the response and effects. *Mar. Environ. Res.* , 32: 79-87.

Giesy, J. P., S. L. Pierens, E. M. Snyder, S. Miles-Richardson, V. J. Kramer, S. A. Snyder, K. M. Nichols, and D. A. Villeneuve. 2000. Effects of 4-Nonylphenol on Fecundity and Biomarkers of Estrogenicity in Fathead Minnows (*Pimephales promelas*). *Environmental Toxicology and Chemistry* 19(5): 1368-1377.

Gill, W. B., F. B. Schumacher, M. Bibbo, F. H. Straus, and H. W. Schoenberg. 1979. Association of diethylstilbestrol exposure in utero with cryptorchidism, testicular hypoplasia and semen abnormalities. *Journal of Urology* 122: 36-39.

Gillesby, B. E., and T. R. Zacharewski. 1998. Exoestrogens: Mechanisms of action and strategies for identification and assessment. *Environmental Toxicology and Chemistry* 17(1): 3-14.

Glassmeyer, S. T., E. T. Furlong, D. W. Kolpin, J. D. Cahill, S. D. Zaugg, S. L. Werner, M. T. Meyer, and D. D. Kryak. 2005. Transport of Chemical and Microbial Compounds from Known Wastewater Discharges: Potential for Use as Indicators of Human Fecal Contamination. *Environmental Science & Technology*, 39: 5157-5169.

Göbel, A., C. S. McArdell, M. J. F. Suter, and W. Giger. 2004. Trace determination of macrolide and sulfonamide antimicrobials, a human sulfonamide metabolite, and trimethoprim in wastewater using liquid chromatography coupled to electrospray tandem mass spectrometry. *Anal. Chem.*, 76: 4756-4764.

Gonzalez-Barreiro, C., M. Lores, M. C. Casais, and R. Cela. 2003. Simultaneous determination of neutral and acidic pharmaceuticals in wastewater by high-performance liquid chromatography-post-column photochemically induced fluorimetry. *Journal of Chromatography A* 993(1-2): 29-37.

Gray, L. E., J. Ostby, V. Wilson, C. Lambright, K. Bobseine, P. Hartig, A. Hotchkiss, C. Wolf, J. Furr, M. Price, L. Parks, R. L. Cooper, T. E. Stoker, S. C. Laws, S. J. Degitz, K. M. Jensen, M. D. Kahl, J. J. Korte, E. A. Makynen, J. E. Tietge, and G. T. Ankley. 2002. Xenoendocrine disrupters-tiered screening and testing Filling key data gaps. *Toxicology* 181-182: 371-382.

Guillette Jr., L. J., T. S. Gross, G. R. Masson, N. M. Matter, H. F. Percival, and A. R. Woodward. 1994. Developmental abnormalities of the gonad and sex hormone concentrations in juvenile alligators from contaminated and control lakes in Florida. *Environmental Health Perspectives* 104: 680-688.

Guillette Jr., L. J., D. B. Pickford, D. A. Crain, A. A. Rooney, and H. F. Percival. 1996. Reduction in penis size and plasma testosterone concentrations in juvenile alligators living in a contaminated environment. *General and Comparative Endocrinology* 101(1): 32-42.

Halden, R. U., and D. H. Paull. 2005. Co-occurrence of triclocarban and triclosan in US water resources. *Environ. Sci. Technol.*, 39(6): 1420-1426.

Halling-Sorensen, B., S. N. Nielsen, P. F. Lanzky, F. Ingerslev, H. C. H. Lutzhoft, and S. E. Jorgensen. 1998. Occurence, Fate and Effects of Pharmaceutical Substances in the Environment - A Review. *Chemosphere* 36(2): 357-393.

Harries, J. E., D. A. Sheahan, S. Jobling, P. Matthiessen, P. Neall, E. J. Routledge, R. Rycroft, J. P. Sumpter, T. Tylor, and N. Zaman. 1996. A survey of estrogenic activity in United Kingdom inland waters. *Environmental Toxicology and Chemistry* 15: 1993-2002.

Hayes, T. B., A. Collins, M. Lee, M. Mendoza, N. Noriega, A. A. Stuart, and A. Vonk. 2002. Hermaphroditic, demasculinized frogs after exposure to the herbicide atrazine at low ecologically relevant doses. *Proceedings of the National Academy of Sciences of the United States of America* 99(8): 5476-5480.

Heberer, T. 2002. Occurrence, fate, and removal of pharmaceutical residues in the aquatic environment: a review of recent research data. *Toxicology Letters* 131(1-2): 5-17.

Heberer, T., and U. Dumnbier. 2000. Polar DDT metabolites as contaminants of surface, ground and drinking water in Berlin. In *National Groundwater Association - Emerging Issues Conference*. Minneapolis, MN: NGWA.

Heberer, T., U. Dunnbier, C. Reilich, and H. J. Stan. 1997. Detection of drugs and drug metabolites in ground water samples of a drinking water treatment plant. *Fresenius Environmental Bulletin*, 6(7-8): 438-443.

Heberer, T., and H.-J. Stan. 1996. Determination of trace levels of dichlorprop,mecoprop,clofibric acid, and naphthylacetic acid in soil by gas chromatography/mass spectrometry with selected-ion monitoring. *Journal of AOAC International* 79(6): 1428-1433.

Herbst, A. L., H. Ulfelder, and D. C. Poskanzer. 1971. Adenocarcinoma of the vagina. Association of maternal stilbesterol therapy with tumor appearance in young women. *New England Journal of Medicine* 284: 878-881.

Hernando, M. D., M. Mezcua, M. J. Gómez, O. Malato, A. Agüeera, and A. R. Fernández-Alba. 2004a. Comparative study of analytical methods involving gas chromatography-mass spectrometry after derivatization and gas chromatography-tandem mass spectrometry for the determination of selected endocrine disrupting compounds in wastewaters. *J. Chromatogr. A*, 1047(1): 129-135.

Hernando, M. D., M. Petrovic, A. R. Fernández-Alba, and D. Barceló. 2004b. Analysis by liquid chromatography-electrospray ionization tandem mass spectrometry and acute toxicity evaluation for β-blockers and lipid-regulating agents in wastewater samples. *J. Chromatogr. A*, 1046(1-2): 133-140.

Hignite, C., and D. L. Azarnoff. 1977. Drugs and drug metabolites as environmental contaminants: chlorophenoxyisobutyrate and salicylic acid in sewage water effluent. *Life Sciences* 20(2): 337-341.

Hilton, M. J., and K. V. Thomas. 2003. Determination of selected human pharmaceutical compounds in effluent and surface water samples by high-performance liquid chromatography-electrospray tandem mass spectrometry. *Journal of Chromatography A* 1015(1-2): 129-141.

Hirsch, R., T. Ternes, K. Haberer, and K.-L. Kratz. 1999. Occurrence of antibiotics in the aquatic environment. *The Science of the Total Environment* 225: 109-118.

Hirsch, R., T. A. Ternes, I. Bobeldijk, and R. A. Weck. 2001. Determination of environmentally relevant compounds using fast GC/TOF-MS. *Chimia* 55(1-2): 19-22.

Hirsch, R., T. A. Ternes, A. Lindart, K. Haberer, and R. D. Wilken. 2000. A sensitive method for the determination of iodine containing diagnostic agents in aqueous matrices using LC-electrospray-tandem-MS detection. *Fresenius' Journal of Analytical Chemistry* 366: 835-841.

Hollert, H., M. Duerr, R. Holtey-Weber, M. Islinger, W. Brack, H. Faerber, L. Erdinger, and T. Braunbeck. 2005. Endocrine disruption of water and sediment extracts in a non-radioactive dot blot/RNAse protection-assay using isolated hepatocytes of rainbow trout. Deficiencies between bioanalytical effectiveness and chemically determined concentrations and how to explain them. *Environmental Science and Pollution Research*, 12: 347-360.

Holm, J. V., K. Rügge, P. J. Bjerg, and T. H. Christensen. 1995. Occurence and distribution of pharmaceutical organic compounds in the groundwater downgradient of a landfill (Grindsted, Denmark). *Environmental Science & Technology* 29(5): 1415-1420.

Huang, C.-H., and D. L. Sedlak. 2001. Analysis of estrogenic hormones in municipal wastewater effluent and surface water using enzyme-linked immunosorbent assay and gas chromatography/tandem mass spectrometry. *Environmental Toxicology and Chemistry* 20(1): 133-139.

Huang, C. H., and D. Sedlak. 2000. Analysis of estrogenic hormones in wastewater and surface water using ELISA and GC/MS/MS. *Abstracts of Papers of the American Chemical Society* 219: 5-ENVR.

Huber, M. M., S. Canonica, G.-Y. Park, and U. von Gunten. 2003. Oxidation of pharmaceuticals during ozonation and advanced oxidation processes. *Environmental Science & Technology* 37(5): 1016-1024.

Huber, M. M., A. Göbel, A. Joss, N. Hermann, D. Löffler, C. S. McArdell, A. Reid, H. Siegrist, T. A. Ternes, and U. Von Gunten. 2005a. Oxidation of pharmaceuticals during ozonation of municipal wastewater effluents: a pilot study. *Environmental Science & Technology* 39(11): 4290-4299.

Huber, M. M., S. Korhonen, T. A. Ternes, and U. von Gunten. 2005b. Oxidation of pharmaceuticals during water treatment with chlorine dioxide. *Water Res.*, 39(15): 3607-3617.

Jobling, S., M. Noylan, C. R. Tyler, G. Brighty, and J. P. Sumpter. 1998. Widespread Sexual Disruption in Wild Fish. *Environmental Science & Technology* 32: 2498-2506.

Khim, J. S., K. Kannan, D. L. Villeneuve, C. H. Koh, and J. P. Giesy. 1999a. Characterization and Distribution of Trace Organic Contaminants in Sediment from Masan Bay, Korea. 1. Instrumental Analyses. *Environmental Science & Technology* 33(23): 4199-4205.

Khim, J. S., D. L. Villeneuve, K. Kannan, and J. P. Giesy. 1999b. Characterization and Distribution of Trace Organic Contaminants in Sediment from Masan Bay, Korea. 2. In Vitro Gene Expression Assays. *Environmental Science & Technology* 33(23): 4206-4211.

Khim, J. S., D. L. Villeneuve, K. Kannan, K. T. Lee, S. A. Snyder, C. H. Koh, and J. P. Giesy. 1999c. Alkylphenols, polycyclic aromatic hydrocarbons, and organochlorines in sediment from Lake Shihwa, Korea: Instrumental and bioanalytical characterization. *Environmental Toxicology and Chemistry* 18(11): 2424-2432.

Kimura, K., G. Amy, J. E. Drewes, T. Heberer, T.-U. Kim, and Y. Watanabe. 2003. Rejection of organic micropollutants (disinfection by-products, endocrine disrupting compounds, and pharmaceutically active compounds) by NF/RO membranes. *Journal of Membrane Science* 227(1-2): 113-121.

King, W. D., L. Dodds, and A. C. Allen. 2000. Relation between stillbirth and specific chlorination by-products in public water supplies. *Environmental Health Perspectives* 108(9): 883-886.

King, W. D., and L. D. Marrett. 1996. Case-control study of bladder cancer and chlorination by-products in treated water (Ontario, Canada). *Cancer Causes and Control* 7: 596-604.

King, W. D., L. D. Marrett, and C. G. Woolcott. 2000. Case-control study of colon and rectal cancers and chlorination by-products in treated waters. *Cancer Epidemiology, Biomarkers, & Prevention* 9: 813-818.

Koistinen, J., M. Soimasuo, K. Tukia, A. Oikari, A. Blankenship, and J. P. Giesy. 1998. Induction of EROD activity in HEPA-1 mouse hepatoma cells and estrogenicity in MCF-7 human breast cancer cells by extracts of pulp mill effluents, sludge, and sediment exposed to effluents. *Environmental Toxicology and Chemistry* 17(8): 1499-1507.

Kolodziej, E. P., J. L. Gray, and D. L. Sedlak. 2003. Quantification of steroid hormones with pheromonal properties in municipal wastewater effluent. *Environmental Toxicology and Chemistry* 22(11): 2622-2629.

Kolpin, D. W., E. T. Furlong, M. T. Meyer, E. M. Thurman, S. D. Zaugg, L. B. Barber, and H. T. Buxton. 2002. Pharmaceuticals, Hormones, and Other Organic Waste Contaminants in U.S. Streams, 1999-2000: A National Reconnaissance. *Environmental Science & Technology* 36(6): 1202-1211.

Kolpin, D. W., M. Skopec, M. T. Meyer, E. T. Furlong, and S. D. Zaugg. 2004. Urban contribution of pharmaceuticals and other organic wastewater contaminants to streams during differing flow conditions. *Science Of The Total Environment* 328(1-3): 119-130.

Koschorreck, J., C. Koch, and I. Ronnefahrt. 2002. Environmental risk assessment of veterinary medicinal products in the EU-a regulatory perspective. *Toxicology Letters* 131(1-2): 117-124.

Kotrikla, A., T. D. Lekkas, and G. Bletsa. 1997. Toxicity of the herbicide atrazine, two of its degradation products and the herbicide metolachlor on photosynthetic microorganisms. *Fresenius Environmental Bulletin*, 6(9-10): 502-507.

Kramer, V. J., S. Miles-Richardson, S. L. Pierens, and J. P. Giesy. 1998. Reproductive impairment and induction of alkaline-labile phosphate, a biomarker of estrogen exposure, in fathead minnows (*Pimephales promelas*) exposed to waterborne 17β-estradiol. *Aquatic Toxicology* 40: 335-360.

Krishnan, V., and S. Safe. 1993. Polychlorinated biphenyls (PCBs), dibenzo-*p*-dioxins (PCDDs), and dibenzofurans (PCDFs) as antiestrogens in MCF-7 human breast cancer cells: Quantitative structure-activity relationships. *Toxicology and Applied Pharmacology* 120: 55-61.

Kubeck, E., and C. G. Naylor. 1990. Trace Analysis of Alkylphenol Ethoxylates. *JAOCS* 67(6): 400-405.

Kuch, H. M., and K. Ballschmiter. 2001. Determination of endocrine-disrupting phenolic compounds and estrogens in surface and drinking water by HRGC-(NCI)-MS in the picogram per liter range. *Environmental Science & Technology* 35(15): 3201-3206.

La Farre, M., I. Ferrer, A. Ginebreda, M. Figueras, L. Olivella, L. Tirapu, M. Vilanova, and D. Barcelo. 2001. Determination of drugs in surface water and wastewater samples by liquid chromatography-mass spectrometry: methods and preliminary results including toxicity studies with Vibrio fischeri. *J. Chromatogr. A*, 938(1-2): 187-197.

Labadie, P., and H. Budzinski. 2005. Determination of Steroidal Hormone Profiles along the Jalle d'Eysines River (near Bordeaux, France). *Environmental Science and Technology*, 39: 5113-5120.

Lagana, A., A. Bacaloni, I. De Leva, A. Faberi, G. Fago, and A. Marino. 2004. Analytical methodologies for determining the occurrence of endocrine disrupting chemicals in sewage treatment plants and natural waters. *Analytica Chimica Acta* 501(1): 79-88.

Lange, R., and D. Dietrich. 2002. Environmental risk assessment of pharmaceutical drug substances-conceptual considerations. *Toxicology Letters* 131(1-2): 97-104.

Lattier, D. L., T. V. Reddy, D. A. Gordon, J. M. Lazorchak, M. E. Smith, D. E. Williams, B. Wiechman, R. W. Flick, A. L. Miracle, and G. P. Totii. 2002. 17.alpha.-ethynyl estradiol-induced vitellogenin gene transcription quantified in livers of adult males, larvae, and gills of fathead minnows (Pimephales promelas). *Environmental Toxicology and Chemistry* 21(11): 2385-2393.

Lee, H.-B., and T. E. Peart. 1998. Determination of 17.beta.-estradiol and its metabolites in sewage effluent by solid-phase extraction and gas chromatography/mass spectrometry. *Journal of AOAC INTERNATIONAL* 81(6): 1209-1216.

Levin, E., J. F. Burns, and V. K. Collins. 1951. Estrogenic, androgenic, and gonadotropic activity in wheat-germ oil. *Endocrinology* 49: 289-301.

Liber, K., J. A. Gangl, T. D. Corry, L. J. Heinis, and F. S. Stay. 1999. Lethality and Bioaccumulation of 4-Nonylphenol in Bluegill Sunfish in Littoral Enclosures. *Environmental Toxicology and Chemistry* 18(3): 394-400.

Liebig, M., J. F. Moltmann, and T. Knacker. 2006. Evaluation of measured and predicted environmental concentrations of selected human pharmaceuticals and personal care products. *Environmental Science and Pollution Research*, 13: 110-119.

Lin, W. C., H. C. Chen, and W. H. Ding. 2005. Determination of pharmaceutical residues in waters by solid-phase extraction and large-volume on-line derivatization with gas chromatography-mass spectrometry. *J. Chromatogr. A*, 1065: 279-285.

Lindqvist, N., T. Tuhkanen, and L. Kronberg. 2005. Occurrence of acidic pharmaceuticals in raw and treated sewages and in receiving waters. *Water Research* 39: 2219-2228.

Lindsey, M. E., M. Meyer, and E. M. Thurman. 2001. Analysis of trace levels of sulfonamide and tetracycline antimicrobials in groundwater and surface water using solid-phase extraction and liquid chromatography/mass spectrometry. *Analytical Chemistry* 73(19): 4640-4646.

Lindstrom, A., I. J. Buerge, T. Poiger, P. A. Bergqvist, M. D. Muller, and H. R. Buser. 2002. Occurrence and environmental behavior of the bactericide triclosan and its methyl derivative in surface waters and in wastewater. *Environmental Science & Technology* 36(11): 2322-2329.

Liu, R., J. L. Zhou, and A. Wilding. 2004. Simultaneous determination of endocrine disrupting phenolic compounds and steroids in water by solid-phase extraction-gas chromatography-mass spectrometry. *Journal of Chromatography A* 1022(1-2): 179-189.

Long, J. L. A., W. A. House, A. Parker, and J. E. Rae. 1998. Micro-organic compounds associated with sediments in the Humber rivers. *The Science of the Total Environment* 210/211: 229-253.

Lopez-Roldan, P., M. J. Lopez de Alda, and D. Barcelo. 2004. Simultaneous determination of selected endocrine disrupters (pesticides, phenols and phthalates) in water by in-field solid-phase extraction (SPE) using the prototype PROFEXS followed by on-line SPE (PROSPEKT) and analysis by liquid chromatography-atmospheric pressure chemical ionisation-mass spectrometry. *Analytical and Bioanalytical Chemistry* 378(3): 599-609.

Loraine, G. A., and M. E. Pettigrove. 2006. Seasonal Variations in Concentrations of Pharmaceuticals and Personal Care Products in Drinking Water and Reclaimed Wastewater in Southern California. *Enivronmental Science & Technology*, 40: 687-695.

Lye, C. M., C. L. J. Frid, M. E. Gill, D. W. Cooper, and D. M. Jones. 1999. Estrogenic Alklphenols in Fish Tissues, Sediments, and Waters from the U.K. Tyne and Tees Estuaries. *Environmental Science & Technology* 33(7): 1009-1014.

Marchese, S., A. Gentili, D. Perret, G. D'Ascenzo, and F. Pastori. 2003a. Quadrupole time-of-flight versus triple-quadrupole mass spectrometry for the determination of non-steroidal antiinflammatory drugs in surface water by liquid chromatography/tandem mass spectrometry. *Rapid Commun. Mass Spectrom.*, 17: 879-886.

Marchese, S., D. Perret, A. Gentili, R. Curini, and F. Pastori. 2003b. Determination of non-steroidal anti-inflammatory drugs in surface water and wastewater by liquid chromatography-tandem mass spectrometry. *Chromatographia*, 58(5-6): 263-269.

McConnell, L. L., J. A. Harman-Fetcho, and J. D. Hagy. 2004. Measured concentrations of herbicides and model predictions of atrazine fate in the Patuxent River estuary. *J. Environ. Qual.*, 33(2): 594-604.

McDowell, D. C., M. M. Huber, M. Wagner, U. Von Gunten, and T. A. Ternes. 2005. Ozonation of Carbamazepine in Drinking Water: Identification and Kinetic Study of Major Oxidation Products. *Environmental Science & Technology*, 39(20): 8014-8022.

Metcalfe, C., B. Koenig, T. Ternes, and R. Hirsch. 2000. Drugs in sewage treatment plant effluents in Canada. *Abstracts of Papers of the American Chemical Society*, 219: U623-U623.

Metzler, M., and E. Pfeiffer. 1995. Effects of estrogens on microtubule polymerization in vitro: correlation with estrogenicity. *Environmental Health Perspectives* 103(Suppl 7): 21-22.

Miles-Richardson, S. R., S. L. Pierens, K. M. Nichols, V. J. Kramer, E. M. Snyder, S. A. Snyder, J. A. Render, S. D. Fitzgerald, and J. P. Giesy. 1999. Effects of Waterborne Exposure to 4-Nonylphenol and Nonylphenol Ethoxylate on Secondary Sex Characteristics and Gonads of Fathead Minnows (*Pimephales promelas*). *Environmental Research Section A* 80: S122-S137.

Millington, A. J., C. M. Francis, and N. R. McKeown. 1964. Wether bioassay of annual pasture legumes. II. The estrogenic activity of nine strains of Trifolium subterraneum L. *Australian J. Agr. Res.* , 15: 527-536.

Mitjans, D., and F. Ventura. 2004. Determination of musks and other fragrance compounds at ng/L levels using CLSA (closed loop stripping analysis) and GC/MS detection. *Water Sci. Technol.*, 50(5): 119-123.

Mitjans, D., and F. Ventura. 2005. Determination of fragrances at ng/L levels using CLSA and GC/MS detection. *Water Sci. Technol.*, 52(10-11): 145-150.

Mottaleb, M. A., D. K. Sarma, S. Sultana, M. M. Husain, S. M. M. Alam, and S. M. Salehuddin. 2003. Determination of normal saturated- and polycyclic aromatic hydrocarbons in the river water of Bangladesh by liquid-liquid extraction and gas chromatography. *Bulletin of the Korean Chemical Society*, 24(1): 99-105.

Mueller, G. C., and U. H. Kim. 1978. Displacement of estradiol from estrogen receptors by simple alkylphenols. *Endocrinology* 102: 1429-1435.

Müller, S., M. Möder, S. Schrader, and P. Popp. 2003. Semi-automated hollow-fibre membrane extraction, a novel enrichment technique for the determination of biologically active compounds in water samples. *J. Chromatogr. A*, 985(1-2): 99-106.

Ng, H. Y., and M. Elimelech. 2004. Influence of colloidal fouling on rejection of trace organic contaminants by reverse osmosis. *J. Membr. Sci.*, 244(1-2): 215-226.

Nghiem, L., A. Schaefer, and M. Elimelech. 2005a. Nanofiltration of Hormone Mimicking Trace Organic Contaminants. *Separation Science and Technology Separation Science and Technology*, 40(13): 2633-2649.

Nghiem, L. D., J. McCutheon, A. I. Schafer, and M. Elimelech. 2004. The role of endocrine disrupters in water recycling: risk or mania? *Water Science and Technology* 50(2): 215-220.

Nghiem, L. D., A. I. Schaefer, and M. Elimelech. 2005b. Pharmaceutical Retention Mechanisms by Nanofiltration Membranes. *Environmental Science & Technology*, 39: 7698-7705.

Öllers, S., H. P. Singer, P. Fassler, and S. R. Muller. 2001. Simultaneous quantification of neutral and acidic pharmaceuticals and pesticides at the low-ng/1 level in surface and waste water. *J. Chromatogr. A*, 911(2): 225-234.

Ouellet, M., J. Bonin, J. Rodrigue, J. J.-L. Desgranges, and S. Lair. 1997. Hindlimb deformities (ectromelia, extrodctyly) in free-living anurans from agricultural habitats. *Journal of Wildlife Disease* 33: 95-104.

Ozaki, H., and H. Li. 2002. Rejection of organic compounds by ultra-low pressure reverse osmosis membrane. *Water Res.*, 36: 123-130.

Pablo, L. J., S.-P. Carmen, G.-J. Carmen, L. Maria, C. Rafael, and G. Mariano. 2004. Solid-phase microextraction-gas chromatography-mass spectrometry for the analysis of selective serotonin reuptake inhibitors in environmental water. *Journal of Chromatography A* 1046(1-2): 241-247.

Panter, G. H., R. S. Thompson, and J. P. Sumpter. 1998. Adverse reproductive effects in male fathead minnows (*Pimephales promelas*) exposed to environmentally relevant concentrations of the natural oestrogens, oestradiol and oestrone. *Aquatic Toxicology* 42: 243-253.

Park, G. Y., J. H. Lee, I. S. Kim, and J. Cho. 2004. Pharmaceutical rejection by membranes for wastewater reclamation and reuse. *Water Sci. Technol.*, 50(2): 239-244.

Peck, A. M., and K. C. Hornbuckle. 2004. Synthetic musk fragrances in Lake Michigan. *Environ. Sci. Technol.*, 38(2): 367-372.

Penalver, A., E. Pocurull, F. Borrull, and R. M. Marce. 2002. Method based on solid-phase microextraction-high-performance liquid chromatography with UV and electrochemical detection to determine estrogenic compounds in water samples. *Journal of Chromatography A* 964(1-2): 153-160.

Petty, J. D., Jones, S.B., Huckins, J.N., Cranor, W.L., Parris, J.T., McTague, T.B., Boyle, T.P. 2000. An Approach for Assessment of Water Quality Using Semipermeable Membrane Devices (SPMDs) and Bioindicator Tests. *Chemosphere* 41(2000): 311-321.

Pitarch, E., F. Hernandez, J. ten Hove, H. Meiring, W. Niesing, E. Dijkman, L. Stolker, and E. A. Hogendoorn. 2004. Potential of capillary-column-switching liquid chromatography-tandem mass spectrometry for the quantitative trace analysis of small molecules. Application to the on-line screening of drugs in water. *Journal of Chromatography A* 1031(1-2): 1-9.

Poiger, T., H.-R. Buser, M. D. Mueller, M. E. Balmer, and I. J. Buerge. 2003. Occurrence and fate of organic micropollutants in the environment: Regional mass balances and source apportioning in surface waters based on laboratory incubation studies in soil and water, monitoring, and computer modeling. *Chimia* 57(9): 492-498.

Pryor, S. W., A. G. Hay, and L. P. Walker. 2002. Nonylphenol in anaerobically digested sewage sludge from New York state. *Environmental Science & Technology* 36(17): 3678-3682.

Psillakis, E., A. Ntelekos, D. Mantzavinos, E. Nikolopoulos, and N. Kalogerakis. 2003. Solid-phase microextraction to monitor the sonochemical degradation of polycyclic aromatic hydrocarbons in water. *J. Environ. Monit.*, 5(1): 135-140.

Purdom, C. E., P. A. Hardiman, V. J. Bye, N. C. Eno, C. R. Tyler, and J. P. Sumpter. 1994. Estrogenic effects of effluents from sewage treatment works. *Chemistry and Ecology* 8: 275-285.

Putschew, A., S. Schittko, and M. Jekel. 2001. Quantification of triiodinated benzene derivatives and X-ray contrast media in water samples by liquid chromatography-electrospray tandem mass spectrometry. *J. Chromatogr. A*, 930(1-2): 127-134.

Quintana, J. B., J. Carpinteiro, I. Rodriguez, R. A. Lorenzo, A. M. Carro, and R. Cela. 2004a. Determination of natural and synthetic estrogens in water by gas chromatography with mass spectrometric detection. *Journal of Chromatography A* 1024(1-2): 177-185.

Quintana, J. B., J. Carpinteiro, I. Rodríguez, R. A. Lorenzo, A. M. Carro, and R. Cela. 2004b. Determination of natural and synthetic estrogens in water by gas chromatography with mass spectrometric detection. *J. Chromatogr. A*, 1024(1-2): 177-185.

Quintana, J. B., and T. Reemtsma. 2004. Sensitive determination of acidic drugs and triclosan in surface and wastewater by ion-pair reverse-phase liquid chromatography/tandem mass spectrometry. *Rapid Commun. Mass Spectrom.*, 18: 765-774.

Quintana, J. B., R. Rodil, and T. Reemtsma. 2004. Suitability of hollow fibre liquid-phase microextraction for the determination of acidic pharmaceuticals in wastewater by liquid chromatography-electrospray tandem mass spectrometry without matrix effects. *J. Chromatogr. A*, 1061: 19-26.

Rebich, R. A., R. H. Coupe, and E. M. Thurman. 2004. Herbicide concentrations in the Mississippi River Basin - the importance of chloroacetanilide herbicide degradates. *Sci. Total Environ.*, 321(1-3): 189-199.

Reddersen, K., and T. Heberer. 2003. Multi-compound methods for the detection of pharmaceutical residues in various waters applying solid phase extraction (SPE) and gas chromatography with mass spectrometric (GC-MS) detection. *J. Sep. Sci.*, 26: 1443-1450.

Rodriguez-Mozaz, S., M. J. Lopez de Alda, and D. Barcelo. 2004. Monitoring of estrogens, pesticides and bisphenol A in natural waters and drinking water treatment plants by solid-phase extraction-liquid chromatography-mass spectrometry. *Journal of Chromatography A* 1045(1-2): 85-92.

Rodríguez, I., J. Carpinteiro, J. B. Quintana, A. M. Carro, R. A. Lorenzo, and R. Cela. 2004. Solid-phase microextraction with on-fiber derivatization for the analysis of anti-inflammatory drugs in water samples. *J. Chromatogr. A*, 1024(1-2): 1-8.

Rodriguez, I., J. B. Quintana, J. Carpinteiro, A. M. Carro, R. A. Lorenzo, and R. Cela. 2003. Determination of acidic drugs in sewage water by gas chromatography-mass spectrometry as tert.-butyldimethylsilyl derivatives. *Journal of Chromatography A* 985(1-2): 265-274.

Romero, J., F. Ventura, and M. Gomez. 2002. Characterization of paint samples used in drinking water reservoirs: Identification of endocrine disruptor compounds. *Journal of Chromatographic Science* 40(4): 191-197.

Rosenfeldt, E., and K. G. Linden. 2004. Degradation of Endocrine Disrupting Chemicals Bisphenol A, Ethinyl Estradiol, and Estradiol during UV Photolysis and Advanced Oxidation Processes. *Environmental Science & Technology*, 38(20): 5476-5483.

Routledge, E. J., D. Sheahan, C. Desbrow, G. C. Brighty, M. Waldock, and J. P. Sumpter. 1998. Identification of estrogenic chemicals in STW effluent. 2. In vivo responses in trout and roach. *Environmental Toxicology and Chemistry* 32(11): 1559-1565.

Routledge, E. J., and J. P. Sumpter. 1997. Structural features of alkylphenolic chemicals associated with estrogenic activity. *Journal of Biological Chemistry* 272(6): 3280-3288.

Rumsey, T. S., and A. C. Hammond. 1990. Effects of intake level on metabolic response to estrogenic growth promoters in beef steers. *Journal of Animal Science* 68(12): 4310-4318.

Saal, F. S. V., C. A. Richter, R. R. Ruhlen, S. C. Nagel, B. G. Timms, and W. V. Welshons. 2005. The importance of appropriate controls, animal feed, and animal models in interpreting results from low-dose studies of bisphenol A. *Birth Defects Research, Part A: Clinical and Molecular Teratology* 73(3): 140-145.

Sabaliunas, D., S. F. Webb, A. Hauk, M. Jacob, and W. S. Eckhoff. 2003. Environmental fate of Triclosan in the River Aire Basin, UK. *Water Res.*, 37(13): 3145-3154.

Sabik, H., B. Rondeau, P. Gagnon, R. Jeannot, and K. Dohrendorf. 2003. Simultaneous filtration and solid-phase extraction combined with large-volume injection in GC/MS for ultra-trace analysis of polar pesticides in surface water. *International Journal of Environmental Analytical Chemistry*, 83(6): 457-468.

Sacher, F., F. T. Lange, H.-J. Brauch, and I. Blankenhorn. 2001. Pharmaceuticals in groundwaters: Analytical methods and results of a monitoring program in Baden-Wuerttemberg, Germany

10th Symposium on Handling of Environmental and Biological Samples in Chromatography, Mainz, Germany, 1-4 April 2001. *Journal of Chromatography A*, 938 199-210.

Safe, S. H., and K. Gaido. 1998. Phytoestrogens and anthropogenic estrogenic compounds. *Environmental Toxicology and Chemistry* 17(1): 119-126.

Schafer, A. I., L. D. Nghiem, and T. D. Waite. 2003. Removal of the natural hormone estrone from aqueous solutions using nanofiltration and reverse osmosis. *Environmental Science & Technology* 37(1): 182-188.

Scheyer, A., S. Morville, P. Mirabel, and M. Millet. 2006. Analysis of trace levels of pesticides in rainwater using SPME and GC-tandem mass spectrometry. *Anal. Bioanal. Chem.*, 384(2): 475-487.

Seiler, R. L., S. D. Zaugg, J. M. Thomas, and D. L. Howcroft. 1999. Caffeine and Pharmaceuticals as Indicators of Waste Water Contamination in Wells. *Ground Water* 37(3): 405-410.

Setchell, K. D. R., S. J. Gosselin, M. B. Welsh, J. O. Johnston, W. F. Balistreri, L. W. Kramer, B. L. Dresser, and M. J. Tarr. 1987. Dietary estrogens - a probable cause of infertility and liver disease in captive cheetahs. *Gastroenterology* 93(2): 225-233.

Sharpe, R. M., and N. E. Skakkebaek. 1993. Are oestrogens involved in falling sperm counts and disorders of the male reproductive tract? *Lancet* 341: 1392-1395.

Shaw, I., and S. McCully. 2002. A review of the potential impact of dietary endocrine disrupters on the consumer. *International Journal of Food Science and Technology* 37(5): 471-476.

Shomar, B. H., G. Muller, and A. Yahya. 2006. Occurrence of pesticides in groundwater and topsoil of the Gaza Strip. *Water Air and Soil Pollution*, 171(1-4): 237-251.

Shore, L. S., M. Gurevitz, and M. Shemesh. 1993. Estrogen as an Environmental Pollutant. *Bulletin of Environmental Contamination and Toxicology* 51: 361-366.

Singer, H., S. Mueller, C. Tixier, and L. Pillonel. 2002. Triclosan: Occurrence and Fate of a Widely Used Biocide in the Aquatic Environment: Field Measurements in Wastewater Treatment Plants, Surface Waters, and Lake Sediments. *Environmental Science & Technology* 36(23): 4998-5004.

Snyder, E. M., S. A. Snyder, K. L. Kelly, T. S. Gross, D. L. Villeneuve, S. D. Fitzgerald, S. A. Villalobos, and J. P. Giesy. 2004. Reproductive responses of common carp (*Cyprinus carpio*) exposed in cages to influent of the Las Vegas Wash in Lake Mead, Nevada from late winter to early spring. *Environmental Science & Technology* 38(23): 6385-6395.

Snyder, S. A., S. Adham, A. M. Redding, F. S. Cannon, J. DeCarolis, J. Oppenheimer, E. C. Wert, and Y. Yoon. 2006a. Role of Membranes and Activated Carbon in the Removal of Endocrine Disruptors and Pharmaceuticals. *Desalination*, 202: 156-181.

Snyder, S. A., T. L. Keith, C. G. Naylor, C. A. Staples, and J. P. Giesy. 2001a. Identification and quantification method for nonylphenol and lower oligomer nonylphenol ethoxylates in fish tissues. *Environmental Toxicology and Chemistry* 20(9): 1870-1873.

Snyder, S. A., T. L. Keith, S. L. Pierens, E. M. Snyder, and J. P. Giesy. 2001b. Bioconcentration of nonylphenol in fathead minnows (*Pimephalas promelas*). *Chemosphere* 44(8): 1697-1702.

Snyder, S. A., T. L. Keith, D. A. Verbrugge, E. M. Snyder, T. S. Gross, K. Kannan, and J. P. Giesy. 1999. Analytical methods for detection of selected estrogenic compounds in aqueous mixtures. *Environmental Science & Technology* 33(16): 2814-2820.

Snyder, S. A., K. L. Kelly, A. H. Grange, G. W. Sovocool, E. M. Snyder, and J. P. Giesy. 2001c. Pharmaceuticals and personal care products in the waters of Lake Mead, Nevada. In

Pharmaceuticals and Personal Care Products in the Environment: Scientific and Regulatory Issues. Edited by C. G. Daughton and T. L. Jones-Lepp. Washington,D.C.: American Chemical Society.

Snyder, S. A., B. J. Vanderford, R. A. Pearson, O. Quinones, and Y. Yoon. 2003a. Analytical methods used to measure endocrine disrupting compounds in water. *Practice Periodical of Hazardous,Toxic, and Radioactive Waste Management* 7(4): 224-234.

Snyder, S. A., D. L. Villeneuve, E. M. Snyder, and J. P. Giesy. 2001d. Identification and quantification of estrogen receptor agonists in wastewater effluents. *Environmental Science & Technology* 35(18): 3620-3625.

Snyder, S. A., E. C. Wert, D. J. Rexing, R. E. Zegers, and D. D. Drury. 2006b. Ozone oxidation of endocrine disruptors and pharmaceuticals in surface water and wastewater. *ozone Science & Engineering*, 28(6): 445-460.

Snyder, S. A., P. Westerhoff, Y. Yoon, and D. L. Sedlak. 2003b. Pharmaceuticals, Personal Care Products, and Endocrine Disruptors in Water: Implications for the Water Industry. *Environmental Engineering Science* 20(5): 449-469.

Soliman, M. A., J. A. Pedersen, and I. H. Suffet. 2004. Rapid gas chromatography-mass spectrometry screening method for human pharmaceuticals, hormones, antioxidants and plasticizers in water. *Journal of Chromatography A* 1029(1-2): 223-237.

Sørensen, L. K., and T. H. Elbæk. 2004. Simultaneous determination of trimethoprim, sulfadiazine, florfenicol and oxolinic acid in surface water by liquid chromatography tandem mass spectrometry. *Chromatographia*, 60: 287-291.

Sower, S. A., C. B. Schreck, and M. Evenson. 1983. Effects of steroids and steroid antagonists on growth, gonadal development, and RNA/DNA ratios in juvenile steelhead trout. *Aquaculture* 32(3-4): 243-254.

Sparling, D. W. 2000. Effects of altosid and abate-4e on deformities and survival in southern leopard frogs under semi-natural conditions. *Journal of Iowa Academic Science* 107(3): 90-91.

Speth, T. F., and R. J. Miltner. 1998. Adsorption capacity of GAC for synthetic organics - part II. *Journal American Water Works Association* 90(4): 171-174.

Stackelberg, P. E., E. T. Furlong, M. T. Meyer, S. D. Zaugg, A. K. Henderson, and D. B. Reissman. 2004. Persistence of pharmaceutical compounds and other organic wastewater contaminants in a conventional drinking-water-treatment plant. *Science Of The Total Environment* 329(1-3): 99-113.

Stahlschmidt-Allner, P., B. Allner, J. Rombke, and T. Knacker. 1997. Endocrine Disrupters in the Aquatic Environment. *Environmental Science and Pollution Research 1* 4(3): 155-162.

Stolker, A. A. M., W. Niesing, E. A. Hogendoorn, J. F. M. Versteegh, R. Fuchs, and U. A. T. Brinkman. 2004. Liquid chromatography with triple-quadrupole or quadrupole-time of flight mass spectrometry for screening and confirmation of residues of pharmaceuticals in water. *Anal. Bioanal. Chem.*, 378(4): 955-963.

Stone, R. 1994. Environmental Estrogens Stir Debate. *Science* 265: 308-310.

Stumm-Zollinger, E., and G. M. Fair. 1965. Biodegradation of steroid hormones. *Journal of the Water Pollution Control Federation* 37: 1506-1510.

Stumpf, M., T. A. Ternes, W. Rolf-Dieter, S. V. Rodrigues, and W. Baumann. 1999. Polar drug residues in sewage and natural waters in the state of Rio de Janeiro, Brazil. *The Science of the Total Environment* 225: 135-141.

Tabak, H. H., R. N. Bloomhuff, and R. L. Bunch. 1981. Steroid hormones as water pollutants II. Studies on the persistence and stability of natural urinary and synthetic ovulation-inhibiting hormones in untreated and treated wastewaters. *Dev. Ind. Microbiol.* , 22: 497-519.

Tabak, H. H., and R. L. Bunch. 1970. Steroid hormones as water pollutants. I. Metabolism of natural and synthetic ovulation-inhibiting hormones by microorganisms of activated sludge and primary settled sewage. *Dev. Ind. Microbiol.* , 11: 367-376.

Ternes, T., M. Bonerz, and T. Schmidt. 2001. Determination of neutral pharmaceuticals in wastewater and rivers by liquid chromatography-electrospray tandem mass spectrometry. *Journal of Chromatography a* 938(1-2): 175-185.

Ternes, T. A. 1998. Occurrence of drugs in German sewage treatment plants and rivers. *Water Research* 32(11): 3245-3260.

Ternes, T. A. 2001. Analytical methods for the determination of pharmaceuticals in aqueous environmental samples. *Trends in Analytical Chemistry* 20(8): 419-434.

Ternes, T. A., R. Hirsch, J. Mueller, and K. Haberer. 1998. Methods for the determination of neutral drugs as well as betablockers and β_2-sympathomimetics in aqueous matrices using GC/MS and LC/MS/MS. *Fresenius' Journal of Analytical Chemistry* 362: 329-340.

Ternes, T. A., M. Meisenheimer, D. Mcdowell, F. Sacher, H.-J. Brauch, B. Haist-Gulde, G. Preuss, U. wilme, and N. Zulei-Seibert. 2002a. Removal of Pharamaceuticals during drinking water treatment. *Environmental Science & Technology* 36(17): 3855-3863.

Ternes, T. A., M. Meisenheimer, D. McDowell, F. Sacher, H. J. Brauch, B. H. Gulde, G. Preuss, U. Wilme, and N. Z. Seibert. 2002b. Removal of pharmaceuticals during drinking water treatment. *Environ. Sci. Technol.*, 36(17): 3855-3863.

Ternes, T. A., J. Stüber, N. Herrmann, D. McDowell, A. Ried, M. Kampmann, and B. Teiser. 2003. Ozonation: a tool for removal of pharmaceuticals, contrast media and musk fragrances from wastewater? *Water Research* 37(8): 1976-1982.

Thomas, P. M., and G. D. Foster. 2004. Determination of nonsteroidal anti-inflammatory drugs, caffeine, and triclosan in wastewater by gas chromatography-mass spectrometry. *J. Environ. Sci. Heal. A*, 39(8): 1969-1978.

Tixier, C., H. P. Singer, S. Canonica, and S. R. Mueller. 2002. Phototransformation of Triclosan in Surface Waters: A Relevant Elimination Process for This Widely Used Biocide-Laboratory Studies, Field Measurements, and Modeling. *Environmental Science & Technology* 36(16): 3482-3489.

Tixier, C., H. P. Singer, S. Oellers, and S. R. Mueller. 2003. Occurrence and Fate of Carbamazepine, Clofibric Acid, Diclofenac, Ibuprofen, Ketoprofen, and Naproxen in Surface Waters. *Environmental Science & Technology* 37(6): 1061-1068.

Tremblay, L., and G. Van Der Kraak. 1999. Comparison between the effects of the phytosterol .beta.-sitosterol and pulp and paper mill effluents on sexually immature rainbow trout. *Environmental Toxicology and Chemistry* 18(2): 329-336.

Trenholm, R. A., B. J. Vanderford, J. C. Holady, D. J. Rexing, and S. A. Snyder. 2006. Broad Range Analysis of Endocrine Disruptors and Pharmaceuticals Using Gas Chromatography and Liquid Chromatography Tandem Mass Spectroscopy. *Chemosphere* 65: 1990-1998.

Trimble, A. J., and M. J. Lydy. 2006. Effects of triazine herbicides on organophosphate insecticide toxicity in Hyalella azteca. *Archives of Environmental Contamination and Toxicology*, 51(1): 29-34.

U.S. Environmental Protection Agency. 2006. 40 CFR Parts 9, 141 and 142. National Primary Drinking Water Regulations: Long Term 2 Surface Water Treatment Rule (LT2 rule); Final Rule. *Fed. Reg. (Part II)*, 71(3): 653-702.

van Aerle, R., N. Pounds, T. H. Hutchinson, S. Maddix, and C. R. Tyler. 2002. Window of Sensitivity for the Estrogenic Effects of Ethinylestradiol in Early Life-Stages of Fathead Minnow, Pimephales promelas. *Ecotoxicology* 11(6): 423-434.

van der Ven, K., W. V. Dongen, B. U. W. Maes, E. L. Esmans, R. Blust, and W. M. D. Coen. 2004. Determination of diazepam in aquatic samples by capillary liquid chromatography-electrospray tandem mass spectrometry. *Chemosphere*, 57: 967-973.

Vanderford, B. J., R. A. Pearson, D. J. Rexing, and S. A. Snyder. 2003. Analysis of Endocrine Disruptors, Pharmaceuticals, and Personal Care Products in Water Using Liquid Chromatography/Tandem Mass Spectrometry. *Analytical Chemistry* 75(22): 6265-6274.

Vieno, N. M., T. Tuhkanen, and L. Kronberg. 2005. Seasonal Variation in the Occurrence of Pharmaceuticals in Effluents from a Sewage Treatment Plant and in the Recipient Water. *Environmental Science & Technology Environmental Science & Technology*, 39(21): 8220-8226.

Villegas-Navarro, A., E. Rosas-L, and J. L. Reyes. 2003. The heart of Daphnia magna: effects of four cardioactive drugs. *Comparative Biochemistry And Physiology C-Toxicology & Pharmacology* 136(2): 127-134.

Vogna, D., R. Marotta, R. Andreozzi, A. Napolitano, and M. d'Ischia. 2004. Kinetic and chemical assessment of the UV/H2O2 treatment of antiepileptic drug carbamazepine. *Chemosphere* 54(4): 497-505.

Walker, B. S., and J. C. Janney. 1930. Estrogenic substances. II. Analysis of plant sources. *Endocrinology* 14: 389-392.

Wang, G., F. Wu, X. Zhang, M. Luo, and N. Deng. 2006. Enhanced TiO_2 photocatalytic degradation of bisphenol A b B-cyclodextrin in suspended solutions. *Journal of Photochemistry and Photobiology A: Chemistry*, 179: 49-56.

Weber, S., M. Gallenkemper, T. Melin, W. Dott, and J. Hollender. 2004. Efficiency of nanofiltration for the elimination of steroids from water. *Water Sci. Technol.*, 50(5): 9-14.

Weigel, S., R. Kallenborn, and H. Huhnerfuss. 2004. Simultaneous solid-phase extraction of acidic, neutral and basic pharmaceuticals from aqueous samples at ambient (neutral) pH and their determination by gas chromatography-mass spectrometry. *Journal of Chromatography A* 1023(2): 183-195.

Weigel, S., J. Kuhlmann, and H. Huhnerfuss. 2002. Drugs and personal care products as ubiquitous pollutants: occurrence and distribution of clofibric acid, caffeine and DEET in the North Sea. *Sci. Total Environ.*, 295(1-3): 131-141.

Welch, R. M., W. Levin, and A. H. Conney. 1969. Estrogenic action of DDT and its analogs. *Toxicology and Applied Pharmacology* 14(2): 358-367.

Westerhoff, P., Y. Yoon, S. Snyder, and E. Wert. 2005. Fate of endocrine-disruptor, pharmaceutical, and personal care product chemicals during simulated drinking water treatment processes *Environmental Science & Technology* 39(17): 6649-6663.

White, R., S. Jobling, S. A. Hoare, J. P. Sumpter, and M. G. Parker. 1994. Environmentally Persistent Alkylphenolic Compounds Are Estrogenic. *Endocrinology* 135(1): 175-182.

Wrenn, T. R., J. R. Wood, G. F. Fries, and J. Bitman. 1970. Tests of Estrogenicity in Rats Fed Low Levels of o,p'DDT. *Bulletin of Environmental Contamination and Toxicology* 5(1): 61-66.

Xu, P., J. E. Drewes, C. Bellona, G. Amy, T. U. Kim, M. Adam, and T. Heberer. 2005. Rejection of emerging organic micropollutants in nanofiltration-reverse osmosis membrane applications. *Water Environment Research*, 77(1): 40-48.

Yang, L., T. Luan, and C. Lan. 2006. Solid-phase microextraction with on-fiber silylation for simultaneous determinations of endocrine disrupting chemicals and steroid hormones by gas chromatography-mass spectrometry. *Journal of Chromatography A*, 1104: 23-32.

Yang, S., and K. H. Carlson. 2004. Solid-phase extraction-high-performance liquid chromatography-ion trap mass spectrometry for analysis of trace concentrations of macrolide antibiotics in natural and waste water matrices. *J. Chromatogr. A*, 1038(1-2): 141-155.

Yang, S., J. Cha, and K. Carlson. 2004. Quantitative determination of trace concentrations of tetracycline and sulfonamide antibiotics in surface water using solid-phase extraction and liquid chromatography/ion trap tandem mass spectrometry. *Rapid Commun. Mass Spectrom.*, 18(18): 2131-2145.

Yook, K.-S., S.-M. Hong, and J.-H. Kim. 1994. Comparison of liquid-liquid extraction and solid-phase extraction coupled with GC/MS for determination of priority pollutants in water. *Analytical Science Technology* 7(4): 441-453.

Yoon, Y., P. Westerhoff, S. A. Snyder, and M. Esparza. 2003. HPLC-fluorescence detection and adsorption of bisphenol A, 17β-estradiol, and 17α-ethynyl estradiol on powdered activated carbon. *Water Research* 37(14): 3530-3537.

Yoon, Y., P. Westerhoff, S. A. Snyder, and E. C. Wert. 2006. Nanofiltration and ultrafiltration of endocrine disrupting compounds,pharmaceuticals and personal care products. *Journal of Membrane Science*, 270: 88-100.

Zhang, Y., and J. L. Zhou. 2005. Removal of estrone and 17β-estradiol from water by adsorption. *Water Research Water Research*, 39(39): 3991-4003.

Zhang, Z. L., J. Huang, G. Yu, and H. S. Hong. 2004. Occurrence of PAHs, PCBs and organochlorine pesticides in the Tonghui River of Beijing, China. *Environmental Pollution*, 130(2): 249-261.

Zuccato, E., D. Calamari, M. Natangelo, and R. Fanelli. 2000. Presence of therapeutic drugs in the environment. *Lancet*, 355: 1789-1790.

Zuehlke, S., U. Duennbier, T. Heberer, and B. Fritz. 2004. Analysis of endocrine disrupting steroids: investigation of their release into the environment and their behavior during back filtration. *Ground Water Monit. R.*, 24: 78-85.

Zwiener, C., and F. H. Frimmel. 2000. Oxidative treatment of pharmaceuticals in water. *Water Research* 34(6): 1881-1885.

Zwiener, C., T. Glauner, and F. H. Frimmel. 2000. Biodegradation of pharmaceutical residue investigated by SPE-GC/ITD-MS and on-line derivatization. *J. High Res. Chromatog.*, 23(7/8): 474-478.

CHAPTER 2
PROJECT DESCRIPTION

OBJECTIVES

The primary goal of this study was to determine the degree to which conventional and advanced water treatment processes were capable of removing a diverse group of organic EDCs and PPCPs from waters of various quality. In order to achieve the primary goal, the project addressed the following objectives:

- Conduct literature reviews regarding EDC/PPCP analysis, occurrence, and treatment
- Select a list of target compounds with a range of physical-chemical properties, usages, and environmental occurrence
- Develop robust and sensitive analytical techniques to identify and quantify the target analytes in a variety of aqueous matrices
- Conduct bench- and pilot-scale treatment studies using a variety of conventional and water treatment processes
- Determine the occurrence and removal of target analytes in full-scale drinking water and water reuse facilities
- Evaluate models that estimate EDC and PPCP physical-chemical properties and removal mechanisms
- Develop recommendations for utilities regarding EDC and PPCP treatment and future research needs

RESEARCH APPROACH

The project consisted of two primary phases. The first phase included the generation of a target suite of EDCs and PPCPs representative of the types of chemicals known or likely to occur in source waters and of a broad range of physical-chemical structure/behavior. This phase included the development of analytical methodology, including sample collection, preservation, extraction, and measurement techniques. This initial phase was critical to the entire project, as the compounds and methods selected here were common throughout the entire investigation. Analytical methods developed were required to perform accurately and precisely in a variety of water matrices, from finished drinking water to raw wastewater. Additionally, methods required high sensitivity in order to detect ng/L levels of target compounds, both for occurrence data and to observe 2-log removals when spiking compounds in the range of environmental relevance.

The second phase of the project involved testing of physical, chemical, and biological treatment processes to determine removal potential of the characteristic EDCs and PPCPs. Process testing was completed at bench-scale, pilot-scale, and full-scale. During most bench and pilot tests, target compounds were spiked into natural waters collected from full-scale water treatment plants across the United States. Chemical dosages and contact times were selected to simulate conditions commonly used during drinking water treatment. Additional detail regarding the treatment processes and operating conditions are presented in Chapter 3. The fate of EDCs and PPCPs at full-scale facilities was conducted using ambient EDC and PPCP levels (no spiking). A flow chart for the sequence of task activities is shown in Figure 2.1.

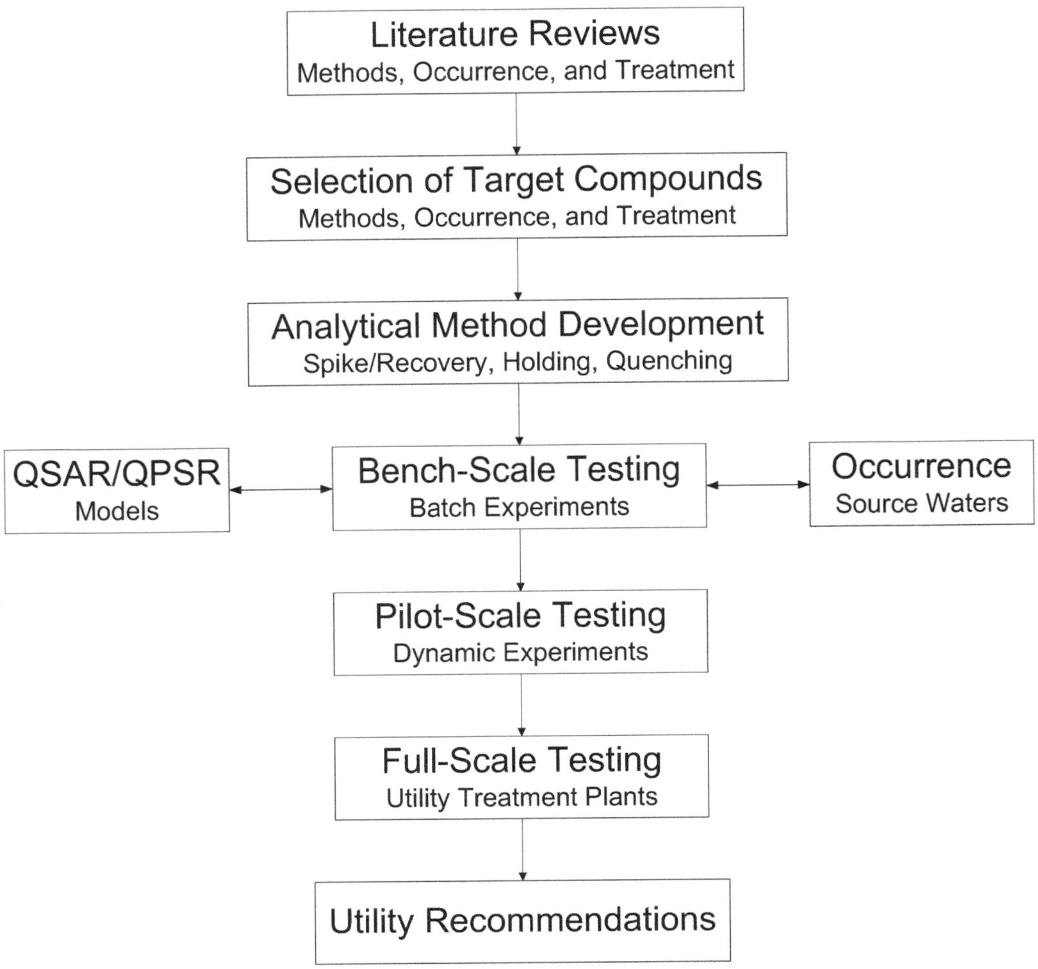

Figure 2.1 Flow chart depicting research approach and primary tasks

SELECTION OF TARGET COMPOUNDS

Target EDCs and PPCPs were selected based on three primary criteria. First, compounds were selected which had been detected in natural waters or had a high likelihood for occurrence. Second, the compounds selected represented a broad diversity of structural and chemical properties (e.g., size, aliphatic, aromatic, acid, base, neutral) as well as a broad range of classifications (e.g., antibiotics, fragrances, steroids). This diversity of chemical structure is extremely important, as it allows predictions of removal for other contaminants based upon molecular structural features. Third, the target compounds had to be amenable to analysis using available equipment, namely GC-MS/MS and LC-MS/MS, with commercially available reference standards. Table 2.1 provides the final target compound list, while Appendix A provides information on additional chemicals that had been considered.

Table 2.1
Target compounds for treatment studies[1]

Compounds	MW – pKa – log K_{ow} – Mol. Volume ($Å^2$) Structural features	Structure
Steroids		
Androstenedione	286.4 – na – 2.75 – 967.7 aliphatic rings, ketones	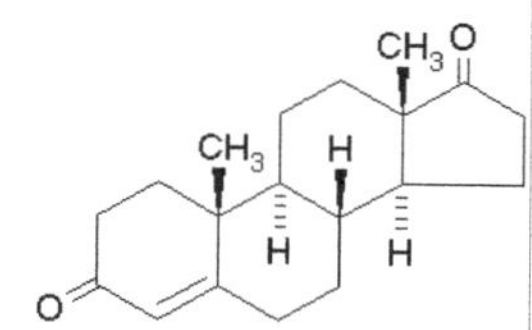
17α-Ethynyl estradiol (oral birth control)	296.4 – 10.4[2] – 3.67 – 982.8 phenol, ethynyl, aliphatic rings, alcohol	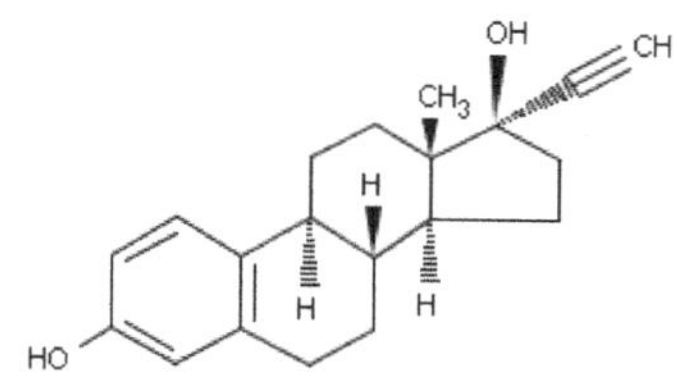
17β-Estradiol	272.2 – 10.4[2] – 4.01 – 907.2 phenol, alcohol, aliphatic rings	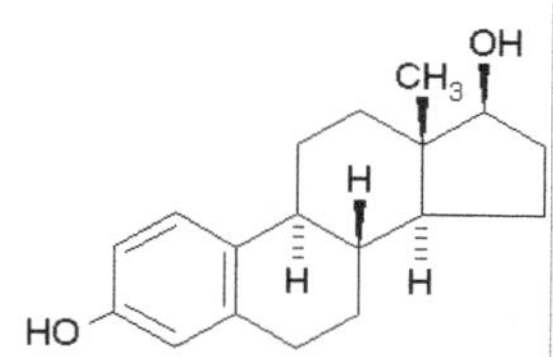
Estriol	288.4 – 10.4[2] – 2.45 – 930.3 phenol, alcohols, aliphatic rings	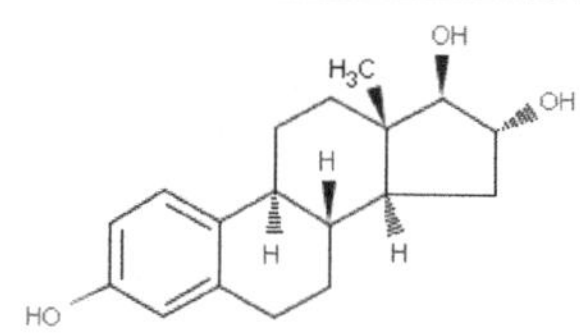
Estrone	270.4 – 10.4[2] – 3.13 – 903.4 phenol, ketone, aliphatic rings	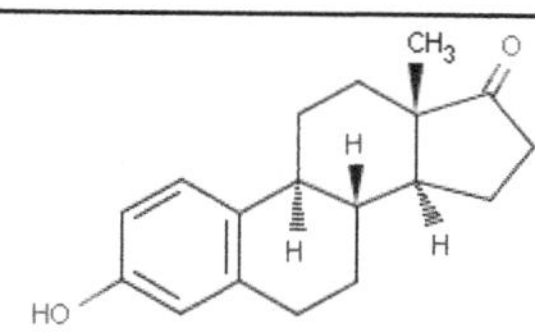
Progesterone	314.5 – na – 3.87 – 1032.4 ketones, aliphatic rings	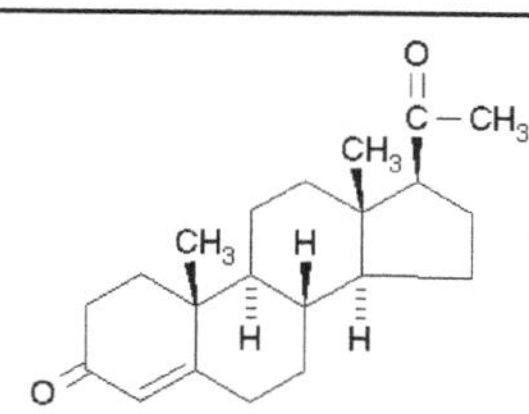
Testosterone	288.4 – na – 3.32 – 960.4 ketone, alcohol, aliphatic rings	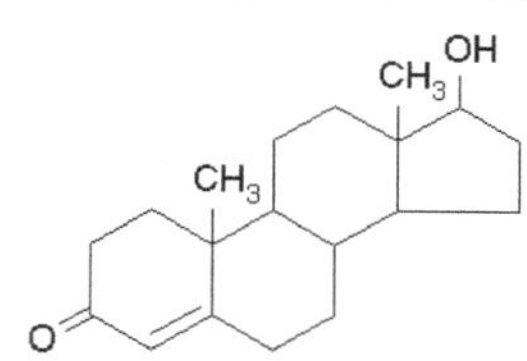

(continued)

Table 2.1 (Continued)

Compounds	MW – pKa – log K_{ow} – Mol. Volume ($Å^2$) Structural features	Structure
Antimicrobials		
Erythromycin (Analyte: Erythromycin-H_2O)	733.5 – 8.88 - 3.06 – 2016.1 complex aliphatic structure, ketone, alcohols, ester, ethers, tertiary amine, heterocyclic rings (oxygen)	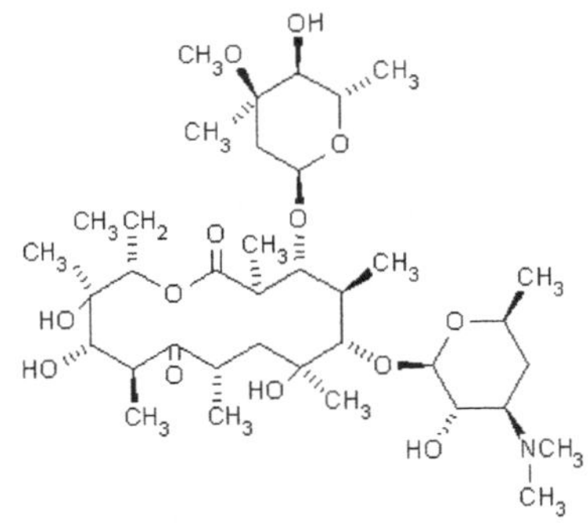
Sulfamethoxazole	253.3 – 5.5[3] – 0.89 – 769.4 sulfone, primary amine, secondary amine, aromatic ring, isoxazole ring	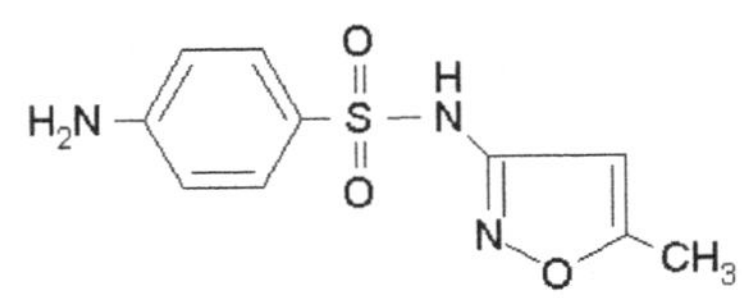
Triclosan	287.9 –7.9[4] – 4.76 – 782.4 aromatic rings, ether, phenol, chlorines	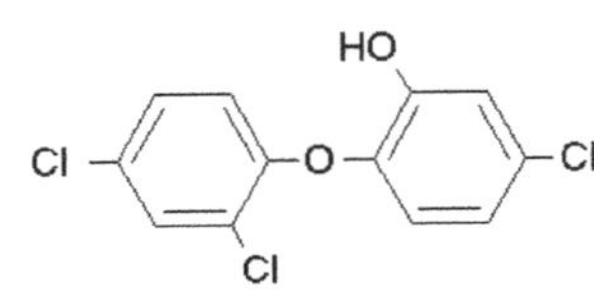
Trimethoprim	290.1 – 7.12 – 0.91 – 922.4 aromatic ring, pyrimidine ring, primary amines, methoxy groups	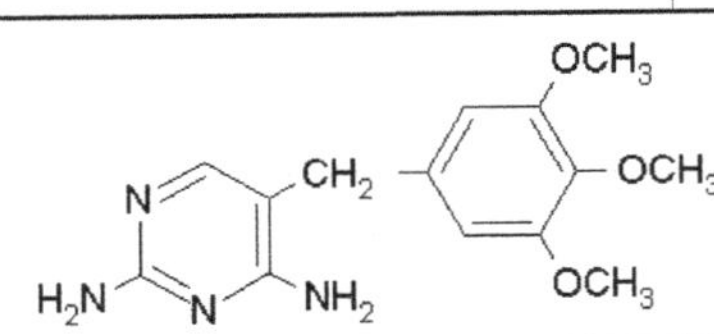
Analgesics		
Acetaminophen (Tylenol)	151.2 – 9.38 – 0.46 – 558.6 phenol, amide	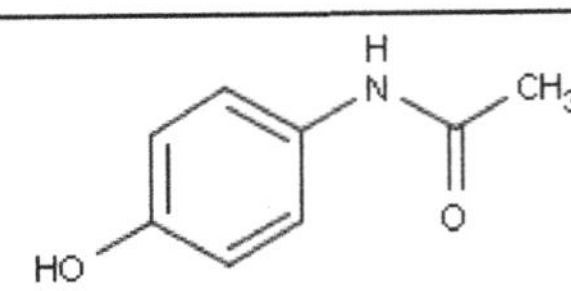
Diclofenac	294.0 – 4.15 – 0.7 – 829.7 carboxylic acid, aromatic rings, secondary amine, chlorines	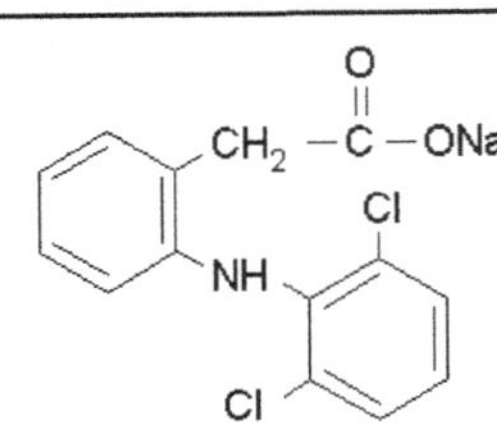

(continued)

Table 2.1 (Continued)

Compounds	MW – pKa – log K_{ow} – Mol. Volume (Å2) Structural features	Structure
Hydrocodone	299.4 – 7.32 (est) – 2.16 – 910.7 aromatic ring, ether, ketone, aliphatic ring, heterocyclic rings (oxygen and tertiary nitrogen)	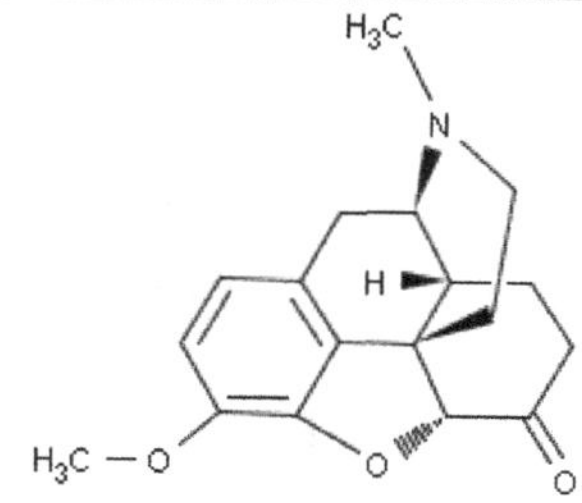
Ibuprofen (Advil)	206.3 – 4.91 – 3.97 – 796.6 aromatic ring, carboxylic acid	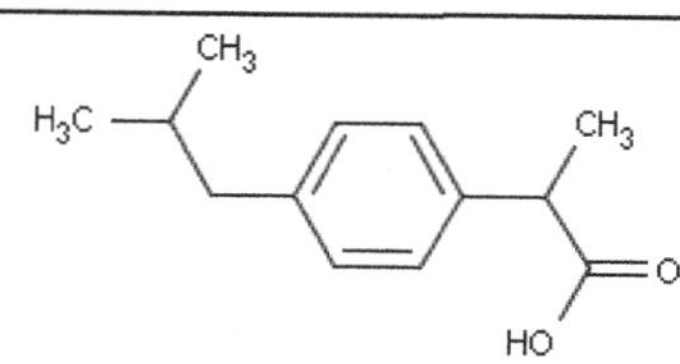
Naproxen	230.1 – 4.15 – 3.18 – 784.7 aromatic rings, carboxylic acid, ether	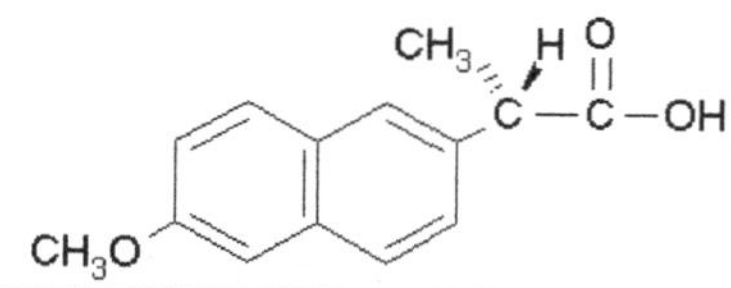
Psychoactive		
Caffeine	194.1 – 10.4 – (-0.07) – 646.3 xanthine ring	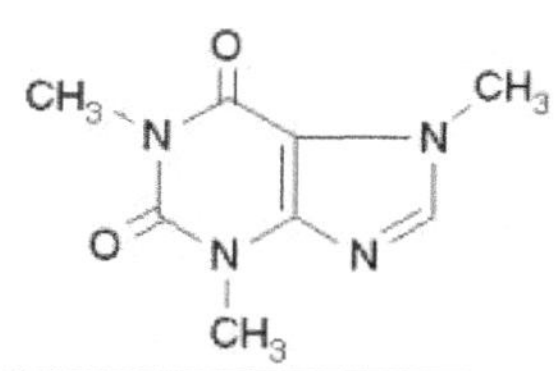
Carbamazepine	236.1 – 13.9[5] – 2.45 – 774.6 aromatic rings, heterocyclic ring (tertiary nitrogen), amide	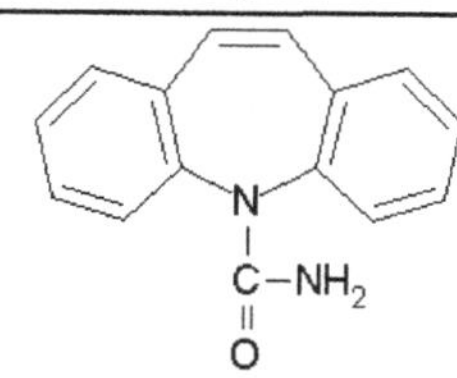
Diazepam (Valium)	284.1 – 3.4 – 2.82 – 888.4 aromatics rings, heterocyclic ring (tertiary nitrogens), cyclic amide	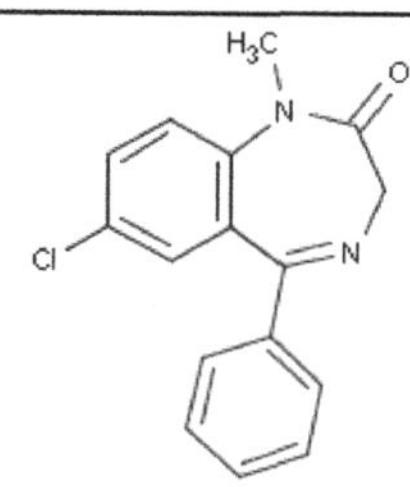

(continued)

Table 2.1 (Continued)

Compounds	MW – pKa – log K_{ow} – Mol. Volume ($Å^2$) Structural features	Structure
Dilantin	252.3 – 8.33 – 2.47 – 781.6 aromatic rings, cyclic amides	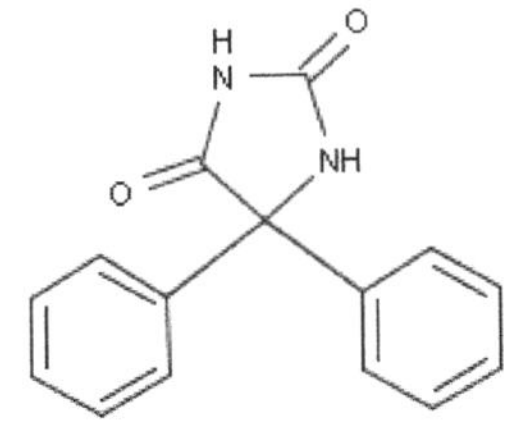
Fluoxetine (Prozac)	309.1– 10.30 (est) – 4.05 – 1007.3 aromatic rings, ether, secondary amine, fluorines	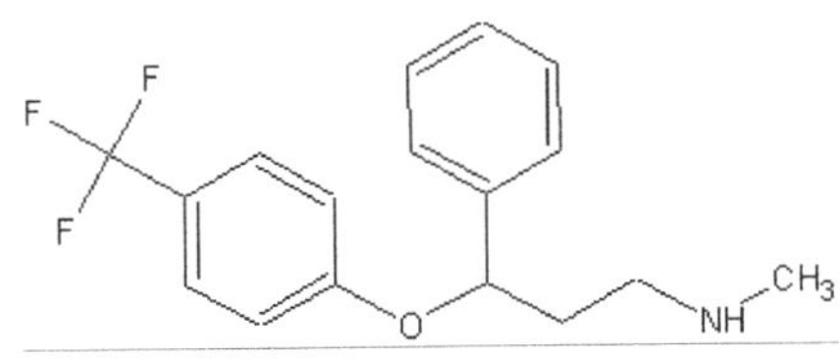
Meprobamate	218.2 – 10.9 (est) – 0.7 – 781.0 aliphatic, carbamates	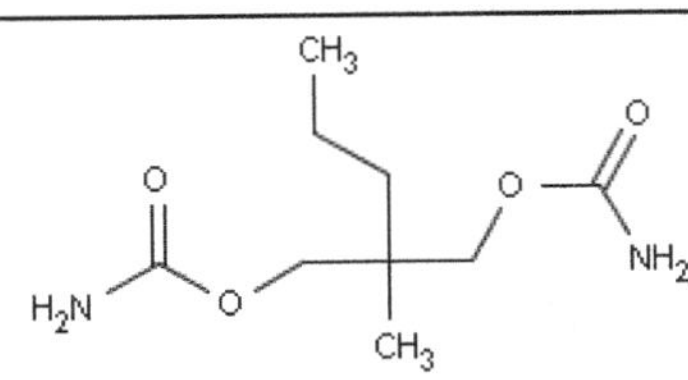
Pesticides		
Atrazine	215.7 – 1.7 – 2.61 – 760.9 triazine ring, secondary amines, chlorine	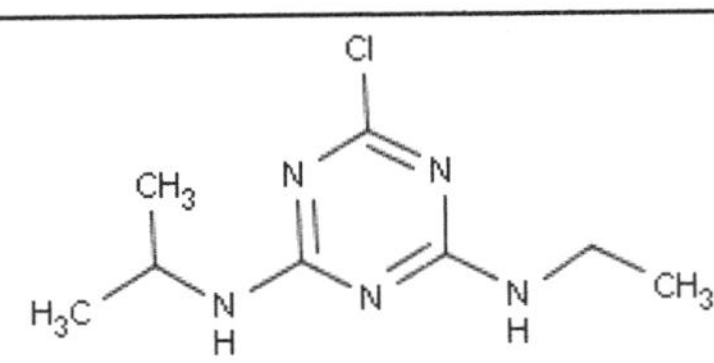
DDT	354.5 – na – 6.91 – 910.2 aromatic rings, chlorines	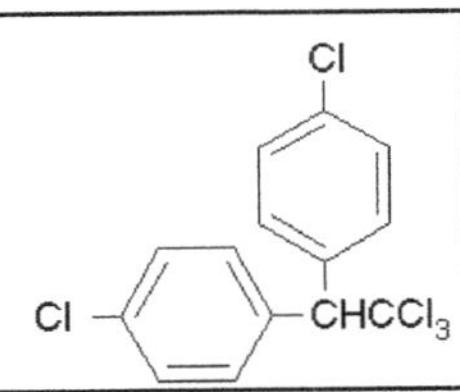
DEET	191.1 – 0.67 (est) - 2.18 – 758.5 aromatic ring, amide	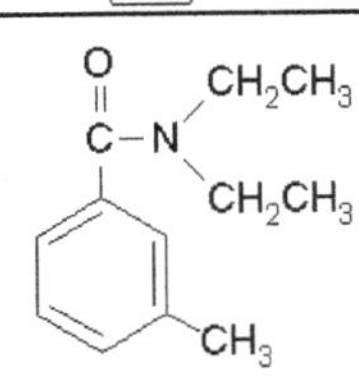

(continued)

Table 2.1 (Continued)

Compounds	MW – pKa – log Kow – Mol. Volume (Å2) Structural features	Structure
Lindane (γ-BHC)	290.8 – na – 3.72 – 669.7 cyclohexane, chlorines	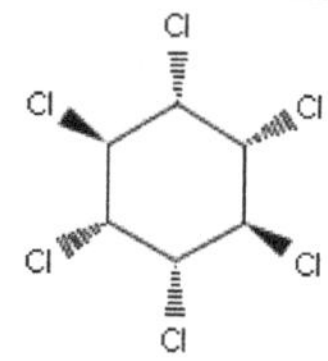
Metolachlor	283.8 – (-1.34) (est) – 3.13 – 948.3 aromatic ring, amide, methoxy, chlorine	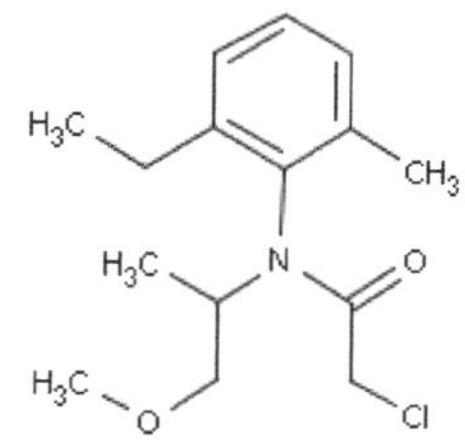
Fragrances		
Galaxolide	258.4 – na – 5.9[6] – 919.3 aromatic ring, aliphatic ring, hetercyclic ring (oxygen)	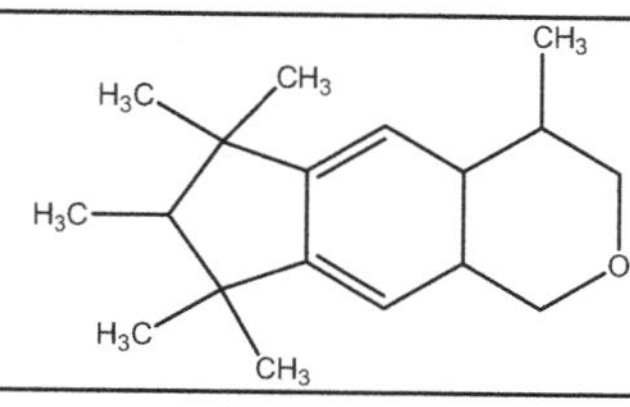
Musk Ketone	294.1 – na – 4.3[7] – 907.3 aromatic ring, ketone, nitro groups	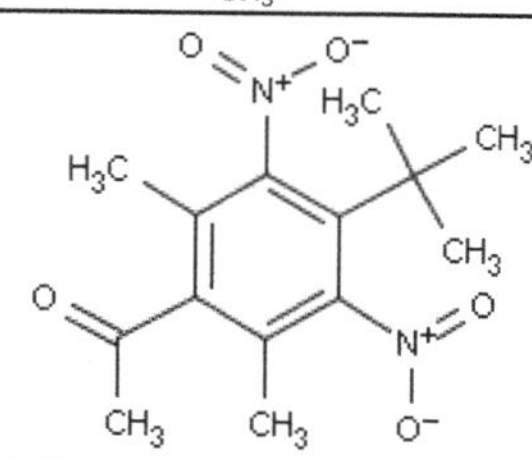
Polycyclic Aromatic Hydrocarbons		
Benzo(a)pyrene	252.3 – na – 6.13 – 824.5 aromatic rings	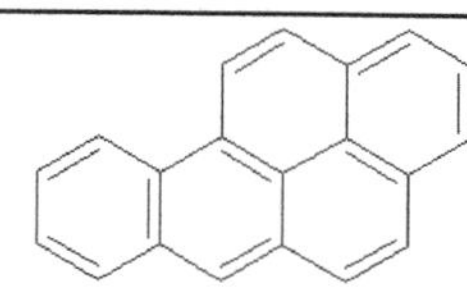
Fluorene	166.2 – na – 4.18 – 624.9 aromatic rings, conjugated ring	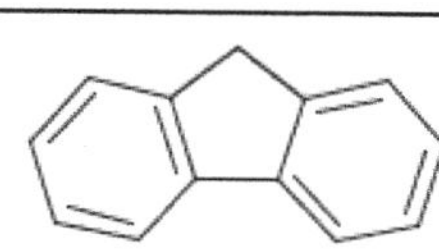
Heart Medications		
Pentoxifylline	278.3 – 0.97 (est) – 0.29 – 949.0 xanthine ring, ketone	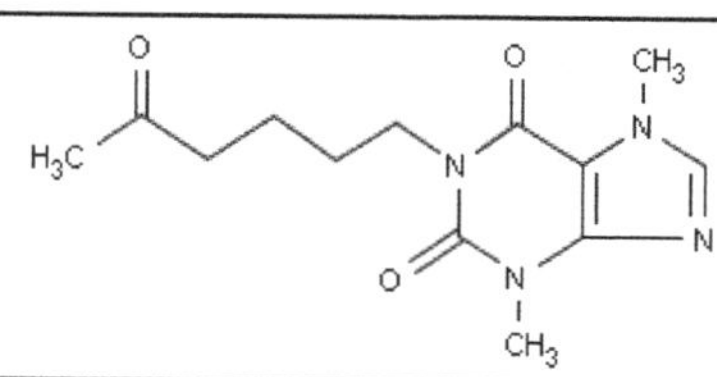

(continued)

Table 2.1 (Continued)

Compounds	MW – pKa – log K_{ow} – Mol. Volume (Å2) Structural features	Structure
Gemfibrozil	250.4 – 4.42 (est) – 4.77 – 945.1 aromatic ring, carboxylic acid, ether	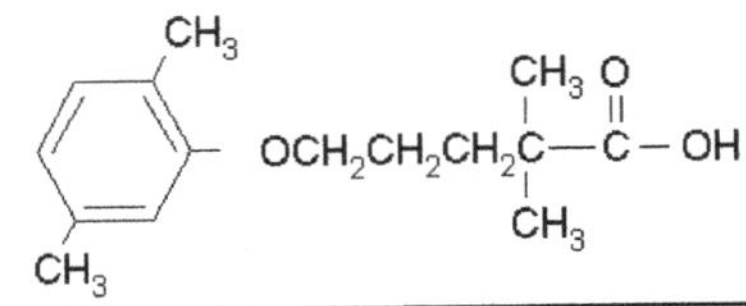
Flame Retardant		
TCEP	285.5 – na – 1.44 – 819.3 phosphate, chlorines	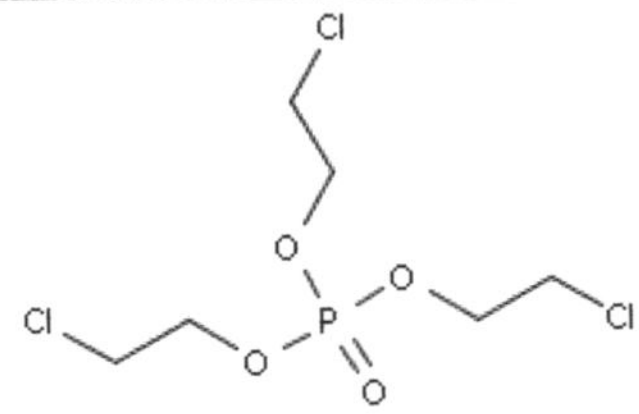
Sun Screen		
Oxybenzone	228.1 – 7.77 (est) – 3.79 – 771.6 aromatic ring, phenol, methoxy, ketone	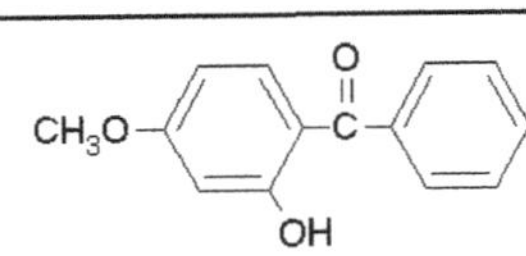
X-ray Contrast Media		
Iopromide	791.1 – 10.24 (est)– (-2.05) – 1461.6 aromatic ring, iodines, alcohols, methoxy, amides	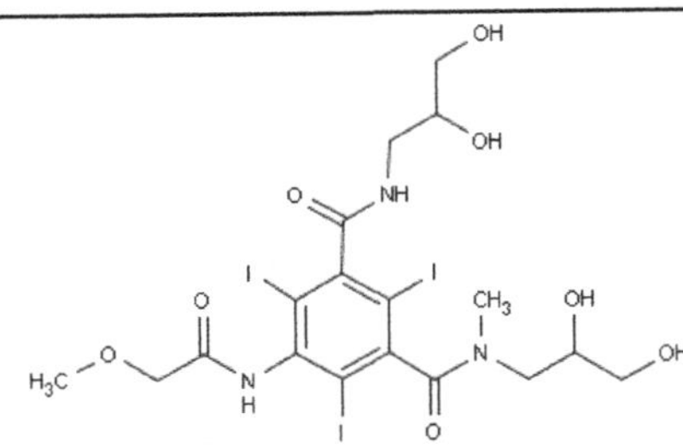

1. Unless indicated,
    a. pKa and logK_{ow} are experimental values obtained from Environmental Science Database SRC PhysProp
    b. Molecular volume is obtained from QikProp, version 2.2, Schrödinger, LLC, New York, NY, 2005
    c. The estimated values are obtained from http://chem.sis.nlm.nih.gov/chemidplus/
2. (Deborde et al. 2005)
3. (Dodd, Buffle, and Von Gunten 2006)
4. (Loftsson and Hreinsdottir 2006)
5. (Jones, Voulvoulis, and Lester 2002)
6. (Balk and Ford 1999)
7. (EC 2003)

This project was initiated in early 2002, thus compounds selected by occurrence were based only upon data available up to that period in time. Conversely, when compounds of similar structure and classification were found to occur, representatives from the class were chosen instead of the entire class (e.g., sulfonamide antibiotics) as similar chemical behavior was expected (e.g., reactions with free chlorine). The USGS published a national reconnaissance study on organic contaminants in "susceptible" surface waters in March of 2002 (Kolpin et al. 2002). These findings were subsequently included in our review of occurrence data. As shown in Table 2.2, our target compounds include several of the most frequently detected compounds in the USGS national survey. However, many compounds were intentionally not included for various reasons, such as coprostanol and cholesterol, which have little human health relevance as chemical toxicants. We also included some contaminants that have enforceable maximum contaminant limits (MCLs) under the Safe Drinking Water Act, such as atrazine (3000 ng/L), lindane (200 ng/L), and benzo(a)pyrene (200 ng/L), which have been implicated as possible endocrine disruptors. Our target compound list was somewhat dynamic, as certain compounds were originally selected, then later dropped due to analytical issues. In addition, not all compounds monitored were summarized in this report, especially those within similar classes, for instance, other polycyclic aromatic hydrocarbons (PAHs) and additional DDT and lindane isomers. These chemicals exhibited similar behavior within the class resulting in redundant information, the inclusion of which would greatly increase the size of this report without contributing significant information.

It was important to this project that selected compounds represent various classes of EDCs and PPCPs. It was critical that compounds come from a broad range of physical-chemical parameters. This study included hydrophilic, hydrophobic, acidic, basic, neutral, large molecular size, small molecular size, semi-volatile, non-volatile, aromatic, and aliphatic target compounds. Structural features are more important than simple class categories (e.g., chlorine reactivity with the phenolic hormone 17β-estradiol versus the ketone hormone progesterone). This project confirmed that compounds with similar structural features often exhibit similar removal patterns during water treatment.

In order to include a particular analyte in our suite of target compounds, suitable analytical standards and methods were required. There was a limit to how many samples the research team could process in the amount of time allotted for this project. Analytical capabilities included four tandem mass spectrometers (MS/MS), two with GC and two with LC interfaces. In order to work at environmentally relevant levels, analytical methods required detection limits in the low ng/L range. Therefore, water samples required extraction and concentration procedures.

Upon reviewing available analytical methodologies (Snyder et al. 2003), including the team's previous and ongoing efforts in this field (Snyder et al. 1999; Snyder et al. 2000; Snyder et al. 2001a; Snyder et al. 2001b; Snyder et al. 2001c), it was determined that only a single solid-phase extraction (SPE) could be used to produce the desired amount of information within the time frame and budgets of this project. Some compromises would be required, for instance the USGS reconnaissance study utilized five different analytical methods for each sample (Kolpin et al. 2002). Another treatment study investigating five pharmaceuticals relied upon four separate analytical methods (Ternes et al. 2002). In order to process the estimated 600 samples generated by this project, the team required a single extraction method followed by both GC-MS/MS and LC-MS/MS analyses. Besides the issue of analytical capability, compounds for which purified standards were not commercially available for a reasonable price could not be included.

Significant quantities of compounds were required in order to conduct the batch and dynamic spiking studies. Some compounds could not be considered due to a lack of, or the cost of, available standards (e.g., certain algal toxins). In order to procure significant amounts of certain controlled pharmaceuticals, the Southern Nevada Water Authority (SNWA) obtained a US Department of Justice - Drug Enforcement Administration (DEA) license to procure and possess the required quantity of schedules 1, 2, 3, 3N, 4, and 5 controlled substances. During method development and the initial bench-scale testing, various sample preservation, handling, and processing issues surfaced. Some of these initial findings required changes in the methodology that resulted in loss of sensitivity for certain analytes while improving the sensitivity of others. For a few compounds, the single SPE and instrumental methods were not robust enough for this project; and those compounds were subsequently dropped from further consideration. Detailed discussion of these findings and issues are beyond the scope of this report. The most significant of discoveries are discussed in the analytical method section of Chapter 3 in this report and in previously published manuscripts (Vanderford et al. 2003; Trenholm et al. 2006).

Table 2.2
Compounds with highest frequency of detection US waters (Calculations based on data from Kolpin *et al.*, 2002)

Compounds	Class	Detection Frequency (%)
Coprostanol	Estrogen Hormone	~80%
Cholesterol	Steroid	~80%
N-N-diethyltoluamide (DEET)	Mosquito repellant	~80%
Caffeine	Stimulant	~75%
Tris(2-chloroethyl)phosphate (TCEP)	Fire retardant	~75%
Triclosan	Antimicrobial	~60%
4-nonylphenol	Surfactant metabolite	~60%
4-nonylphenol monoethoxylate	Surfactant	~50%
Ethanol, 2-butoxy-phosphate	Plasticizer	~45%
4-octylphenol monoethoxylate	Surfactant	~45%
Bisphenol-A	Plasticizer	~45%
Cotinine	Nicotine metabolite	~35%
4-nonylphenol diethoxylate	Surfactant	~35%
5-methyl-1H-benzotriazole	Antioxidant	~30%
Fluoranthene	PAH	~30%
1,7-dimethylxanthine	Caffeine metabolite	~30%
Pyrene	PAH	~25%
Trimethoprim	Antibiotic	~25%
1,4-dichlorobenzene	Deodorizer/Insecticide	~25%
Acetaminophen	Analgesic	~25%
Terachloroethylene	Solvent	~20%
4-octylphenol diethoxylate	Surfactant	~20%
Erythromycin-H_2O	Antibiotic	~20%
Estriol	Estrogen	~20%
Lincomycin	Antibiotic	~15%
Sulfamethoxazole	Antibiotic	~15%
Phthalic anhydride	Plasticizer	~15%
Carbaryl	Insecticide	~15%

DATA ANALYSIS

The reduction in target compound concentration through a treatment process is shown as percent removal throughout this report, while noting that removal through oxidation is likely a transformation of the particular compound rather than mineralization to carbon dioxide. Appendices are included which provide additional information, including measured concentrations. For experiments using water spiked with target compounds, the initial concentrations generally ranged from 100 - 300 ng/L for each compound. Method reporting limits (MRLs) for the target compounds ranged from 1-10 ng/L. From this relatively narrow working range, percent removal was highly consistent among the experiments conducted for a particular process. In full-scale and certain pilot investigations, only detectable compounds could be evaluated. Some target compounds were not detected in any full-scale drinking water plants at the established MRLs; therefore, only data for detectable compounds are shown in these cases.

All analytical data are subject to a degree of variability dependent upon the methodology applied, the expertise of the technicians and chemists performing the analyses, and instrumental fluctuation. Experimental variability arises from compound spiking, volume and mass measurements, chemical doses, and holding times. While not all variability can be eliminated, the research team took careful steps to minimize experimental and analytical variability. Even with these protective measures in place, some variability from the experiments and analytical methods is evident. In Chapter 3, several replicates of both complete experiments and of analytical measurements are shown. From these, the researchers determined that analytical variability (i.e., the measurement of multiple samples of the same water) was generally less than 5% for LC-MS/MS and less than 20% for GC-MS/MS analyses. GC-MS/MS is inherently more variable due to losses during the sample clean-up procedure (including two evaporation/concentration steps), where more volatile compounds can be lost. In LC-MS/MS analyses, matrix effects can cause erratic quantitation through ion suppression or enhancement (see Analytical Methods in Chapter 3). These effects were reduced in the experiments through use of isotopically-labeled surrogates and dilution of extracts when matrix effects were observed. In consideration of experimental and analytical variation, the researchers believe that differences of less than 10% for LC-MS/MS analytes and 20% for GC-MS/MS analytes may not be significant.

REFERENCES

Balk, F., and R. A. Ford. 1999. Environmental risk assessment for the polycyclic musks AHTN and HHCB in the EU - I. Fate and exposure assessment. *Toxicol. Lett.*, 111(1-2): 57-79.

Deborde, M., S. Rabouan, J.-P. Duguet, and B. Legube. 2005. Kinetics of aqueous ozone-induced oxidation of some endocrine disruptors. *Environ. Sci. Technol.*, 39(16): 6086-6092.

Dodd, M. C., M. O. Buffle, and U. Von Gunten. 2006. Oxidation of antibacterial molecules by aqueous ozone: Moiety-specific reaction kinetics and application to ozone-based wastewater treatment. *Environ. Sci. Technol.*, 40(6): 1969-1977.

EC. 2003. Risk assessment musk ketone. European Union Risk Assessment Report. Final Draft June 2003.

Jones, O. A. H., N. Voulvoulis, and J. N. Lester. 2002. Aquatic environmental assessment of the top 25 English prescription pharmaceuticals. *Water Res.*, 36: 5013-5022.

Kolpin, D. W., E. T. Furlong, M. T. Meyer, E. M. Thurman, S. D. Zaugg, L. B. Barber, and H. T. Buxton. 2002. Pharmaceuticals, hormones, and other organic waste contaminants in U.S. streams, 1999-2000: A national reconnaissance. *Environ. Sci. Technol.*, 36(6): 1202-1211.

Loftsson, T., and D. Hreinsdottir. 2006. Determination of aqueous solubility by heating and equilibration: A technical note. *AAPS PharmSciTech*, 7(1): E1-E4.

Snyder, S. A., T. L. Keith, C. G. Naylor, C. A. Staples, and J. P. Giesy. 2001a. Identification and quantification method for nonylphenol and lower oligomer nonylphenol ethoxylates in fish tissues. *Environ. Toxicol. Chem.*, 20(9): 1870-1873.

Snyder, S. A., T. L. Keith, D. A. Verbrugge, E. M. Snyder, T. S. Gross, K. Kannan, and J. P. Giesy. 1999. Analytical methods for detection of selected estrogenic compounds in aqueous mixtures. *Environ. Sci. Technol.*, 33(16): 2814-2820.

Snyder, S. A., K. L. Kelly, A. H. Grange, G. W. Sovocool, E. M. Snyder, and J. P. Giesy. 2001b. Pharmaceuticals and personal care products in the waters of Lake Mead, Nevada. In *Pharmaceuticals and Personal Care Products in the Environment: Scientific and Regulatory Issues*. Edited by C. G. Daughton and T. L. Jones-Lepp. Washington, D.C.: American Chemical Society.

Snyder, S. A., E. Snyder, D. Villeneuve, K. Kurunthachalam, A. Villalobos, A. Blankenship, and J. Giesy. 2000. Instrumental and Bioanalytical Measures of Endocrine Disruptors in Water. In *Analysis of Environmental Endocrine Disruptors*. Edited by L. H. Keith, T. L. Jones-Lepp and L. L. Needham. Washington,DC: American Chemical Society.

Snyder, S. A., B. J. Vanderford, R. A. Pearson, O. Quinones, and Y. Yoon. 2003. Analytical methods used to measure endocrine disrupting compounds in water. *Pract. Periodical of Haz., Toxic, and Radioactive Waste Mgmt.*, 7(4): 224-234.

Snyder, S. A., D. L. Villeneuve, E. M. Snyder, and J. P. Giesy. 2001c. Identification and quantification of estrogen receptor agonists in wastewater effluents. *Environ. Sci. Technol.*, 35(18): 3620-3625.

Ternes, T. A., M. Meisenheimer, D. Mcdowell, F. Sacher, H.-J. Brauch, B. Haist-Gulde, G. Preuss, U. wilme, and N. Zulei-Seibert. 2002. Removal of pharmaceuticals during drinking water treatment. *Environ. Sci. Technol.*, 36(17): 3855-3863.

Trenholm, R. A., B. J. Vanderford, J. C. Holady, D. J. Rexing, and S. A. Snyder. 2006. Broad Range Analysis of Endocrine Disruptors and Pharmaceuticals Using Gas Chromatography and Liquid Chromatography Tandem Mass Spectroscopy. *Chemosphere* 65: 1990-1998.

Vanderford, B. J., R. A. Pearson, D. J. Rexing, and S. A. Snyder. 2003. Analysis of Endocrine Disruptors, Pharmaceuticals, and Personal Care Products in Water Using Liquid Chromatography/Tandem Mass Spectrometry. *Analytical Chemistry* 75(22): 6265-6274.

CHAPTER 3
EXPERIMENTAL AND ANALYTICAL METHODS

SUMMARY OF TREATMENT PROCESSES EVALUATED

This chapter describes, in detail, the experimental and analytical methods used throughout this project. Results are provided in individual chapters; however, the experimental details are shown here in order to keep the results from unit processes as concise as possible. This project involved the following processes:

- Physico-chemical processes: metal salt coagulation, chemical softening, activated carbon adsorption, granular media filtration, membrane filtration, ion exchange
- Oxidation processes: chlorination, chloramination, ozonation (with and without hydrogen peroxide), ultraviolet light irradiation (with and without hydrogen peroxide)
- Biological processes: biological filtration, riverbank filtration, soil aquifer treatment, aquifer storage and recovery, and membrane bioreactor

Bench-scale experiments were designed using chemical doses, contact times, and natural waters in order to mimic conditions at actual water treatment facilities. Bench-scale experiments were conducted as static experiments (jar tests) using various waters fortified with target contaminants. Pilot-scale evaluations were dynamic (flow-through) experiments, designed to investigate the operating conditions associated with contaminant removal observed during bench-scale experiments. Full-scale treatment facilities were investigated to validate bench- and pilot-scale experiments under a broad range of water qualities and treatment process combinations.

RESEARCH FACILITIES UTILIZED

Bench-scale and pilot-scale experiments were performed at several locations depending on facility capabilities and technical expertise as shown in Table 3.1. Bench-scale studies were performed at SNWA, Arizona State University (ASU), Metropolitan Water District of Southern California (MWD), Pennsylvania State University (PSU), and Trojan UV Technologies. Pilot-scale studies were performed at SNWA, Louisville Water Company (LWC), Trojan UV Technologies, and several anonymous water reclamation facilities. Anonymously located pilot facilities are listed by the engineering firm, namely Black & Veatch (B&V) or Montgomery Watson Harza (MWH), who were contracted to design and/or operate the pilot systems. Schematics for pilot facilities, where available, are provided in Appendix B. The subsequent sections will provide a description and the role of each facility used during the project.

Several full-scale drinking water and water recycling facilities also participated in the research effort allowing bench- and pilot-scale data to be compared to removals observed at operating water treatment facilities. Table 3.2 shows the range of processes evaluated at full-scale drinking water treatment plants, while Table 3.3 provides information on full-scale water reuse facilities tested. It is important to note that several facilities had multiple processes that were not evaluated during this study, due to redundancy, sample timing, and/or difficulties in isolating the particular process. Several of the treatment plants requested anonymity; therefore we have coded each system numerically. The treatment plants evaluated come from a wide geographical dispersion throughout the US; therefore, many different water qualities are represented. Interestingly, as will be seen in later chapters of this report, water quality did not play a large role in compound removal for most treatment processes when chemical doses were adjusted to compensate for water quality (e.g., chlorine doses to achieve similar free chlorine residuals).

For off-site pilot- and full-scale evaluations, bottles were prepared by adding the appropriate amount of preservative and quenching agent. These bottles were shipped in ice-packed, well-protected coolers to the participating facility. Detailed instructions were provided to ensure that samplers at the facility were capable of collecting representative samples while minimizing chances of contamination. When possible, utilities were urged to collect samples between processes following a plug flow using the hydraulic retention time within each process. This was often not feasible due to logistical restraints. Likewise, it was sometimes difficult to isolate processes. For instance, coagulation often took place in conjunction with pre-oxidation (e.g., chlorine and permanganate).

Table 3.1
Processes evaluated and facilities utilized

	Bench Scale Testing Facilities					Pilot Scale Testing Facilities				
	ASU	MWD	PSU	SNWA	Trojan	LWC	SNWA	Trojan	B&V	MWH
Coagulation/Flocculation/Sedimentation	X			X		X	X			
Softening/Sedimentation	X			X		X				
Chlorination	X			X						
Chloramination				X		X				
Ozone/ Ozone w/H_2O_2	X			X			X			
UV/ UV w/H_2O_2		X			X			X		X
Granular Activated Carbon (GAC)			X							
Biologically Active Carbon (BAC)							X			
Biologically Active Filtration (BAF)							X			
Powder Activated Carbon (PAC)	X			X						
Magnetic Ion Exchange Resin (MIEX)				X			X		X	
Microfiltration (MF)										X
Ultrafiltration (UF)	X			X						X
Nanofiltration (NF)	X			X						
Reverse Osmosis (RO)									X	X
Membrane Bioreactor (MBR)										X
Electrodialysis Reversal (EDR)										X
River Bank Filtration (RBF)						X				

Table 3.2
Full-scale drinking water treatment facilities investigated

	Drinking Water Processes Evaluated																			
	#1	#2	#3	#4	#5	#6	#7	#8	#9	#10	#11	#12	#13	#14	#15	#16	#17	#18	#19	#20
Coagulation/Flocculation/Sedimentation	X	X	X	X			X	X	X	X		X					X	X	X	X
Softening/Sedimentation								X												
Chlorination	X	X	X	X		X	X		X	X	X	X			X	X				X
Chloramination								X									X	X		
Ozone/ Ozone w/H_2O_2													X	X					X	
UV/ UV w/H_2O_2					X	X	X													
River Bank Filtration (RBF)																		X		
Granular Activated Carbon (GAC)	X			X															X	

Table 3.3
Full-scale water reuse facilities investigated

	Water Reuse Processes					
	#1	#2	#3	#4	#5	#6
Slow Sand Filtration				X		X
Activated Sludge			X	X	X	
Microfiltration (MF)	X	X				
Ultrafiltration (UF)						
Nanofiltration (NF)						
Reverse Osmosis (RO)	X	X				
Granular Activated Carbon (GAC)						X
Ultraviolet Irridiation (UV)		X	X			X
Chloramination					X	
Chlorination				X		

BENCH-SCALE MATERIALS AND METHODS

All water treatment standards and chemicals were at least reagent grade and/or of the highest purity commercially available. Target compounds were obtained from Sigma-Aldrich, except atrazine and DEET (Accustandard, New Haven, CT, USA); fluoxetine and iopromide (United States Pharmacopeia, Rockville, MD, USA); and hydrocodone (Cerilliant, Round Rock, TX, USA). All glassware and supplies were solvent rinsed three times each using trace-residue grade acetone, hexane, and methanol obtained from Burdick & Jackson (Muskegon, MI, USA) or Sigma-Aldrich (Milwaukee, WI, USA).

Waters and Conditions Evaluated

Five different water qualities were evaluated for bench-scale experiments. Table 3.4 describes the water qualities and collection locations for the various waters evaluated. All natural waters were collected at the location in glass or stainless steel containers and shipped immediately to the testing locations. Most natural waters had detectable levels of some target compounds; however, these concentrations were generally less than 10 ng/L and were accounted for in the measurement of the raw spiked water prior to testing. The model water was prepared by adding Suwannee River Reverse Osmosis (RO) isolated natural organic matter (NOM), purchased from International Humic Substances Society (St. Paul, MN, USA), to DI water with sodium bicarbonate added as a pH buffer. Chemical doses were adjusted for each water quality evaluated. These doses can be found in Table 3.5.

Spiking Procedures

All compounds were spiked into bench-scale waters using a methanol carrier, with the exception of the biological degradation experiment. The EDC/PPCP spiking solution cocktail was prepared in methanol with individual compound concentrations between 10 and 250 mg/L. It was quickly determined that the volume of methanol must be kept to a minimum as methanol can react with hydroxyl radicals and ozone. However, a water-soluble carrier was required as many EDC/PPCP compounds were only available dissolved in solvents rather than as neat standards, or required a solvent carrier due to low water solubility.

Table 3.4
Waters evaluated at bench-scale

Water	Code	pH	DOC (mg/L)	Alkalinity (mg/L as $CaCO_3$)
Colorado River	CR	8.2	2.5	140
Passaic River	PV	6.8	3.4	52
Suwannee Model	SR	7.5	4.0	400
Ohio River	OR	7.9	3.5	79
Lake Huron	LH	7.4	1.3	100

For bench-scale oxidation, membrane, and powdered activated carbon (PAC) experiments, waters were prefiltered using a 0.7 μm glass-fiber filter (Whatman GF/F) prior to the addition of target compounds. For all other experiments (e.g., coagulation, softening, and MIEX®), unfiltered water was spiked. For all waters except the Lake Huron water used for the second round of UV bench-scale, spiking occurred by injecting 70 – 1000 μL of stock solution into 25 L of water contained in a pre-cleaned stainless steel barrel and thoroughly mixed. From each water, samples of unspiked and spiked raw waters were analyzed for the suite of target compounds. When spiked waters were used over more than one day, samples of the spiked water were analyzed each day in order to determine any losses of target analytes.

Coagulation

Two coagulants, aluminum sulfate (alum, $Al_2(SO_4)_3{\cdot}14H_2O$) and ferric chloride (iron, $FeCl_3{\cdot}6H_2O$) (Fisher Scientific, USA), were evaluated using a six-place programmable jar tester (Phipps and Bird, Richmond, VA, USA) and 2 L glass beakers filled with 1.5 L of the source water. Stock solutions of alum and iron were initially prepared in deionized (DI) water at 10,000 mg/L. Coagulant doses and mixing conditions are provided in Table 3.5. The alum dose was determined by multiplying the DOC by a factor of ten. The ferric chloride dose was determined using approximately 1 Al^{3+}:1 Fe^{3+} meq stoichiometric conversion. All experiments were conducted at a room temperature of approximately 20°C. Samples were carefully collected and filtered through a 0.7 μm (GF/F) glass-fiber filter prior to instrumental analysis.

Chemical Softening

Chemical softening experiments using calcium hydroxide (Ca(OH)2) and soda ash (Na2CO3) were conducted with a six-place programmable jar tester (Phipps and Bird, Richmond, VA, USA) and 2 L glass beakers filled with 1.5 L of the source water. Stock solutions of calcium hydroxide and soda ash were initially prepared in DI water at 50,000 mg/L. Softening doses and mixing conditions for each water are provided in Table 3.5. The applied/theoretical lime and soda ash doses were calculated based upon initial pH, alkalinity, and carbonic acid concentration for excess-lime softening conditions. After lime and soda ash addition, the pH was adjusted to 11.3 ± 0.2 by adding sodium hydroxide solution for removal of magnesium and calcium. Similar sampling and analysis processes were conducted as described for the coagulation tests.

Table 3.5
Chemical and UV energy doses evaluated for bench-scale experiments

Treatment Technology	Reaction Time	Test #	Source Water: CRW	PRW	ORW	SRW	LHW
Coagulation/ Softening	2 min @100 rpm, 20 min @ 30 rpm, 1 hr Settling	1	30 mg/L $Al_2(SO_4)_3$	40 mg/L $Al_2(SO_4)_3$	35 mg/L $Al_2(SO_4)_3$ (pH=5.5)	40 mg/L $Al_2(SO_4)_3$ (pH=5.5)	
		2	28.4 mg/L $FeCl_3$	37.9 mg/L $FeCl_3$	35 mg/L $Al_2(SO_4)_3$	40 mg/L $Al_2(SO_4)_3$	
		3			8.3 mg/L $FeCl_3$	9.45 mg/L $FeCl_3$	
		4	238 mg/L as $Ca(OH)_2$, 177 mg/L as Na_2CO_3	133 mg/L as $Ca(OH)_2$, 32 mg/L as Na_2CO_3	135 mg/L as $Ca(OH)_2$, 41 mg/L as Na_2CO_3	53 mg/L as $Ca(OH)_2$, 0 mg/L as Na_2CO_3	
Powdered Activated Carbon	4 hrs @100 rpm 1 hr Settling	1	5 mg/L PAC WPM	5 mg/L PAC WPM	5 mg/L PAC WPM	1 mg/L PAC WPM	
		2			5 mg/L PAC AC800	5 mg/L PAC WPM	
		3				20 mg/L PAC WPM	
		4				5 mg/L PAC AC800	
Ozone/AOP	5 minutes	1	2.5 mg/L	3.0 mg/L O_3	3.5 mg/L O_3	4.0 mg/L O_3	
		2			7.0 mg/L O_3	8.0 mg/L O_3	
		3	2.5 mg/L O_3, 0.0625 mg/L H_2O_2	3.0 mg/L O_3, 0.075 mg/L H_2O_2	3.5 mg/L O_3, 0.0875 mg/L H_2O_2	8.0 mg/L O_3, 0.20 mg/L H_2O_2	
UV		1	43.6 mJ/cm^2	39 mJ/cm^2	41 mJ/cm^2	41 mJ/cm^2	39.5 mJ/cm^2
		2	1020 mJ/cm^2	955 mJ/cm^2	949 mJ/cm^2	968 mJ/cm^2	439 mJ/cm^2
		3					787 mJ/cm^2
		4					1318 mJ/cm^2
UV-AOP		1	1022 mJ/cm^2, 5 mg/L H_2O_2	955 mJ/cm^2, 5 mg/L H_2O_2	950 mJ/cm^2, 5 mg/L H_2O_2	968 mJ/cm^2, 5 mg/L H_2O_2	372 mJ/cm^2, 5 mg/L H_2O_2
		2					671 mJ/cm^2, 5 mg/L H_2O_2
		3					969 mJ/cm^2, 5 mg/L H_2O_2
		4					371 mJ/cm^2, 7.5 mg/L H_2O_2
		5					668 mJ/cm^2, 7.5 mg/L H_2O_2
		6					966 mJ/cm^2, 7.5 mg/L H_2O_2
Chlorination	24 hours	1	3.5 mg/L Cl_2 (pH=5.5)	3.8 mg/L Cl_2 (pH=5.5)	2.8 mg/L Cl_2 (pH=5.5)	6.75 mg/L Cl_2 (pH=5.5)	
		2	3.5 mg/L Cl_2	3.8 mg/L Cl_2	2.8 mg/L Cl_2	6.75 mg/L Cl_2	
		3	3 mg/L Cl_2				
	168 hours	4	3 mg/L Cl_2				
Chloramination	24 Hours	1	2 mg/L Cl_2				
		2	3 mg/L Cl_2				
	168 hours	3	2 mg/L Cl_2				
		4	3 mg/L Cl_2				
MIEX®	10 minutes	1	5 mL/L				
		2	15 mL/L				
		3	20 mL/L				
	20 minutes	4	5 mL/L				
		5	10 mL/L				
		6	15 mL/L				
		7	20 mL/L				

Granular Activated Carbon Adsorption

Bench-scale testing was conducted using rapid, small-scale column tests (RSSCTs) to predict GAC performance. The RSSCT design used constant diffusivity similitude, in which the ratio of empty bed contact time is proportional to the square of the ratio of the grain diameters (Crittenden et al. 1999). Constant diffusivity similitude was selected because Crittenden and others have found this to be more appropriate than proportional diffusivity for "smaller" molecular weight discrete molecules that are not highly charged (Crittenden et al. 1999). Proportional diffusivity has been found most appropriate for highly charged natural organic matter with an average molecular weight of ≥ 1000 Daltons and for highly charged inorganic species such as perchlorate (Na, Cannon, and Hagerup 2002). Moreover, constant diffusivity is the more "conservative" similitude to use, i.e., if RSSCT shows that an organic molecule will be removed within a certain bed life when using constant diffusivity, then the predicted bed life at full scale will be the same or longer.

Tests were performed to compare two lignite-based GACs manufactured by Norit Americas: Hydrodarco 4000 (HD4000) and Norit Superdarco (Superdarco). Superdarco has been adapted to have considerably more micropore and mesopore volume than the HD4000. It also hosts a more stable surface that will not become as oxygenated and acidified as compared to HD4000 and other commercial carbons. The Superdarco has been manufactured at full-scale in Marshall, Texas, to adapt key features of the tailored activated carbons that have been developed in bench-scale research (Nowack, Cannon, and Mazyck 2004; Rangel-Mendez and Cannon 2005). In accordance with constant diffusivity similitude for the target compounds, a column with 0.307 cm^3 empty bed volume was utilized with the test carbons crushed and wet-sieved to 75-90 μm particle size (170 x 200 US mesh). With this design, RSSCTs simulated a full-scale column that operates at a 7.6 minute empty bed contact time (EBCT). Colorado River water was concurrently spiked with 100-200 ng/L of the target compounds; however only LC-MS compounds were analyzed. The test column was maintained at room temperature and therefore adsorption occurred at 20-25°C.

Powdered Activated Carbon Adsorption

The bench-scale activated carbon adsorption (PAC) studies used two commercially available brands of PAC. AC800 (AC800, Acticarb, Dunnellon, FL, USA) and WPM (PAC form of F400, Calgon Carbon Corp., Pittsburgh, PA, USA) were selected because they exhibited different removal capability for bisphenol A (common plasticizer), 17β-estradiol (natural estrogen), and 17α-ethynyl estradiol (synthetic estrogen) in screening tests with 6 different PAC brands (Yoon et al. 2003).

The EDC/PPCP cocktail solution was spiked into filtered (GF/F) test waters and stored in a 28 L SS container. Filtered water, as opposed to the unfiltered water used in chemical coagulation and softening studies, reduced the potential of competitive partitioning of EDCs/PPCPs between suspended materials (i.e., turbidity) and PAC. PAC was hydrated for 24 hours in DI water prior to use and added to the samples as a slurry (1000 mg/L). Actual PAC doses and mixing conditions are provided in Table 3.5. PAC experiments were performed in a six-place jar tester (mixing at 100 rpm) using 2 L glass beakers filled with 1.5 L source water that served as reactors. The doses and contact times were based upon full-scale WTPs that frequently use PAC contact times of 1 to 5 hours and PAC doses of 5 to 50 mg/L. Sampling and filtration procedures were the same as described for the coagulation and chemical softening experiments. Supernatant was collected and filtered (GF/F) to remove any residual PAC.

The removal of tritium-labeled 17β-estradiol by WPM PAC (1.0 mg/L and 4 h contact time) was conducted using both CRW and distilled water to investigate the effect of initial estradiol concentration. 17β-estradiol [2,4-^{3}H] was purchased from Sigma-Aldrich (St. Louis, MO, USA) with an activity of 16.8 Ci/mmol at a concentration of 1.0 mCi/mL in ethanol. This concentration corresponds to 16.2 μg/mL 17β-estradiol. Radioactivity was measured by a liquid scintillation counter (GMI, Beckman LS6500, Albertville, MN, USA). For scintillation measurements, 4 mL of scintillation liquid (ScintiSafe Plus, Fisher Scientific) were pipetted into a standard glass vial to which 100 μL of the sample were added. The method detection limit was 0.005 nM (1.4 ng/L).

Nanofiltration and Ultrafiltration

EDC and PPCP rejection at bench scale was tested using commercially available nanofiltration (NF) and ultrafiltration (UF) membranes. The NF (ESNA, Hydranautics, U.S.A.) and UF (GM, Desal/Osmonics, U.S.A.) membranes were thin film composites made of aromatic polyamide (ESNA) or sulfonated polyethersulfone (PES) coating an ultrathin polyimide (GM). The membranes also had different pore sizes based upon their nominal molecular weight cut-offs (MWCOs). The MWCO was 600 Daltons and 8,000 Daltons for the NF and UF membranes, respectively. These membranes had functional groups, such as carboxylic acids, that were ionizable based upon their zeta potential values. Membrane filtration experiments were performed at pH 7.5 buffered with 1 mM phosphate for SR and ambient pH for CR and OR.

A commercial dead-end stirred-cell membrane unit (SEPA® ST, Osmonics, Minnetonka, MN, U.S.A.) was used to evaluate flat-sheet membranes. A stainless steel membrane filtration unit was employed to minimize adsorption of the compounds onto the cell. All of the experiments were performed at a constant initial pure water flux. The membranes were pre-compacted for five to seven hours at the same pressures with NaCl solutions. When the flux remained constant, the feed water was exchanged. In these experiments, an initial volume of 300 mL of a given sample were passed through the membrane until 200 mL of permeate were obtained. The corresponding retentate was also collected. The experiments were repeated until a total volume of 1,000 mL of permeate were obtained.

Chlorination

Chlorination experiments were performed using liquid sodium hypochlorite (NaOCl, Fisher Scientific, USA). Stock solutions of chlorine were initially prepared in DI water at 1,200 mg/L. Chlorine was dosed directly into 1 L bottles containing the source water. Chlorine doses were determined based upon the chlorine demand in order to achieve a residual goal of approximately 0.5 mg/L after 24 hours. Free chlorine residuals were measured by the DPD Method using a Hach DR4000 spectrophotometer (USEPA-approved Hach Method #8021, Hach Company, Loveland, CO, USA). Most experiments were conducted at ambient pH and at a suppressed pH of 5.5. After the 24 hour contact time, residual chlorine was quenched with 50 mg/L of ascorbic acid. Analytical surrogates sensitive to chlorine were added just prior to solid-phase extraction; therefore, analytical recovery of these surrogates provided definitive information on whether the residual chlorine was quenched.

Chloramination

Chloramine experiments were performed by first adding ammonia to a 1 L sample of raw water followed by sodium hypochlorite. Stock solutions of each chemical were created from 29% ammonium hydroxide (J.T.Baker, USA) and 5% sodium hypochlorite (NaOCl, Fisher Scientific, USA). A chlorine:ammonia ratio of 4:1 was targeted because this is commonly used in drinking water treatment. This sequence of chemical addition was selected over preformed monochloramine solution to closely simulate actual full-scale plant conditions while minimizing exposure to free chlorine. As a result, dichloramine and monochloramine are likely formed, resulting in higher target compound removal. In addition, a brief free chlorine exposure may result in false compound removal by chloramine during cases in which free chlorine reacts faster with the target compounds than with ammonia. The experiments were conducted at room temperature (approximately 20°C) and ambient pH. The chloramine residual was quenched with ascorbic acid using a ratio of 20:1 ascorbic acid:chlorine.

Mixed Oxidants

Mixed oxidant experiments were conducted using a Brine Pump System (BPS) (MIOX Corporation, USA). Using a 3% brine feed solution, the BPS produced around 230 mg/L of free chlorine from the combined anode and cathode product stream. Bench scale testing using free chlorine and chloramines were performed at room temperature (20°C) and ambient pH. Freshly generated mixed oxidant solution was transferred from the BPS into the sample within 10 seconds of collection to achieve the desired dose. EDC samples were collected after 24 hours of exposure. The chlorine residual was quenched with ascorbic acid after sample collection.

Ozonation

Dissolved ozone stock solutions were made by dissolving a high concentration of gaseous ozone into deionized water at 2°C. Dissolved ozone stock solution concentrations were measured spectrophotometrically (258 nm; ε: 3150 m^{-1} s^{-1}) after a 2:1 dilution with a few drops of 1.0 N phosphoric acid, or by using the indigo method (Standard Methods 4500-O_3) with dilution. Ozone experiments were conducted by injecting a high concentration of dissolved ozone into a 1 L sample bottle containing the source water. All dissolved ozone residuals were measured using the indigo method (Standard Methods 4500-O_3). Ozone doses were determined using an ozone demand/decay curve generated for each water type.

UV Irradiation

UV bench-scale experiments were conducted using two different collimated beam systems, each equipped with medium pressure lamps. Experiments conducted at MWD used a collimated beam apparatus (Calgon Carbon Corporation, Pittsburgh, PA) using a medium pressure lamp. Aliquots of 500 mL were irradiated in a 600 mL glass beaker on a magnetic stir plate that had been centered with respect to the light beam. Experiments using the Trojan Technologies collimated beam system utilized a medium pressure mercury arc lamp with a known spectral power distribution. Aliquots of 220 mL were tested in a crystallization dish placed on a magnetic stir plate that had been centered with respect to the light beam.

At both locations, samples were irradiated for pre-determined periods of time corresponding to selected UV fluences (UV dose). Irradiance measurements were collected with a 1700 International Light Research Radiometer with SED40 detector, calibrated at each wavelength within a 200-340 nm range. UV fluences were calculated using the incident irradiance, sample geometry, and water absorption spectrum, as described elsewhere (Bolton and Linden 2003). UV fluence information is provided in Table 3.5. During advanced oxidation experiments, hydrogen peroxide was added 1 minute prior to exposure, and residuals were quenched with 0.2 mg/L bovine catalase to prevent any dark oxidation of the target compounds.

Biological Degradation and Soil Sorption

Batch Biodegradation Experiments

The following descriptions and field investigations were previously published in Snyder et al. 2004.

Batch experiments using biologically-active sand (BAS) were used to assess biodegradation of EDCs, PPCPs, and dissolved organic carbon (DOC). An adaptation of existing methods to determine biologically-degradable organic carbon (BDOC) was used (Allgeier et al. 1996; Escobar and Randall 1999). Colorado River water freshly collected from Lake Mead, NV (pH 8) was filtered (ashed GF/F 0.7 μM nominal pore size) and then spiked with the target compounds, without methanol, to prepare a solution between 10 and 100 ng/L. The DOC of the CRW (2.52 ± 0.07 mg/L) was expected to serve as a primary substrate for the BAS; therefore, the addition of methanol would not have been a representative primary substrate and hence was eliminated (by nitrogen evaporation) from the target compound spike solution. It is important to note that some compounds were not completely recovered after evaporation to dryness, most likely due to adsorption to vessel walls or loss due to volatility. The reactor consisted of a 1 L amber glass bottle containing 100 mL of BAS sealed with a Teflon-lined cap to prevent atmospheric contamination. The sand used in these experiments was acclimated for a period of at least one month with CRW in order to establish a mixed population of microorganisms on the media. Exactly 550 mL of spiked CRW were added to the reactor, leaving approximately 400 mL of headspace (air) and incubated at room temperature ($20 \pm 1^{\circ}$C). The reactors were agitated on a shaker table in order to increase the mass transfer of oxygen and DOC. After two and five days, duplicate reactors were dismantled. For each reported sample, supernatant from two reactors were combined to reach the required 1 L sample volume, filtered (using GF/F filters), acidified, and analyzed for target EDCs and PPCPs. Approximately 50 mL of water from each reactor were retained for DOC analysis.

Two different duplicate control samples were prepared. One control sample included CRW containing spiked target compounds without sand, which was immediately acidified with sulfuric acid to pH 2.0 to minimize biodegradation. The second control sample used pre-acidification of the spiked CRW prior to addition to the reactor containing BAS. It should be noted that this was not a perfect abiotic control since the shift in pH may have changed the adsorption behavior of polar compounds. However, it was reasonable to assume that the acidified CRW would inactivate the biomass; therefore, any removal observed would be attributed to partitioning onto the BAS. Separate studies indicated that acidification minimized loss due to biodegradation (Ternes et al. 1998; Ternes 2001; Vanderford et al. 2003). The difference between the pre-acidified samples with the inactivated biomass and the samples collected after incubation with the BAS should represent EDC and PPCP loss due solely to biodegradation.

DOC was measured using a combustion/non-dispersive infrared gas analysis method (Shimadzu Model TOC-5050A) immediately after inoculation of the sample with BAS (DOC0) and over the two and five day incubation periods (DOC2 and DOC5). A previous study showed that limited additional DOC degradation of surface waters was observed after a five-day incubation period (Allgeier et al. 1996).

Soil/Aquifer Treatment Simulation

Three 2.8 m long stainless steel columns were connected in series. The diameter of the columns increased stepwise from 4.4 cm in column 1 to 8.8 cm in column 3. This simulated flow from an aquifer recharge well in which linear velocities are expected to decrease radially (Figure B.1 in Appendix B). Linear flow in each column was vertically upward against gravity. Column 1 was fed from a collapsible 20 L HDPE reservoir using a piston pump fitted with a stainless steel cylinder head. Effluent from the top of each column was fed into the bottom of the next using 1/8-inch stainless steel tubing. The columns were filled with graded sand to a density of between 1.60 – 1.65 g/mL. A series of sample ports were located at different locations along each column and samples were collected using a syringe fitted with a luer connection. The columns were in continuous use for over a year, initially fed with tertiary effluent to establish a biological colony, then fed with dechlorinated CRW for over six months.

Field Investigations

Sites within the Las Vegas Valley (Valley), situated in Clark County, in southeastern Nevada, United States were used as field sites for this investigation (see Appendix G, Figure G.1). Within the Valley are the cities of Las Vegas, Henderson, and North Las Vegas, in addition to extensive unincorporated urbanized areas and Nellis Air Force Base. Turf irrigation with reuse water is one of many conservation activities that are encouraged by the Southern Nevada Water Authority (SNWA) and municipal water purveyors including the cities of Henderson and Las Vegas. Three sites were examined for the present investigation: the Desert Rose Golf Course in Las Vegas, the Wildhorse Golf Course in Henderson, and the Henderson wastewater treatment facility (WWTF). Both golf courses irrigate with reuse water supplied by different WWTFs. At Desert Rose, reuse water is supplied by a City of Las Vegas facility. This water is used for coolant at a nearby power plant before being piped to the golf course for irrigation supply. Wildhorse receives reuse water from a Clark County WWTF. The Henderson WWTF, in addition to supplying reuse water to other golf courses, also discharges to a series of ponds for disposal via evaporation and percolation. In combination, data from all three sites enabled some constraints to be placed on residence times and transport distances for selected EDCs and PPCPs within shallow groundwater. Note that water provided by WWTFs contains a chloramine disinfectant residual, which may alter the chemical makeup of the effluent.

Field Investigagtion Site 1: Desert Rose Golf Course

Treated wastewater from a nearby power plant is pumped through a supply pipe to a pond, which serves as an on-site irrigation reservoir. The irrigation supply pipe and three monitoring wells were sampled at Desert Rose (Figure G.2). Well depths range from ~20 to ~50 ft (~6 – 15 m) and water levels from ~12 to ~20 ft (~3.6 – 6 m) below land surface (BLS). Major ion chemistry and previous investigations suggest that shallow groundwater in this area derives principally from the leaching fraction associated with irrigation application (Zikmund

1996; Leising 2003). Black and Veatch (1999) estimated hydraulic conductivity values between ~5 and ~400 ft/day (~1.5 – 150 m/d) for lithologically similar sites within ~2 miles (3.2 km) of Desert Rose. Based on other investigations by the authors, a hydraulic conductivity of ~50 ft/day (~15 m/d) is representative for the area. Groundwater gradients are 0.022 ft/ft northeast from the pond toward well DROSE. Within limits imposed by the hydraulic conductivity estimates, data from this well constrain travel times within the saturated zone. Data from the remaining wells indicate the extent to which soil processes concentrate and attenuate EDCs and PPCPs in the unsaturated (vadose) zone above the water table.

Field Investigation Site 2: Wildhorse Golf Course

Sampling at Wildhorse Golf Course involved one vadose zone monitoring station on a fairway and three down gradient monitoring wells (Figure G.3). At the vadose zone station, soil water was obtained from tensiometers placed at ~0.5 ft (15 cm), ~1.5 ft (45 cm), and ~3.4 ft (105 cm). Monitor well depths ranged between ~8 and ~11 ft (~2.4 – 3.3 m) and saturated zone water levels between 4.6 and ~7 ft (~1.4 – 2.1 m). At this site, groundwater flows east-northeast with a gradient of 0.02 ft/ft toward the monitor wells. Hydraulic conductivity values reported by (Black/Eagle 1999) range between 1 and 15 ft/d (~0.3 – 4.5 m/d) with those nearest the site at ~6 ft/d (~1.8 m/d). Combined with inferences as to the composition of the irrigation water, these data help provide upper limits to aquifer residence times and transit distances.

Field Investigation Site 3: Henderson WWTF

Two sites each consisting of an evaporation/percolation pond and an adjoining monitoring well were examined and sampled at the Henderson WWTF (Figure G.4). These particular basins, Pond 4 and Pond 6, receive and retain treated wastewater prior to reuse or surface discharge. Monitor well HMW-14 is situated ~40 ft (~12 m) N of Pond 4 and is ~40 ft (~12 m) deep. On the sampling date, the water level at HMW-14 was ~17 ft (~ 5.2 m) and ~11 ft (~3.3 m) below that of Pond 4, indicating a steep northward groundwater gradient of ~0.28 ft/ft. Well PC-105 lies ~50 ft (~15.2 m) NE from Pond 6 and is ~51 ft (~12 – 15.5 m) deep. Its water level was ~19 ft (~5.8 m) BLS and ~20 ft (~6 m) below the pond surface, implying a very steep groundwater gradient of ~0.4 ft/ft toward the northeast. As part of a previous investigation, Kerr-McGee Chemical Corp. (KMG) conducted tracer and pumping tests in the area (KMG 2001). Their calculated horizontal hydraulic conductivity was between ~160 and ~280 ft/d (~49 – 85 m/d). SNWA determinations based on solute mass fluxes in the area are somewhat lower at 60 – 80 ft/d (~18 – 24 m/d). Estimates based on water balances within the Henderson WWTF are lower: ~4 – 40 ft/d (~1.2 – 12 m/d) depending on hydraulic gradient. To resolve these discrepancies is a legitimate subject for subsequent investigation. In the present analysis, a figure of 20 ft/d (~6 m/d) was selected.

Sampling methods used at all sites were the same. For the monitoring wells, a minimum of three well volumes were purged using a 12-volt, 1.5 inch (3.8 cm) diameter Whale No. 921 submersible pump attached to clear Tygon® tubing. Prior to the next sampling event, the pump and tubing were submerged and flushed with ~3 – 4 gallons (~11 – 15 L) of tap water containing Simple Green® detergent, followed by a ~2 – 3 gallon (~7.5 – 11 L) tap water rinse. Samples were taken using a new disposable bailer at each well. The collected water was stored in a 1 L salinized amber glass bottle that was immediately placed on ice. Prior to securing the sample, the bottle was rinsed three times with water from the bailer. Disposable Nalgene® gloves were worn at all times during sampling. The protocols at the feed pipe and ponds were the same, except that the glass bottle was directly submerged both to rinse and to acquire the sample. Soil water samples were pumped from tensiometers into 100 mL glass bottles, and from these decanted into 10 mL salinized amber glass bottles, which were then placed on ice.

PILOT-SCALE MATERIALS AND METHODS

Southern Nevada Water Authority

Three pilot testing studies were conducted using Colorado River water at the Southern Nevada Water Authority (SNWA) in Las Vegas, NV. A micro-pilot plant with a flow rate of 0.26 gpm was used to evaluate ozonation. A conventional pilot plant with a flow rate of 6 gpm was used to investigate ozonation, coagulation, biologically active filtration with anthracite, and biologically active filtration with GAC. A 7.5 gpm pilot plant skid was also on-site to evaluate the magnetic ion exchange resin (MIEX®) process.

Bench-Top Pilot Plant Description

A process flow schematic of the bench-top pilot plant (BTPP) is provided in Figure B.2 of Appendix B. The 1 L/min bench-top pilot plant (BTPP) consists of a continuous-flow ozone contactor constructed using inert materials such as glass, fluorocarbon polymers, and stainless steel that do not leach or adsorb target analytes. A 55 gallon drum was filled with 45 gallons of Colorado River water and spiked with the target compounds. A peristaltic pump was used to control the flow rate of 1 L/min. Prior to entering the ozone contactor, chemical feed ports allowed the injection of hydrogen peroxide into the process stream followed by two static mixers. The ozone contactor consisted of 12 glass chambers each providing 2 minutes of contact time for a total of 24 minutes. Each glass chamber was equipped with Teflon sample ports. Ozone feed gas was produced from oxygen with a laboratory-scale ozone generator. Ozone was added in the first contactor chamber with counter-current flow through a glass-fritted diffuser with bubble size of 0.1 μm. An ozone feed gas flow rotameter and a feed gas concentration analyzer were used to calculate the ozone dose. The off-gas was directed from the top of each cell into one central manifold and sent to an ozone destruction unit containing manganese dioxide destruct catalyst. The effluent was discharged to sewer or contained in another 55 gallon drum.

Conventional Pilot Plant Description

The entire pilot plant process scheme (Figure B.3 in Appendix B) includes ozonation, coagulation, flocculation, and two dual media filters. Raw water from Lake Mead was supplied to the 22.7 L/min (6 gpm) SNWA pilot plant prior to any chemical addition. A syringe pump was used to introduce the target compounds into the process stream. Two static mixers followed the injection to ensure adequate mixing. The ozone contactor consisted of 12 PVC cells to provide approximately 24 minutes of contact at the design flow rate of 22.7 L/min (6 gpm). Each contactor cell had sample ports located at the top, middle, and bottom to allow for sampling after various contact times. Ambient air supplied to the ozone generator (model SGC21, Pacific Ozone Technology, Benicia, CA) was used to produce the ozone feed gas. The ozone feed gas concentration was measured (model HI-X, IN USA Inc., Needham, MA), controlled by a gas rotameter, and injected counter-currently through a porous stone diffuser mounted horizontally near the bottom of cell 1. Contactor cells 3, 5, and 9 were equipped with dissolved ozone monitors used to calculate disinfection levels and other ozone parameters such as half-life, CT, demand, and decay rate. The dissolved ozone monitors were calibrated using the indigo method (Standard methods 4500-O3). The off-gas from each contactor cell was directed into a central manifold. A sample line from the manifold supplied a sample conditioning system and an off-gas concentration monitor. A stainless-steel rapid mix chamber and 4 stage stainless steel flocculation basin followed the ozone contactor. Two dual media, stainless steel filters followed the flocculation basin. Filter #1 was loaded with 20" of anthracite and 10" of sand. Filter #2 was loaded with F100 GAC (Calgon Carbon Corp. Pittsburgh, PA) and 10" of sand.

Magnetic Ion Exchange (MIEX®) Pilot Plant Skid

A 7.5 gpm MIEX® pilot plant (Figure B.4 in Appendix B) was installed and operated at the SNWA using raw Colorado River water. A diaphragm pump was used to introduce the target compounds into the process stream. MIEX® resin (Orica Watercare Inc., Watkins, CO) was added to the spiked water using two peristaltic pumps. The resin feed was composed of 10 percent freshly regenerated resin and 90 percent recycled resin from previous testing. Two high-density polyethylene (HDPE) MIEX® contactor cells in series provided a total of 15 minutes of contact time for the anion exchange process. Each contactor cell was mixed with a variable speed, low shear mixer at 70 to 100 rpm. The contactor effluent entered a HDPE conical gravity settler near the bottom. Gravity separation of the resin is very efficient because of the "magnetically" enhanced agglomeration of individual resin beads that yields resin settling against the rapidly rising water in the settler. A peristaltic pump transferred the settled resin (~189 mL/min) into the HDPE accumulator tank, which stores resin during the regeneration process.

Louisville Water Company

The pilot testing studies at the Louisville Water Company (LWC) were conducted in Louisville, KY using Ohio River water. A conventional pilot plant with a flow rate of 7.57 L/min (2 gpm) was used to evaluate metal-salt coagulation, PAC addition, softening, chlorination and chloramination. A second pilot plant with a flow rate of 0.01 gpm was used to investigate river bank filtration.

Conventional Pilot Plant Description

The LWC pilot plant arrangement includes coagulation, flocculation, sedimentation and two dual-media clear PVC filters. A schematic for this pilot is shown in Figure B.5 in Appendix B. Raw water from the Ohio River was supplied to the 7.57 L/min (2 gpm) LWC pilot plant prior to any chemical addition. A positive-displacement diaphragm pump was used to introduce the target compounds into the process stream. A static mixer followed the injection to ensure adequate mixing. Coagulant chemicals were injected upstream of a second static mixer before entering the flocculation basin. Hydrated lime slurry and PAC slurry were fed at the flocculation basin influent. The flocculated water flowed to a plate settler for sedimentation. Following the sedimentation basin, injectors for liquid chlorine and ammonia were located immediately upstream of each of the two dual media filters. Both filters were filled with 16" of anthracite and 12" of sand.

Riverbank Filtration Pilot Plant Description

The riverbank filtration (RBF) pilot plant consisted of three river-sand filter columns operated in series. A schematic of the RBF pilot is shown in Figure B.6 in Appendix B. A peristaltic pump fed Ohio River raw water from an overflow tank to the top of Filter #1 at a flow rate of 500 mL/min. A Dionex eluent pump was used to introduce the target compounds into the raw water stream at a flow rate of 1.0 mL/min. The injection was followed by a static mixer to assure adequate homogenization. The effluent of Filter #1 was transferred to Filter #2 using a peristaltic pump at a flow rate of 60 mL/min. Filter #2 was sealed and operated in an up-flow orientation, followed by Filter #3, which was of the same configuration. The filtrate flow from Filter #3 was maintained at a flow rate of 60 mL/min (0.05 ft/hour). Sampling points were located at the influent and effluent of each filter column.

Trojan UV Technologies

UV pilot-scale experiments were performed at Trojan UV Technologies' research facility in London, Ontario, Canada. The UV system was comprised of a medium pressure UV reactor operated through an electronic ballast, static mixers, an electronic flowmeter, ports for spiking hydrogen peroxide and other solutions, sampling ports, and metering peristaltic pumps. The pilot setup contained two 41,600 L tanks and tests could be conducted in either flow-through or re-circulation mode. A photograph of the pilot-plant is shown in Figure B.7 of Appendix B.

Dechlorinated drinking water (30,000 L) from Lake Huron (LHW) was spiked with 100-200 ng/L of the target compounds in a large feed tank and recirculated for one hour at 2840 L/min (750 gpm) to ensure thorough mixing. UV energies delivered to the water were varied by adjusting the flow rate through the UV reactor. The disinfection UV dose was achieved by running the system with two lamps at 50% power and a flow rate of 2690 L/min (710 gpm), whereas the high UV fluence required for contaminant oxidation were delivered using four lamps at 60% power level and flow rates from 416 to 1400 L/min (110 to 370 gpm). Three different test conditions were selected: UV alone and UV light in the presence of either ~5 or ~7.5 mg/L of hydrogen peroxide. Hydrogen peroxide was injected directly into the pipeline through one of the chemical injection ports using a peristaltic metering pump. The working H_2O_2 solution (11 mg/mL) was made from a 35% stabilizer-free H_2O_2 concentrated solution.

Black & Veatch Reverse Osmosis Pilot

The process schematic of the B&V RO Pilot using a shallow saline aquifer is shown in Figure B.8 and Figure B.9 in Appendix B. A well was drilled to supply groundwater for the pilot plant. Feedwater was pumped from the equalization basin by a low-pressure pump and dosed with anti-scalant ahead of a static mixer. Initial testing for EDC and pharmaceutical target compounds indicated that only one or two compounds were present in the saline groundwater. Therefore, the equalization basin was used as a head tank in which target contaminants were spiked and thoroughly mixed before pumping the spiked water though the RO system. All flows emanating from the pilot plant were discharged to the trickling filter of a wastewater treatment plant.

The pilot skid contained two stages of pressure vessels in a 2:2:1:1 array. Each vessel contained three 4" diameter by 40" long RO elements. The elements were Koch model TFC-HR, thin film composite (TFC) polyamide elements. The first stage included the first two pairs of pressure vessels in series and the second stage comprised the last two vessels in series. EDC/PPCP rejection was evaluated by testing spiked raw water, permeate water, and brine water. Samples were also collected before and after anti-scale chemical addition.

Montgomery Watson Harza Pilot Plants

Water Recycling Membrane Bioreactor and Ultrafiltration Pilot

Membrane bioreactor (MBR) and ultrafiltration (UF) processes were evaluated separately at the pilot-scale to assess their ability to prevent EDC/PPCP passage. The pilot-testing site for the study was a full-scale tertiary treatment water reuse facility located in Nevada. Tertiary treatment at the full-scale facility consisted of influent screening, grit removal, chemical coagulation, flocculation, primary sedimentation, aeration, secondary clarification, dual media filtration and UV disinfection. The MBR pilot system utilized during this testing was provided by US Filter. A general process flow schematic of the system is provided in Figure B.10 of Appendix B. The pilot used 3 mm pre-screen, influent holding tank and pumps, membrane tank and backwash tank. The system was also equipped with a self-priming pump to provide mixed liquor from the full–scale plant aeration basin to the membrane pilot unit. The membranes, under a light suction, filtered a portion of the mixed liquor from the membrane tank, with the remaining mixed liquor overflowed back to the feed tank. This operation resulted in a cross flow velocity across the membrane surface to prevent fouling. Four membrane modules were submerged in the membrane tank. Each membrane module was comprised of hollow fibers with a nominal pore size of 0.2 microns. During operation, air and mixed liquor were continuously pumped into the membrane tank to scrub and shake the membrane fibers. Technical information on the membranes is provided in Table B.1 of Appendix B.

The second phase of testing was conducted with two UF systems provided by Zenon and Ionics. Feed water to the pilot units was secondary effluent from the full scale tertiary plant described above. The Zenon UF pilot system contained three hollow fiber UF membrane elements (nominal pore size = 0.04 micron) submerged inside a tall rectangular stainless steel process tank. A schematic of the Zenon UF pilot is provided in Figure B.11 of Appendix B. The Ionics UF pilot used two membrane modules housed horizontally in a standard 8" diameter pressure vessel. Each of the two membrane modules contains approximately 10,000 hollow fibers with a fiber length of 5 feet. During filtration, the flow pattern through the membranes is from the inside to the outside and the nominal pore size of the membrane fibers is 0.05 micron. A schematic of the Ionics UF pilot is shown in Figure B.12 of Appendix B.

Ultrafiltration Followed by Reverse Osmosis Water Recycling Pilot

A pilot system consisting of UF followed by RO was evaluated for its ability to reject EDCs/PPCPs. A process flow diagram of the pilot treatment train employed during the study is provided in Figure B.13 of Appendix B. Tertiary wastewater was used as feed water for this pilot system. Values of general water quality parameters measured at various locations in the pilot train are presented in Table 3.6.

A Zenon ZeeWeed® 1000 UF pilot system was used during the study to provide pretreatment to downstream RO membranes. Specifications for the ZW 1000 membrane are provided in Table B.2 in Appendix B. A photo of the UF pilot is provided as Figure B.14 in Appendix B. The UF pilot system included the following components: feed pump, 3 membrane elements, vacuum pump used for filtrate production and backwash, filtrate storage tank used for backwash water supply, air compressor for air scour during backwash, and computerized control and data storage system.

A multi-stage RO pilot system was utilized during this study. The system was configured as a 2-1 array. Stage 1 consisted of four pressure vessels arranged as two parallel 1-1 arrays. Stage 2 contained two single vessels arranged in series. Each vessel housed three RO elements with nominal dimensions of 4" x 40". The stages were arranged in series, to allow concentrate from Stage 1 to serve as feed water for Stage 2. A photo of the RO pilot system is provided as Figure B.15 in Appendix B. Operating conditions for the UF and RO pilot systems are shown in Table B.3 in Appendix B.

Membrane Bioreactor followed by Reverse Osmosis Pilot

The ability of a pilot scale MBR followed by RO to reduce EDC/PPCP concentrations was evaluated at one treatment facility in Southern California and one facility in New Mexico. Table 3.7 summarizes the influent source water and pilot equipment installed at each participating location.

The advanced primary treatment in the Southern California full-scale facility consisted of influent screening, grit removal, chemical coagulation, flocculation, and sedimentation. Feed water for the MBR pilot was pumped from effluent of the sedimentation basin.

A schematic of the Zenon and Mitsubishi MBR pilot units is shown in Figures B.16 and B.17 of Appendix B, respectively. The Zenon MBR pilot unit was equipped with one membrane cassette, composed of three submerged UF membrane elements. An anoxic tank for denitrification was included in the process train installed in New Mexico. Only an aerobic tank for nitrification was included in the process train in Southern California. Parallel testing of a Mitsubishi MBR and a Zenon MBR was conducted in Southern California as shown in Figure B.18 of Appendix B. The MBR samples were obtained for the following range of MBR operating conditions: flux range of 25.5 – 37.4 $Lm^{-2}h^{-1}$ (15-22 gal/ft/d); HRT of 2-6 hours and MLSS range of 4,000-13,000 mg/L.

The RO membranes tested were provided by GE/Osmonics and Saehan Industries. Each membrane is made of polyamide thin-film composite material with similar spiral wound configurations as shown in the specifications provided in Table B.4 in Appendix B. The RO systems were operated in a 2-1 array at a flux range of 15.0-20.4 $Lm^{-2}h^{-1}$ (10-12 gal/ft/d) and a 75% feed water recovery rate. Samples were collected from the MBR influent, MBR effluent, and RO effluent and analyzed for the full suite of EDC and PPCP compounds.

Table 3.6
Water quality measured at UF/RO pilot

			3/25/2005			4/13/2005	
Parameter	**Unit**	**Tertiary**	**UF effluent**	**RO permeate**	**Tertiary**	**UF effluent**	**RO permeate**
Aluminum	µg/L	16.4	13.3	ND	10.7	7.89	ND
Alkalinity total	mg/L	146	147	9.01	145	143	8.49
Arsenic	µg/L	2.03	1.93	ND	1.99	1.99	ND
Boron	µg/L	365	353	283	363	367	301
Chloride	mg/L	268	276	7.93	248	248	7.02
Conductivity	µmho/cm	2420	2550	0.91	1640	1660	65
Hardness	mg/L	352	356	4.4	342	338	5.32
Manganese	µg/L	119	115	ND	80	74.4	1.47
Silica	mg/L	18.4	18.2	0.714	15.7	15.9	0.667
Sulfate	mg/L	233	234	0.769	215	214	0.679
TOC	mg/L	8.13	7.8	ND	7.98	7.51	ND
TDS	mg/L	934	1020	77	960	956	64
Total Coliform	MPN/100mL		<2	4		8	ND
Turbidity	NTU		0.25	0.1		0.4	0.1

Table 3.7
Source water and treatment systems for pilot installation locations

	Southern California	**New Mexico**
Source water	**Advanced Primary**	**Raw wastewater**
MBR System	Zenon ZW 500D and Mitsubishi Sterapore HF	Zenon ZW 500C
RO System	Saehan 4040 FL Dual stage operated at a 75% feed water recovery rate	Osmonics AK4040 Dual stage operated at a 75% feed water recovery rate

Microfiltration followed by Reverse Osmosis and Electrodialysis Reversal

Treatment trains consisting of microfiltration (MF) followed by RO and MF followed by electrodialysis reversal (EDR) were evaluated at the pilot scale to remove EDC/PPCP compounds present in tertiary treated wastewater. The tertiary treated wastewater used in this study was characterized by relatively high levels of total dissolved solids, hardness and alkalinity, with moderate levels of organic material and low turbidity. Table 3.8 presents typical feed water quality at the pilot site.

The trial equipment consisted of a modified US Filter "H" series RO, model number ROSLH 3180. A photograph of this pilot system is provided as Figure B.19 in Appendix B. The pilot system was capable of utilizing up to 12 vessels, but for the purposes of this study, three vessels were used, configured in a 2:1 array with a feed water recovery (FWR) of 50 – 75% (15.0 – 25.5 $Lm^{-2}h^{-1}$ permeate flow). The membrane elements were Dow FilmTec BW30-4040 brackish water membranes. Each vessel housed four RO membrane elements.

Table 3.8
Pilot feed water quality (tertiary treated wastewater)

Parameter		
Cl^- (Chloride)	188	mg/L
NO_3-N	7.1	mg/L
SO_4	96	mg/L
Ca	59.1	mg/L
Cr (Total)	<0.002	mg/L
Fe	0.07	mg/L
Mg	31.7	mg/L
SiO_2	24.0	mg/L
Na	156	mg/L
Conductivity	1200	μmhos/cm
pH	7.3	SU
TOC	9	mg/L
Turbidity	0.7	NTU
Hardness, total ($CaCO_3$)	250	mg/L
Alkalinity, total ($CaCO_3$)	190	mg/L
TDS	720	mg/L
UV-254	0.109	Abs/cm

Membrane pretreated water was continuously fed to the RO pilot system at a flow rate of approximately 75.7 L/min (20 gpm). The pilot system utilized DOW FilmTec BW30-4040 membrane elements. Four elements were placed in series in each of the three pressure vessels in a two-stage, 2:1 array configuration. Sodium bisulfite and anti-scalant were added to the MF pretreated water to control RO membrane fouling and protect the membrane elements from chemical damage due to free chlorine or chloramines.

The EDR equipment consisted of an Aquamite V with a bipolar membrane stack. The capacity of the Aquamite V was 56,800 – 132,500 L/day. The maximum feed flow for this unit was 227,000 L/day. The EDR operated at a range of flows (83-102 L/min) to continually produce demineralized water without constant chemical addition during normal operation. Current was supplied at 2-4 amps depending on the specific water quality goals to be achieved. Membrane fouling and scaling was controlled by using electrical polarity reversal every fifteen minutes. Photographs of the EDR pilot-plant are provided as Figure B.20 in Appendix B.

Typically, EDR systems are configured using multiple stages to provide the maximum membrane surface area and retention time to remove a specified fraction of salt from the demineralized stream. Two types of staging are used: hydraulic and electrical. For this study, the Aquamite V pilot unit operated as a single stack with two electrical stages that could be independently controlled to achieve a desired water quality. Electrical staging was accomplished by inserting additional electrode pairs into the membrane stack to provide maximum salt removal rates while avoiding polarization and hydraulic pressure limitations.

Ultraviolet and Ultraviolet Peroxide Oxidation Pilot

The ability of UV with and without AOP to destroy EDCs/PPCPs was evaluated at the pilot scale level. Feedwater supply for this study was tertiary wastewater from a full-scale reclamation facility which had undergone pretreatment by MF and RO.

Testing was conducted using a low-pressure high-output (LPHO) amalgam UV lamp system, Trojan UV Logic8AL30. The LPHO system has eight lamps arranged parallel to the direction of flow. The lamps are rated 250 watts per lamp as driven from a ballast and the lowest percentage lamp power set point is 60%. The various power setting points and the capability of switching individual lamps on/off allow the pilot system to deliver different doses. The reactor chamber is made of stainless steel with a diameter of 30 cm, a length of 160 cm, and a total reactor volume of 110 L. A complete list of specifications for the UV system is provided in Table B.5 in Appendix B.

The pilot testing was performed in order to evaluate the electrical energy per order parameter, which describes the electrical energy efficiency of a treatment process. By definition, for a first order kinetics decay of a chemical pollutant, the Electrical Energy per Order (E_{EO}) is the electrical energy (kWh) required to reduce the concentration of a pollutant by one order of magnitude in 1000 L of water (Bolton et al. 2001). This figure-of-merit can be expressed in units of kWh/m^3/order or kWh/kgal/order. The E_{EO} expression for a flow-through system is given as:

$$E_{EO} = \frac{P}{F \times \log\left(c_i / c_f\right)} \qquad (3.1)$$

where P is the rated power (kW), F is the flow rate (m^3/h or kgal/h), and c_i and c_f are the influent and the effluent concentration of chemical pollutant. The pilot system was operated to provide a delivered dose necessary to achieve 1-log NDMA destruction. Results from the spiking experiments conducted with NDMA showed the EEO of the pilot system to be approximately 0.17 kWh/1000 gal/order.

Since feedwater used during this study was tertiary wastewater which had undergone pretreatment by UF and RO, essentially all EDC/PPCP compounds were already removed prior to the UV system. Therefore, a solution of selected EDCs/PPCPs was infused upstream of the UV. A list of the chemicals spiked is provided in Table 3.9 with the same spiking procedures described in bench-scale experiments. For hydrogen peroxide addition in the pilot tests, the chemical was dosed to yield a concentration of 5 mg/L in the RO permeate ahead of the UV reactor. A static mixer was installed in the inlet pipe to the UV reactor to ensure adequate mixing of the hydrogen peroxide.

Table 3.9
Specifications of chemicals used in spiking experiment

Chemical	Mass available	Concentration	Form
TCEP	25 g	97.0%	Liquid
Oxybenzone	5 g	98.0%	Powder
Caffeine	25 g	99.0%	Powder
Triclosan (Irgasan)	5 g	97.0%	Powder
DEET	250 mg	97.3%	Liquid
Ibuprofen	1 g	98.0%	Powder
Estrone	1 g	99.0%	Powder
Iopromide	400 mg	97.9%	Powder

FULL-SCALE MATERIALS AND METHODS

As shown in Tables 3.2 and 3.3, twenty full-scale drinking water and six full-scale water reuse systems were tested for contaminant removal. In each case, the facility was asked to collect samples between each unit process and immediately quench oxidant residual with ascorbic acid. The utilities were also directed to follow a plug-flow through the system, with sample collections corresponding to unit hydraulic retention time. In addition, each utility was shipped salinized glass sampling bottles with instructions on proper sampling technique (including following the hydraulic retention times) and a water quality worksheet to incorporate water quality and treatment parameters.

Analytical Methods

Background

To evaluate the removal of trace levels of EDCs and pharmaceuticals, sensitive and robust analytical methods using equipment available to the project team were required. This team has extensive experience in the analyses of these types of contaminants using a variety of analytical and bioanalytical tools (Snyder et al. 1999; Snyder et al. 2000; Snyder et al. 2001a; Snyder et al. 2001b; Snyder et al. 2001c; Snyder et al. 2003). Since EDCs and pharmaceuticals encompass a wide range of chemical structures and properties, analytical methods required a great deal of flexibility in the analysis of these diverse compounds. Due to the number of proposed experiments needed to evaluate many processes at various scales, the analytical methods developed needed to be capable of analyzing large numbers of samples within the project timeframe. Therefore, it was determined that a single solid-phase extraction followed by gas chromatography with tandem mass spectrometry (GC-MS/MS) and liquid chromatography with tandem mass spectrometry (LC-MS/MS) would provide the maximum diversity in chemical analyses within the allotted project timeframe while avoiding time-consuming and variable derivatizations and/or immunoassays. Likewise, significant sensitivity was required, yet sample volumes of no greater than one liter were desired. By utilizing automated solid-phase extraction and highly sensitive and selective MS/MS instruments, highly robust and sensitive methods were developed.

Sample Collection and Preservation

Eagle-Picher 1 L ultra clean, pre-salinized, amber glass sample bottles were provided to utilities for sample collection. For bench scale tests, each sample bottle was preserved after collection with sulfuric acid to pH < 2. For pilot and full-scale experiments, 1 g of sodium azide was added prior to collection to each sample bottle for preservation due to concerns over the handling of sulfuric acid. For full-scale experiments, 50 mg of ascorbic acid were added to sample bottles prior to collection to quench residual oxidants. Samples were kept on ice during transportation back to the laboratory. Once received, samples were stored at 4ºC until solid-phase extraction (SPE). All samples were extracted within 14 days of collection.

Extraction and Concentration

Complete sample extraction and LC-MS/MS and GC-MS/MS analysis details have been published previously (Vanderford et al. 2003; Trenholm et al. 2006). A flow chart of the extraction and analysis procedures is shown in Figure 3.1. In brief, 1 L samples were extracted in batches of six using hydrophilic-lipophilic balance (HLB) cartridges from Waters Corp. All extractions were performed using a Zymark Autotrace automated SPE system. Prior to extraction, the samples were spiked with GC-MS/MS and LC-MS/MS carbon-13 and deuterium labeled surrogate standards (Table 3.10). The HLB cartridges were sequentially preconditioned with dichloromethane (DCM), tert-butyl methyl ether (MTBE), methanol, and reagent water. The samples were then loaded onto the SPE cartridges at 15 mL/min. After sample loading, the SPE cartridges were rinsed with reagent water and dried with a stream of nitrogen. The SPE cartridges were eluted in two separate fractions. The first fraction (F1) was eluted with a solution of 10% methanol/90% MTBE. The second fraction (F2) was eluted using DCM. F1 was concentrated with a gentle stream of nitrogen to a final volume of 1 mL of methanol. Five hundred μL was removed from Fraction 1 and placed into a 500 μL autosampler vial insert for LC-MS/MS analysis, while the remaining 500 μL was retained for a liquid/liquid extraction (LLE) and solvent transfer. F2 was concentrated to 3 mL of DCM with a gentle stream of nitrogen.

Prior to GC-MS/MS analysis, a LLE was performed on the 500 μL aliquot from F1 to remove polar interferences and to exchange the solvent to DCM, after which it was combined with F2. Sodium chloride (2.5 mL of a 25% w/v solution) was added to F1 in order to increase the ionic strength of the solution. A solution of 10% DCM in hexane (3.5 mL) was added and the mixture shaken vigorously and allowed to settle. The top layer (organic) was removed and combined with F2. This procedure was performed on F1 two more times and the resulting extracts were combined with F2. The final combined organic extract was then concentrated with a gentle stream of nitrogen to 1 mL. Two mL of iso-octane were added and the extract concentrated to ~450 μL. A solution of carbon-13 and deuterium labeled internal standards was now added. The extract was adjusted to a final volume of 500 μL using iso-octane and transferred to a 500 μL autosampler vial insert for GC-MS/MS analysis.

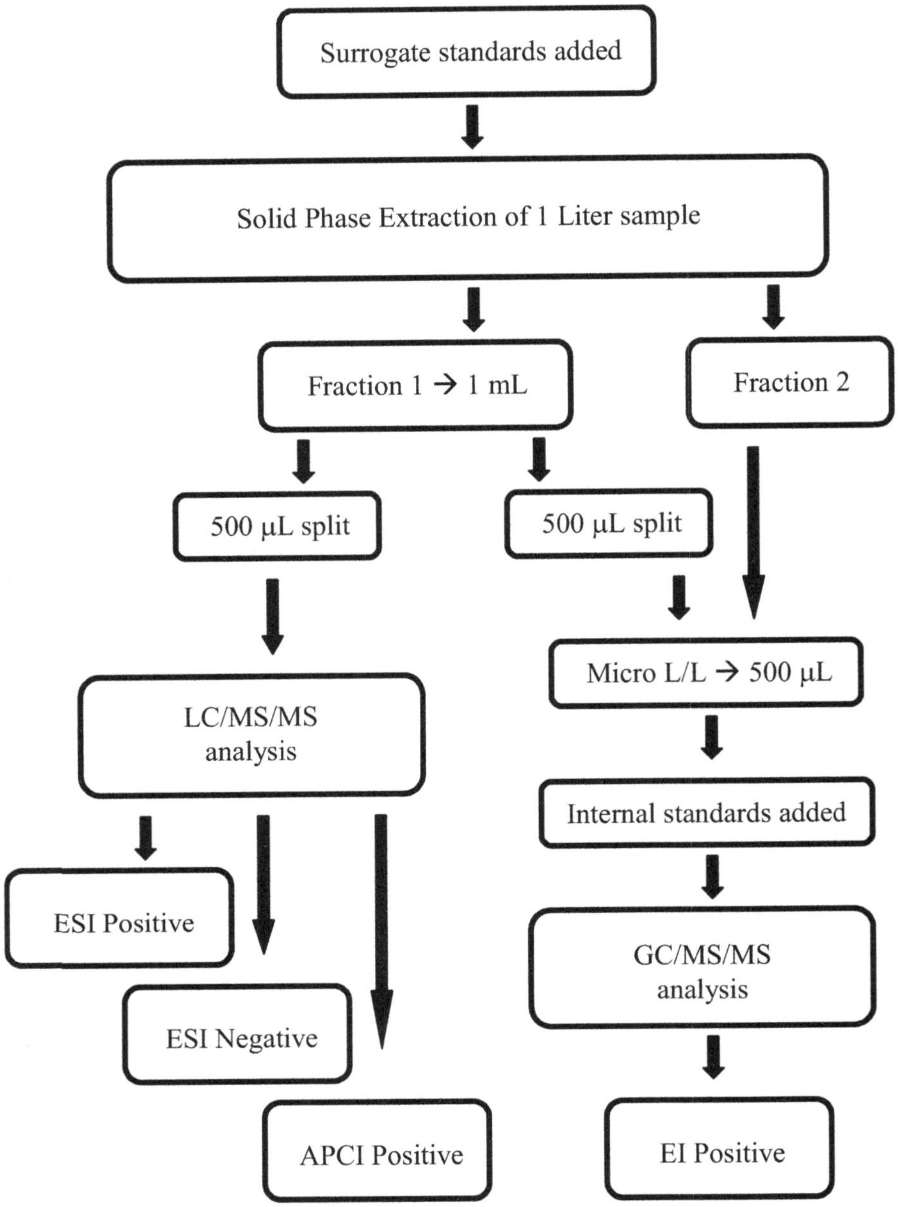

Figure 3.1 Analytical method for target EDCs and pharmaceuticals

Source: Trenholm et al. 2006. Reprinted from *Chemosphere,* 65, Trenholm, R. A., B. J. Vanderford, J. C. Holady, D. J. Rexing, and S. A. Snyder, Broad Range Analysis of Endocrine Disruptors and Pharmaceuticals Using Gas Chromatography and Liquid Chromatography Tandem Mass Spectroscopy, 1990-1998., Copyright 2006, with permission from Elsevier.

Table 3.10
Analytical methods for target compounds, surrogates, and internal standards

LC/MS/MS			GC/MS/MS
ESI – Positive	**ESI – Negative**	**APCI – Positive**	**EI – Positive**
Acetaminophen	Diclofenac	Androstenedione	Benzo[a]pyrene
Atrazine	Gemfibrozil	Estradiol	DDT
Caffeine	Ibuprofen	Estriol	Fluorene
Carbamazepine	Iopromide	Estrone	Galaxolide
DEET	Naproxen	Ethynyl estradiol	Metolachlor
Diazepam	Triclosan	Progesterone	Musk Ketone
Dilantin		Testosterone	Lindane
Erythromycin-H_2O			
Fluoxetine	$^{13}C_3$-Ibuprofen – SU	$^{13}C_2$-Estradiol – SU	Fluorene-d_{10} – SU
Hydrocodone	$^{13}C_{12}$-Triclosan – SU	Testosterone-d_3 – SU	Perylene-d_{12} – SU
Meprobamate			Anthracene-d_{10} – SU
Oxybenzone			$^{13}C_{10}$-Heptachlor – SU
Pentoxifylline			$^{13}C_{10}$-trans-Chlordane – SU
Sulfamethoxazole			$^{13}C_6$-alpha-BHC – IS
TCEP			DDD-d_8 – IS
Trimethoprim			Acenaphthylene-d_8 – IS
$^{13}C_3$-Caffeine – SU			
$^{13}C_3$-Atrazine – SU			
Diazepam-d_5 – SU			
SU – Surrogate; IS – Internal Standard			

Liquid Chromatography with Tandem Mass Spectrometry

The complete LC-MS/MS method for the target analytes has been published previously (Vanderford et al. 2003). Briefly, an Agilent binary pump and a CTC autosampler were used for all analyses. Analytes were separated using a Phenomenex 250 x 4.6 mm Synergi Max-RP C12 column. A binary gradient consisting of 0.1% formic acid (v/v) in water (A) and 100% methanol (B) at a flow rate of 700 µL/min was used. A nine min equilibration step at 5% B was also used at the beginning of each run. An injection volume of 10 µL was used for all analyses .

Mass spectrometry was performed using an Applied Biosystems API 4000 triple-quadrupole mass spectrometer. All analyses were performed using multiple reaction monitoring (MRM) with electrospray ionization (ESI) in both positive and negative modes and atmospheric pressure chemical ionization (APCI) in positive mode. A table showing the analysis method of each of the target compounds is shown in Table 3.10.

Gas Chromatography with Tandem Mass Spectrometry

A detailed description of the GC-MS/MS methodology used in this project has been published previously (Trenholm et al. 2006). Briefly, a Varian CP-3800 gas chromatograph and autosampler were used for all analyses. Analytes were separated using an Agilent 30 m x 0.25 mm ID x 0.25 µm DB5-MS column with a 1.0 mL/min helium flow rate and an initial pressure pulse of 45 PSI for 0.85 min. The temperature program was as follows: 90°C, hold for 2.0 min; 90-150°C @ 20°C/min; 150-280°C @ 3°C/min, hold for 5.0 min; 280-315°C @ 30°C/min, hold for 2.5 min. An injection volume of 2 μL was used for all analyses.

Mass spectrometry was performed using a Varian 2200 ion trap mass spectrometer. All analyses were performed using MRM in positive electron impact (EI) mode. Helium was used for the collision gas.

Quality Assurance and Quality Control

Details of the determination of the instrument detection limits, method detection limits, and other instrument related information can be found in previously published papers (Vanderford et al. 2003; Trenholm et al. 2006). Reporting limits were 1.0 ng/L and 10 ng/L for LC-MS/MS and GC-MS/MS compounds, respectively. During the project, a study of extracted blanks was performed to determine whether any compounds were contaminating the analyses. The data from 24 blanks were plotted and it was determined that caffeine and TCEP were possibly contaminating samples in the range of 1 - 5 ng/L; therefore, reporting limits for caffeine and TCEP were raised to 10 ng/L.

During instrumental analysis, calibration was performed using a minimum of five calibration points, with resulting regression coefficients in excess of 0.990. LC-MS/MS analyses used external calibration and GC/MS/MS analyses used internal calibration using the internal standards shown in Table 3.10. Continuing calibration verifications were analyzed at regular intervals to verify the performance of the instruments.

Analyte recoveries were determined by adding an appropriate amount of a stock solution of the target compounds to one liter of deionized water and Colorado River water and then extracted using the procedure described previously. Table 3.11 shows the mean recovery and percent relative standard deviation (%RSD) for both LC-MS/MS and GC-MS/MS compounds. Within each batch of six extracted samples a deionized water blank and spike were included to monitor the reliability of the extraction process. As discussed above, surrogate standards were added to each sample to monitor the sample extraction and any matrix effects that may have been present. During the project, it was found that the carbon-13 labeled surrogate standards $^{13}C_3$-caffeine and $^{13}C_2$-estradiol proved useful in detecting whether residual oxidants had been properly quenched. Due to the susceptibility of the two compounds to residual oxidant degradation, their addition to the samples prior to analysis and their subsequent absence from the final extract was indicative of unquenched oxidant residual.

Table 3.11
Percent recovery and relative standard deviation for target analytes

Compound	Percent Recovery	Percent RSD
ESI Positive	n=17	
Hydrocodone	112	7
Trimethoprim	91	9
Acetaminophen	41	14
Caffeine	93	10
Erythromycin – H_2O	71	10
Sulfamethoxazole	72	16
Fluoxetine	80	7
Pentoxifylline	84	15
Meprobamate	96	11
Dilantin	86	11
TCEP	75	18
Carbamazepine	91	12
DEET	81	19
Atrazine	89	8
Diazepam	80	19
Oxybenzone	68	19
ESI Negative		
Iopromide	91	9
Naproxen	91	9
Ibuprofen	96	15
Diclofenac	83	11
Triclosan	79	17
Gemfibrozil	94	10
APCI Positive		
Ethynyl estradiol	92	11
Androstenedione	96	12
Estradiol	92	14
Testosterone	83	9
Progesterone	90	12
Estriol	90	12
Estrone	89	10
EI Positive	n=21	
Benzo[a]pyrene	67	9
DDT	71	11
Fluorene	72	8
Galaxolide	30	12
Metolachlor	75	17
Musk Ketone	72	16
Lindane	81	11

Data from Vanderford et al. 2003

As discussed by Vanderford, et al. (Vanderford et al. 2003), it was discovered that the recoveries of androstenedione, testosterone, progesterone, trimethoprim, acetaminophen, and fluoxetine spiked into unpreserved surface waters would quickly decline over time. During initial bench-scale treatment studies, the concentrations of these compounds appeared to increase after chlorination, ozonation, and chloramination. It was determined that these compounds were highly susceptible to biodegradation, hence natural water samples disinfected using chlorine, ozone, or chloramine hindered degradation of these compounds while the spiked raw water controls continued to degrade. Therefore, it appeared that androstenedione, testosterone, progesterone, trimethoprim, acetaminophen, and fluoxetine were "created" during disinfection. However, this observation was actually the loss of these compounds in non-disinfected raw water due to biodegradation.

Since low pH is commonly used to preserve samples from biodegradation, sulfuric acid preservation at pH < 2 was tested to determine whether it would prevent degradation of the aforementioned compounds and not collaterally affect the recoveries of other compounds. Eight surface water samples from the Boulder Basin of Lake Mead, Nevada were collected in 1L salinized amber glass bottles. Four sets of samples, each containing one sample and one duplicate, were prepared. Two sets were fortified at 100 ng/L with the target compounds and left unpreserved and the other two sets were fortified at the same concentration and preserved to pH 2 with concentrated sulfuric acid. One set of preserved/unpreserved samples was held for seven days at 4°C and the duplicate set was held for 14 days. The samples were then extracted and analyzed as described previously. The sulfuric acid preservation both prevented the degradation of the affected compounds and did not adversely affect the recoveries of the other compounds being studied. Table 3.12 shows the results of the compounds that degraded in surface water and several other unaffected compounds for reference. All compounds not shown in Table 3.12 had recoveries that did not change in preserved versus unpreserved samples (Vanderford et al. 2003).

Since pilot and full-scale sampling required that pre-preserved bottles be shipped to various testing locations, it was agreed that shipment and addition of the concentrated sulfuric acid preservative discussed above was impractical. Therefore, sodium azide, which had been previously found to have minimal impact on target analyte recovery, was used as a preservative in all pilot and full-scale samples.

Quenching of oxidants was also investigated for possible impacts on the analyses of target analytes. Three common quenching agents were compared: sodium thiosulfate ($Na_2S_2O_3$), ammonium chloride (NH_4Cl), and ascorbic acid. Tests were initially performed on spiked deionized water samples to determine the effects of each quenching agent on compound recoveries. Sodium thiosulfate resulted in significant loss of trimethoprim, erythromycin, fluoxetine, atrazine, diazepam, progesterone and diclofenac. Ascorbic acid resulted in a 50% loss of only erythromycin, while ammonium chloride did not affect compound recoveries. Even with the loss of erythromycin, ascorbic acid was chosen as the preferred quenching agent as it was most compatible with the sulfuric acid preservation and also had little impact on analyte recovery. Selected compound recoveries for each quenching agent are shown in Figure 3.2.

Table 3.12
Sulfuric acid preservation study

Compound	Mean Surface Water Recovery - Unpreserved	Mean Surface Water Recovery - Sulfuric acid preserved *
Degradation Observed		
Trimethoprim	70%	101%
Fluoxetine	49%	76%
Androstenedione	56%	97%
Testosterone	50%	97%
Progesterone	22%	93%
Unaffected Compounds		
Caffeine	71%	72%
Atrazine	70%	73%
Ibuprofen	65%	64%
Diazepam	60%	62%
Ethynyl estradiol	98%	100%

* Samples were preserved for 7 and 14 days respectively and the average was reported

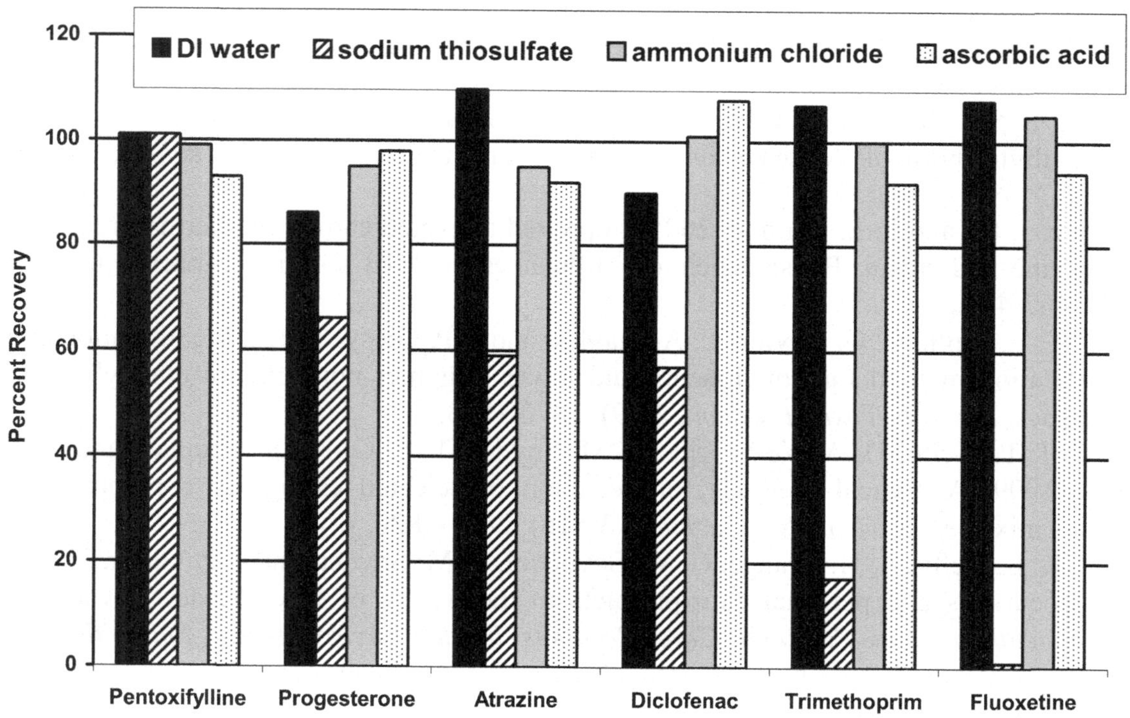

Figure 3.2 Selected compound recoveries from quenching agents in spiked deionized water samples

REFERENCES

Allgeier, S. C., S. R. Summers, J. G. Jacangelo, V. A. Hatcher, D. M. Moll, S. M. Hooper, J. W. Swertfeger, and R. B. Green. 1996. A simplified and rapid method for biodegradable dissolved organic carbon measurement. In *AWWA WQTC*: AWWA.

Black/Eagle.1999. Phase III Shallow Ground Water Investigation, Volumes I and II. Report prepared for the City of Henderson by Black Eagle Consulting.

Bolton, J. R., K. G. Bircher, W. Tumas, and C. A. Tolman. 2001. Figures-of-merit for the technical development and application of advanced oxidation technologies for both electric- and solar-driven systems - (IUPAC Technical Report). *Pure and Applied Chemistry*, 73(4): 627-637.

Bolton, J. R., and K. G. Linden. 2003. Standardization of methods for fluence (UV dose) determination in bench-scale experiments. *Journal of Environmental Engineering* 129(3): 209-216.

Crittenden, J. C., S. Sanongraj, J. L. Bulloch, D. W. Hand, T. N. Rogers, T. F. Speth, and M. Ulmer. 1999. Correlation of aqueous-phase adsorption isotherms *Environ. Sci. Technol.*, 33: 2926-2933.

Escobar, I. C., and A. A. Randall. 1999. Influence of NF on distribution system biostability. *Journal of American Water Works Association* 91(6): 76-89.

KMG.2001. Seep Area Groundwater Characterization Report. Report prepared for the Nevada Division of Environmental Protection by Kerr-McGee Chemical, LLC

Leising, J. F.2003. Chemical Conditions in the Primary Producing Aquifers and Portions of the Shallow Groundwater System in the Las Vegas Valley in 2000. Southern Nevada Water Authority Report

Na, C., F. S. Cannon, and B. Hagerup. 2002. Perchlorate removal via iron-reloaded GAC and borohydride regeneration. *Journal American Water Works Association* 94(11): 90-102.

Nowack, K. O., F. S. Cannon, and D. W. Mazyck. 2004. Enhancing activated carhon adsorption of 2-methylisoborneol: Methane and steam treatments. *Environ. Sci. Technol.*, 38(1): 276-284.

Rangel-Mendez, J. R., and F. S. Cannon. 2005. Improved activated carbon by thermal treatment in methane and steam: Physicochemical influences on MIB sorption capacity. *Carbon* 43(3): 467-479.

Snyder, S. A., T. L. Keith, C. G. Naylor, C. A. Staples, and J. P. Giesy. 2001a. Identification and quantification method for nonylphenol and lower oligomer nonylphenol ethoxylates in fish tissues. *Environ. Toxicol. Chem.*, 20(9): 1870-1873.

Snyder, S. A., T. L. Keith, D. A. Verbrugge, E. M. Snyder, T. S. Gross, K. Kannan, and J. P. Giesy. 1999. Analytical methods for detection of selected estrogenic compounds in aqueous mixtures. *Environ. Sci. Technol.*, 33(16): 2814-2820.

Snyder, S. A., K. L. Kelly, A. H. Grange, G. W. Sovocool, E. M. Snyder, and J. P. Giesy. 2001b. Pharmaceuticals and personal care products in the waters of Lake Mead, Nevada. In *Pharmaceuticals and Personal Care Products in the Environment: Scientific and Regulatory Issues*. Edited by C. G. Daughton and T. L. Jones-Lepp. Washington,D.C.: American Chemical Society.

Snyder, S. A., E. Snyder, D. Villeneuve, K. Kurunthachalam, A. Villalobos, A. Blankenship, and J. Giesy. 2000. Instrumental and Bioanalytical Measures of Endocrine Disruptors in Water. In *Analysis of Environmental Endocrine Disruptors*. Edited by L. H. Keith, T. L. Jones-Lepp and L. L. Needham. Washington,DC: American Chemical Society.

Snyder, S. A., B. J. Vanderford, R. A. Pearson, O. Quinones, and Y. Yoon. 2003. Analytical methods used to measure endocrine disrupting compounds in water. *Practice Periodical of Hazardous,Toxic, and Radioactive Waste Management* 7(4): 224-234.

Snyder, S. A., D. L. Villeneuve, E. M. Snyder, and J. P. Giesy. 2001c. Identification and quantification of estrogen receptor agonists in wastewater effluents. *Environ. Sci. Technol.*, 35(18): 3620-3625.

Ternes, T. A. 2001. Analytical methods for the determination of pharmaceuticals in aqueous environmental samples. *Trends in Analytical Chemistry* 20(8): 419-434.

Ternes, T. A., R. Hirsch, J. Mueller, and K. Haberer. 1998. Methods for the determination of neutral drugs as well as betablockers and β_2-sympathomimetics in aqueous matrices using GC/MS and LC/MS/MS. *Fresenius' Journal of Analytical Chemistry* 362: 329-340.

Trenholm, R. A., B. J. Vanderford, J. C. Holady, D. J. Rexing, and S. A. Snyder. 2006. Broad Range Analysis of Endocrine Disruptors and Pharmaceuticals Using Gas Chromatography and Liquid Chromatography Tandem Mass Spectroscopy. *Chemosphere* 65: 1990-1998.

Vanderford, B. J., R. A. Pearson, D. J. Rexing, and S. A. Snyder. 2003. Analysis of Endocrine Disruptors, Pharmaceuticals, and Personal Care Products in Water Using Liquid Chromatography/Tandem Mass Spectrometry. *Anal. Chem.*, 75(22): 6265-6274.

Yoon, Y., P. Westerhoff, S. A. Snyder, and M. Esparza. 2003. HPLC-fluorescence detection and adsorption of bisphenol A, 17β-estradiol, and 17α-ethynyl estradiol on powdered activated carbon. *Water Res.*, 37(14): 3530-3537.

Zikmund, K. S.1996. Extent and Potential Use of the Shallow Aquifer and Wash Flow in the Las Vegas Valley, Nevada. Southern Nevada Water Authority Report

CHAPTER 4
COAGULATION, FLOCCULATION AND CHEMICAL SOFTENING

BASIC THEORY

Chemical coagulation using aluminum or iron based salts will precipitate metal hydroxides. Chemical softening removes dissolved calcium and magnesium using lime and soda-ash to precipitate calcium carbonate (pH > 9.0) and magnesium hydroxide (pH > 11). Organic compounds can co-precipitate with the metal hydroxides and carbonates or adsorb onto surfaces after precipitation. Chemical coagulation and softening aid in removing suspended solids (i.e., turbidity), colloids, and some dissolved organic carbon (DOC) from water; therefore, organic compounds associated with the surfaces of suspended solids (e.g., organic debris, clays, etc.), colloids, and DOC can also be removed. Sedimentation and filtration follow coagulation or chemical softening to remove these newly formed particles.

BENCH-SCALE EVALUATIONS

Two coagulants, aluminum sulfate (alum, $Al_2(SO_4)_3$) and ferric chloride (ferric, $FeCl_3$), were used in jar tests simulating coagulation, flocculation, and sedimentation (see Chapter 3, section of Bench-Scale Materials and Methods). Calcium hydroxide ($Ca(OH)_2$) and soda ash (Na_2CO_3) were used to evaluate chemical softening treatments in separate jar tests.

During coagulation experiments, the average percent removal for the four waters studied, namely Colorado River Water, Passaic River Water, Ohio River Water, and Suwannee River Water (CRW, PRW, ORW, SRW), was less than 15% for 34 of the 36 target compounds. The complete data set can be found in Appendix C. There was no significant difference in percent removal when using stoichiometrically equivalent dosages of ferric chloride or alum. In addition, percent removals did not appear to be enhanced when using a lower operating pH (pH=5.5) during alum coagulation compared to ambient pH. The remaining two compounds, DDT and benzo(a)pyrene, showed percent removals of 31% and 70%, respectively. Removal likely occurred partitioning onto the particulate matter initially present in the source water or partitioning/adsorbing onto metal (hydr)oxide or carbonate precipitates formed during coagulation. Higher removals for these compounds were expected since they are the most hydrophobic compounds (log $K_{ow} > 6.0$) within the target compound list. When considering the log K_{ow} of the compounds evaluated versus percent removal, a general increase in removal is observed with log K_{ow}, even for those compounds experiencing less than 10% removal (Figure 4.1). However, the overall octanol-water partition coefficient (D_{ow}) should be considered for those compounds that are present in anionic form in the range of pH conditions evaluated. The values for D_{ow} depend on both K_{ow} and pKa, with the latter not readily available for most target analytes. Therefore, diclofenac, gemfibrozil, musk ketone, naproxen, and triclosan were all removed from Figure 4.1 since they exist as anions at ambient pH. The log K_{ow} provides a rough indication of percent removal by coagulation. From these data, a log K_{ow} of greater than five is required in order to observe any appreciable removal by aluminum or iron salt coagulation. However, other factors not related to log K_{ow} may occur during surface adsorption that may lead to apparent contaminant removal (Tolls 2001).

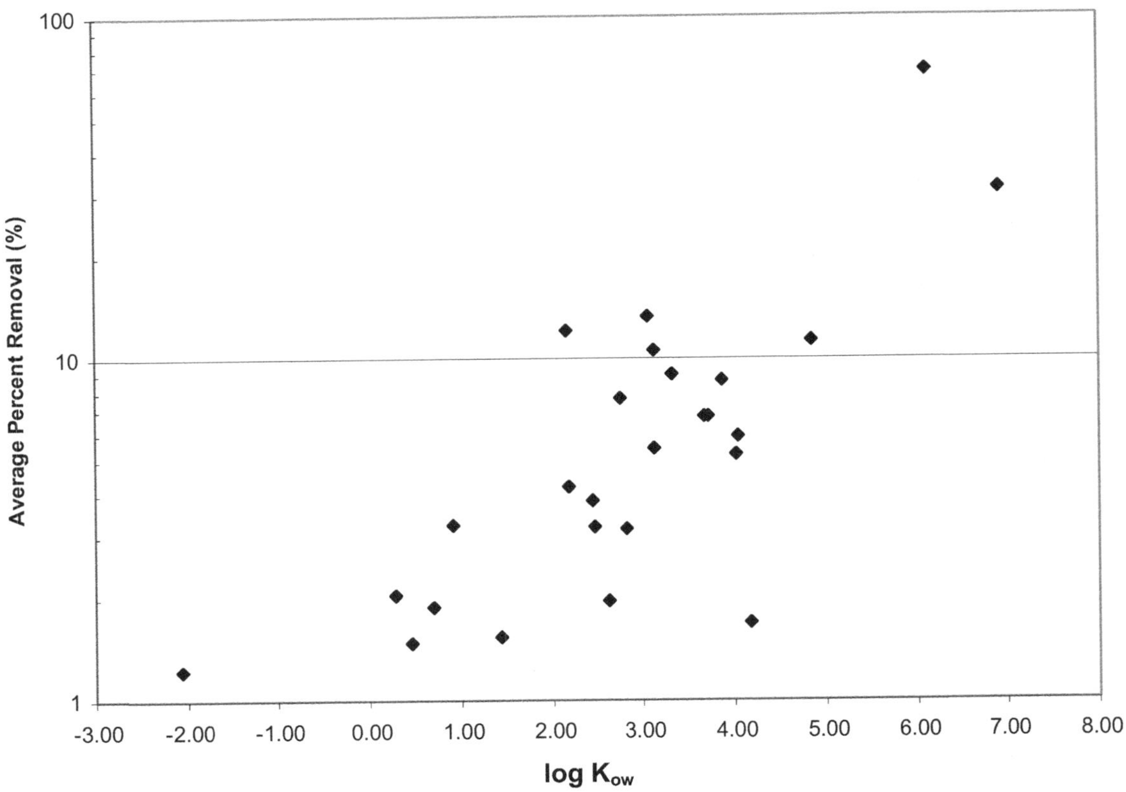

Figure 4.1 Relationship between average percent removal and log K_{ow} during bench-scale coagulation experiments with ferric chloride and alum

During chemical softening experiments, the average percent removal for the four waters studied was again less than 15% for 34 of the 36 target compounds. Once again, DDT and benzo(a)pyrene experienced significant removals at 34% and 74%, respectively. The removal mechanisms are likely the same for chemical softening as coagulation. Overall, chemical precipitation processes achieved minimal removal of the target compounds examined.

PILOT-SCALE EVALUTATIONS

Louisville Water Company (LWC)

The LWC pilot-plant was operated using enhanced coagulation with sedimentation. During pilot coagulation testing, 29 of the 36 target compounds were spiked into the raw Ohio River water. Ferric chloride dosages of 8 mg/L and 26.5 mg/L resulted in less than 30% removal for each target compound. An alum dosage of 35 mg/L produced less than 15% removal for each target compound. The complete data set can be found in Appendix D. Overall, these results agreed with bench scale findings indicating relatively poor (<30%) contaminant removal by coagulation.

Southern Nevada Water Authority (SNWA)

The SNWA treatment plant operates as a direct filtration plant without sedimentation. Since the CRW is low in turbidity and particles, a small ferric chloride dose of 0.2-0.3 mg/L is applied during daily operation. Pilot-scale experiments were conducted with CRW using a 6-gpm pilot-plant operated at a ferric dosage of 0.25 mg/L. As expected from bench-scale studies, coagulation was ineffective at removing the target compounds (Table D.2 in Appendix D). No target compound was removed by greater than 20%. The primary purpose of the post-coagulation sampling was to provide influent concentration data for subsequent biological filtration studies discussed in Chapter 11.

FULL-SCALE OBSERVATIONS

As described in Chapter 3 (Table 3.2), contaminant removal by coagulation was evaluated at 13 of the 20 full-scale facilities investigated. One facility incorporated coagulation followed by lime softening and sedimentation. Since the occurrence of the target compounds varied by facility, Table 4.1 presents only the target compounds detected and the average percent removal. Although the percent removals appear to be high in some cases, these compounds were often were detected at concentrations near the MRL. Additionally, some utilities used pre-oxidants which likely increased removal. While utilities were encouraged to sample a "plug" of water in accordance with the hydraulic retention time (HRT), proper sample timing and/or accurate knowledge of the HRT may have been problematic. If a utility detected a target compound pre-coagulation and not post-coagulation, then the percent removal was calculated based upon the minimum reporting level (1.0 ng/L).

RESULT SUMMARY AND DISCUSSION

Coagulation and chemical softening processes remove particles during water treatment. Therefore, compounds bound to particles can be removed by these processes. Likewise, compounds dissolved in water are not expected to be substantially removed by coagulation, flocculation, and/or softening. Table 4.2 summarizes the removal categories for the target compounds. The experiments conducted were unable to provide significant removal of the target compounds investigated with the exception of DDT and benzo(a)pyrene. These results have confirmed previous finding about the inefficiency of coagulation to remove diclofenac, carbamazepine and sulfamethoxazole (Adams et al. 2002; Ternes et al. 2002). Contaminant removal appeared to be related to the log K_{ow} of the individual compounds. If the compound of concern has a log K_{ow} greater than 5.0, then some removal could be expected due to particle binding and subsequent settling/filtering. However, these highly hydrophobic contaminants are most likely to enter a facility already bound to particles and may not be detected in the analytical protocol is not suitable for particle analysis (i.e., pre-filtration). Although coagulation and softening processes were relatively ineffective for the removal of target compounds, these processes will play a major role in the removal of contaminants that are bound to particles entering a water treatment plant.

Table 4.1
Summary of full-scale occurrence and percent removal

Target Compound	Full-Scale Utilities Detected at 2*MRL (n)	Degree of Removal Observed			
		< 20%	20-50%	50-80%	> 80%
Acetaminophen	1				1
Atrazine	11	11			
Caffeine	14	12	1	1	
Carbamazepine	11	8	1	1	1
DEET	14	12	2		
Dilantin	11	8	3		
Erythromycin	3	1		2	
Gemfibrozil	7	4	1	2	
Ibuprofen	10	10			
Iopromide	11	8	3		
Meprobamate	9	9			
Metolachlor	6	6			
Naproxen	5	3			2
Oxybenzone	1			1	
Sulfamethoxazole	11	5	1		5
TCEP	14	6	3	3	2
Triclosan	3	1	1	1	
Trimethoprim	2			2	

Table 4.2
Summary of percent removal by coagulation and chemical softening

Coagulation with Alum or Ferric Chloride			
> 80% Removal	50-80% Removal	20-50% Removal	< 20% Removal
	Benzo(a)pyrene	DDT	Acetaminophen (Tylenol)
			Androstenedione
			Atrazine
			Caffeine
			Carbamazepine
			DEET
			Diazepam (Valium)
			Diclofenac
			Dilantin
			Erythromycin
			Estradiol
			Estriol
			Estrone
			Ethinyl Estradiol
			Fluorene
			Fluoxetine (Prozac)
			Galaxolide
			Gemfibrozil
			Hydrocodone
			Ibuprofen (Advil)
			Iopromide
			Lindane (γ-BHC)
			Meprobamate
			Metolachlor
			Musk Ketone
			Naproxen
			Oxybenzone
			Pentoxifylline
			Progesterone
			Sulfamethoxazole
			TCEP
			Testosterone
			Triclosan
			Trimethoprim

REFERENCES

Adams, C., Y. Wang, K. Loftin, and M. Meyer. 2002. Removal of antibiotics from surface and distilled water in conventional water treatment processes. *Journal of Environmental Engineering* 128(3): 253-260.

Ternes, T. A., M. Meisenheimer, D. Mcdowell, F. Sacher, H.-J. Brauch, B. Haist-Gulde, G. Preuss, U. wilme, and N. Zulei-Seibert. 2002. Removal of pharmaceuticals during drinking water treatment. *Environ. Sci. Technol.*, 36(17): 3855-3863.

Tolls, J. 2001. Sorption of veterinary pharmaceuticals in soils: A review. *Environ. Sci. Technol.*, 35(17): 3397-3406.

CHAPTER 5
ACTIVATED CARBON

BASIC THEORY

The activation process for the manufacture of activated carbon creates highly porous materials with a distribution of pore sizes and surface areas. Depending upon the activation process and base material, activated carbons also have a distribution of surface functional groups that influence their adsorption characteristics. Activated carbon primarily removes natural organic matter (NOM) and synthetic organic compounds (e.g., pesticides and chlorinated solvents) from water through hydrophobic interactions between the organics and the activated carbon surface. Contaminants are adsorbed by attraction to and accumulation within the activated carbon. Pore structure, as defined by the distribution of pore sizes, has a large influence on both adsorption capacity and adsorption kinetics. An appropriate distribution of mesopores (20-500 Å) and micropores (<20Å) will increase the adsorption capacity for organic contaminants (Rangel-Mendez and Cannon 2005). Those contaminants with low aqueous solubility, and a size conducive to fitting within the pore structure, are most readily adsorbed by activated carbon. In drinking water treatment, the concentration and type of NOM in the source water will compete with contaminants for surface adsorption sites, thus NOM can reduce the effectiveness of activated carbon for contaminant removal (Knappe et al. 1998; Li et al. 2003).

POWDERED ACTIVATED CARBON

Bench-Scale Evaluations

Powdered Activated Carbon

While the EDC/PPCP spiked concentrations varied slightly among experiments with the four source waters, the percentage removal of EDC/PPCPs has been reported to be independent of initial concentration (Westerhoff et al. 2005). Table 5.1 summarizes the average percentage removal for all four waters. The standard deviation of the percentage removal among the four waters was generally less than 10. The percentage removals were consistently higher in SRW experiments. This may be related to the lack of preservation during SRW experiments (see Analysis section in Chapter 3) or the result of differences in organic matter characteristics between the high specific UV absorbance (SUVA) of SRW and as compared to the lower SUVA of the other waters. Increasing SUVA correlates with higher aromatic carbon content, increased hydrophobicity, and higher molecular weight, all of which can affect removal of contaminants by PAC (Graham et al. 2000).

Octanol-water partition coefficients (K_{ow}) of the neutral compounds were well correlated with the percentage EDC/PPCP removal by PAC (Westerhoff et al. 2005). This is anticipated, as adsorption of neutral compounds on activated carbon is dominated by hydrophobic interactions for which K_{ow} is a good indicator. However, K_{ow} was generally not a good indicator for deprotonated acids or protonated bases, since reported K_{ow} values generally represent the neutral form of the molecules. For these species, the overall octanol-water partititon coefficient (D_{ow}) is more appropriate as it relates contaminant hydrophobicity to the particular water pH. Since the experimental pKa of many emerging contaminants is not readily available, it is difficult to conclude definitively about the relevance of D_{ow} in contaminant removal.

Table 5.1
Bench-scale average removal by PAC (5 mg/L WPM, 4 hour contact time)

	Average Spiked Concentration (ng/L)	Average Percent Removal (%)	log K_{ow}
Acetaminophen	41	78	0.46
Androstenedione	59	70	2.75
Atrazine	62	63	2.61
Benzo(a)pyrene	110	72	6.13
Caffeine	59	59	-0.07
Carbamazepine	71	72	2.45
DDT	89	70	6.91
DEET	59	54	2.18
Diazepam	51	67	2.82
Diclofenac	29	49	4.51
Dilantin	25	56	2.47
Erythromycin	33	52	3.06
Estradiol	67	55	4.01
Estriol	90	58	2.45
Estrone	100	77	3.13
Ethinyl Estradiol	91	70	3.67
Fluorene	172	94	4.18
Fluoxetine	40	91	4.05
Galaxolide	130	59	5.3
Gemfibrozil	40	38	4.77
Hydrocodone	79	72	2.16
Ibuprofen	47	26	3.97
Iopromide	59	31	-2.05
Lindane (γ-BHC)	134	70	3.72
Meprobamate	50	36	0.7
Metolachlor	221	50	3.13
Musk Ketone	161	69	4.3
Naproxen	25	60	3.18
Oxybenzone	28	92	3.79
Pentoxifylline	40	71	0.29
Progesterone	46	84	3.87
Sulfamethoxazole	19	43	0.89
TCEP	33	60	1.44
Testosterone	73	71	3.32
Triclosan	47	90	4.76
Trimethoprim	48	69	0.91

In order to evaluate the impact of initial contaminant concentration on removal, adsorption experiments using estradiol (E2) at concentrations ranging from 0.025 nM (6.8 ng/L) to 5 nM (1362 ng/L) were performed (Figure 5.1). The effect of initial E2 concentration within the range investigated was minimal. E2 removal ranged from 70 to 85% in reagent water and from 40 to 50% in CRW. The results were consistent with a previous study where the removal of micropollutants (2-methylisoborneol and geosmin) by PAC was independent of initial contaminant concentrations (Knappe et al. 1998). These data suggest that percent E2 removal at lower concentrations would be equivalent to the percent removals observed at higher levels.

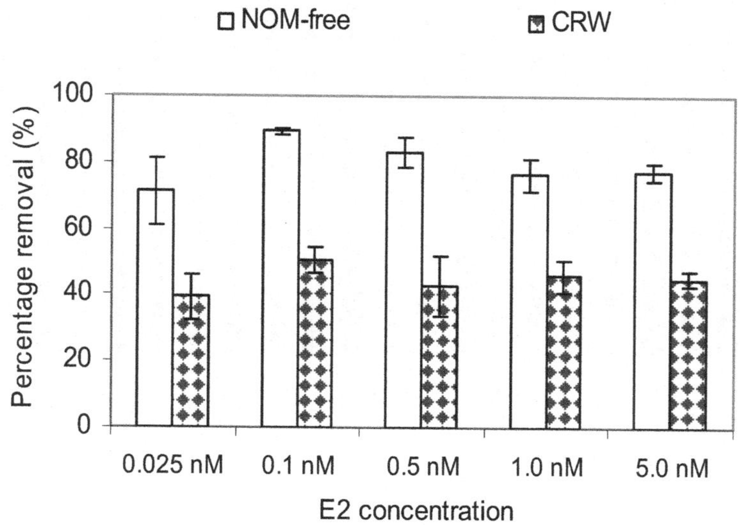

Figure 5.1 Effect of initial concentration on E2 removal from reagent water and CRW using PAC

Pilot-Scale Evaluations

PAC was evaluated at pilot-scale using a dynamic spike with flow-through conditions using Ohio River water with an ambient pH of 7.9. A positive-displacement diaphragm pump was used to introduce the target compounds into the process stream. Compounds were spiked to achieve nominal doses of 100-300 ng/L per compound. A static mixer followed the injection to provide mixing. Coagulant chemicals were injected upstream of a second static mixer before entering the flocculation basin. Norit Hydrodarco B, which is produced by steam-activation of lignite based coal, was used for all pilot experiments. The PAC slurry was fed to the flocculation basin influent at concentrations of both 5 and 35 mg/L. Samples for analysis were collected from the spiked raw water and after 5 hours and 20 hours of PAC contact time. Only minimal differences in removal were observed between the 5 and 20 hour contact times; therefore, results are shown for the 5 hour contact time only (Figure 5.2). Although GC-MS compounds were spiked, γ-BHC, DDT, galaxolide, and benzo[a]pyrene were not detectable in spiked raw water. The rapid loss of certain spiked compounds from raw water was mostly likely due to binding to suspended materials in the relatively turbid Ohio River raw water.

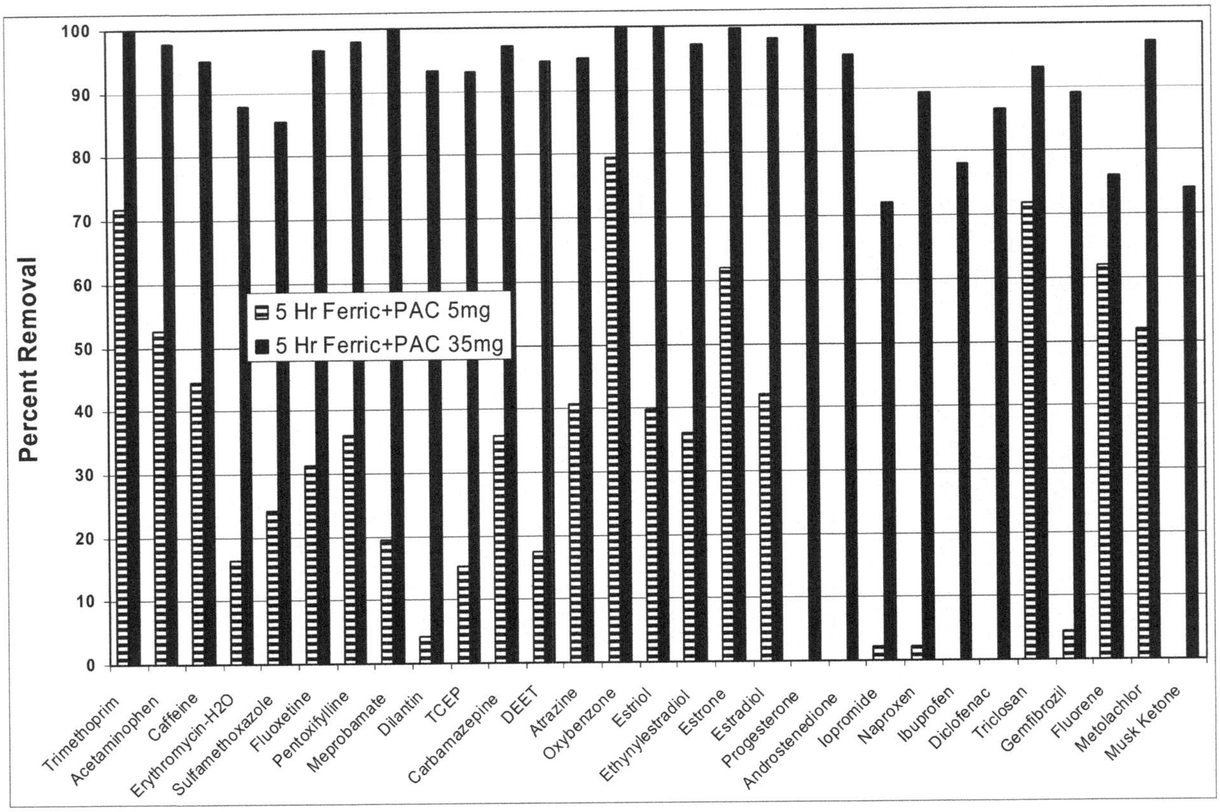

Figure 5.2 Pilot-scale removal by PAC (5 hour contact time)

GRANULAR ACTIVATED CARBON

Bench-Scale Evaluations

Table 5.2 contains results for rapid small-scale column tests (RSSCTs) with GAC. Constant diffusivity similitude (described in chapter 3) was used in these tests. Data presented in Table 5.2 depict the improvement that Superdarco carbon provides relative to HD4000. On average, the Superdarco offered 3.3 times longer bed life to a given extent of breakthrough than did the HD4000. Compared to the Superdarco GAC, the HD4000 GAC had 36% less micropore volume and 18% less mesopore volume. The higher adsorption capacity of Superdarco implies that greater pore volume and the correct distribution of pore volume is necessary for compound removal. Figures 5.3-5.8 compare the breakthrough curves for target compounds when using the two carbons.

Table 5.2
Rapid small-scale column tests with GAC

	Bed Volumes Treated						Improve-ment Factor*	Avg. Influent Conc. (ng/L)
	5% Breakthrough		10% Breakthrough		20% Breakthrough			
	HD4000	Superdarco	HD4000	Superdarco	HD4000	Superdarco		
Acetaminophen	19500	36000	28100	45100	40400	57300	1.6	148
Androstenedione	37200	89600	58000	>90000	>55000	>90000	2.4	209
Atrazine	13000	41600	18400	49600	29300	66300	2.7	227
Caffeine	19700	50900	23600	56800	34200	62900	2.3	234
Carbamazepine	16400	54100	23700	67100	34000	78600	2.8	181
DEET	10900	36800	14300	40600	23200	51000	2.8	210
Diazepam	11700	65300	26100	73000	36000	88400	3.6	153
Diclofenac	9810	37900	12800	47800	18100	63400	3.7	146
Dilantin	7900	32800	11300	38900	17100	54200	3.6	170
Erythromycin	18000	141000	32100	>90000	38500	>90000	7.8	14
Estradiol	41900	82800	51200	101000	>55000	>90000	2.0	200
Estriol	32000	66100	38200	73800	48200	89300	2.0	227
Estrone	36900	70900	58000	85200	>55000	>90000	1.9	228
Ethynylestradiol	31800	69600	44000	83000	65500	107000	1.9	209
Fluoxetine	37700	94300	>55000	>90000	>55000	>90000	2.5	117
Gemfibrozil	10200	40300	13200	49300	18400	63500	3.7	186
Hydrocodone	19100	57900	27500	69500	39700	82100	2.5	204
Ibuprofen	4280	28900	6880	30400	10100	33200	4.8	179
Iopromide	2290	28600	4580	29600	8260	31700	7.6	215
Meprobamate	5960	28600	8070	29800	12000	32100	3.7	198
Naproxen	11500	42200	14800	52100	18400	68200	3.6	173
Oxybenzone	>55000	>90000	>55000	>90000	>55000	>90000	--	6
Pentoxifylline	18100	59100	26300	70100	36800	84300	2.7	177
Progesterone	>55000	>90000	>55000	>90000	>55000	>90000	--	179
Sulfamethoxazole	3090	28300	6030	29100	9370	30700	5.8	164
TCEP	11900	37100	13000	39300	15200	43600	3.0	178
Testosterone	35100	87700	51400	110000	>55000	>90000	2.5	201
Triclosan	>55000	93100	>55000	>90000	>55000	>90000	--	95
Trimethoprim	23300	65200	31600	72700	47400	87600	2.3	199

*Improvement Factor (IF) = (ratios of 5% break through + ratios of 10% BT + ratios of 20% BT)/3

$$IF = \frac{(BVT)_{\left[C/C_0=5\%\right]} + (BVT)_{\left[C/C_0=10\%\right]} + (BVT)_{\left[C/C_0=20\%\right]}}{3} \qquad (5.1)$$

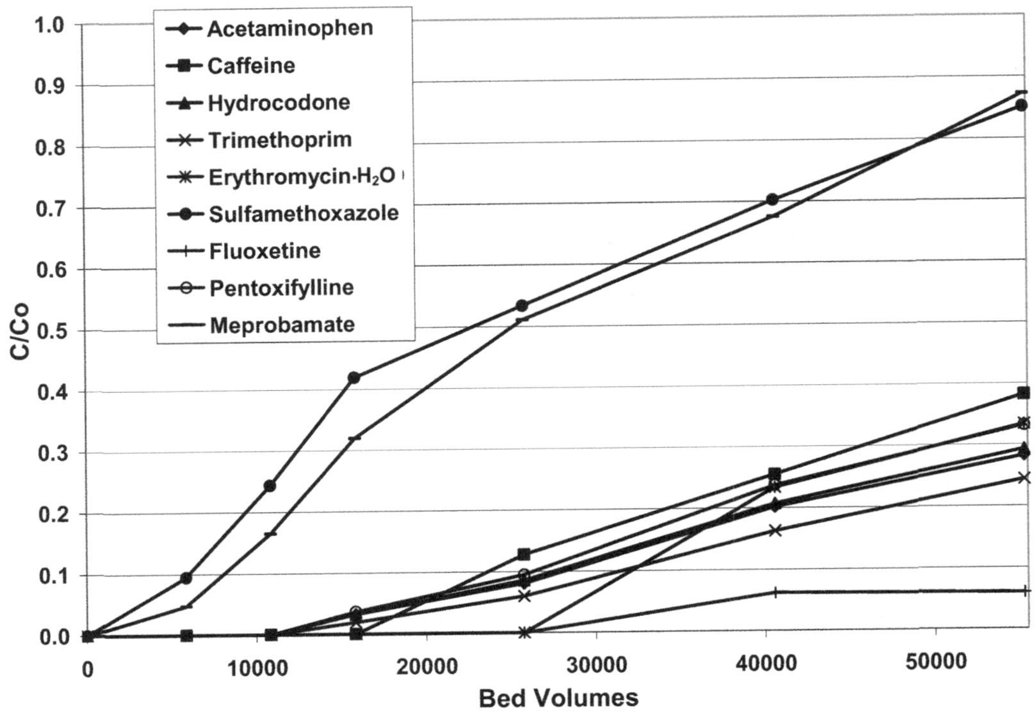

Figure 5.3 Results of select target analytes RSSCT using HD4000 GAC – Part 1 (Data from Snyder et al. 2006a)

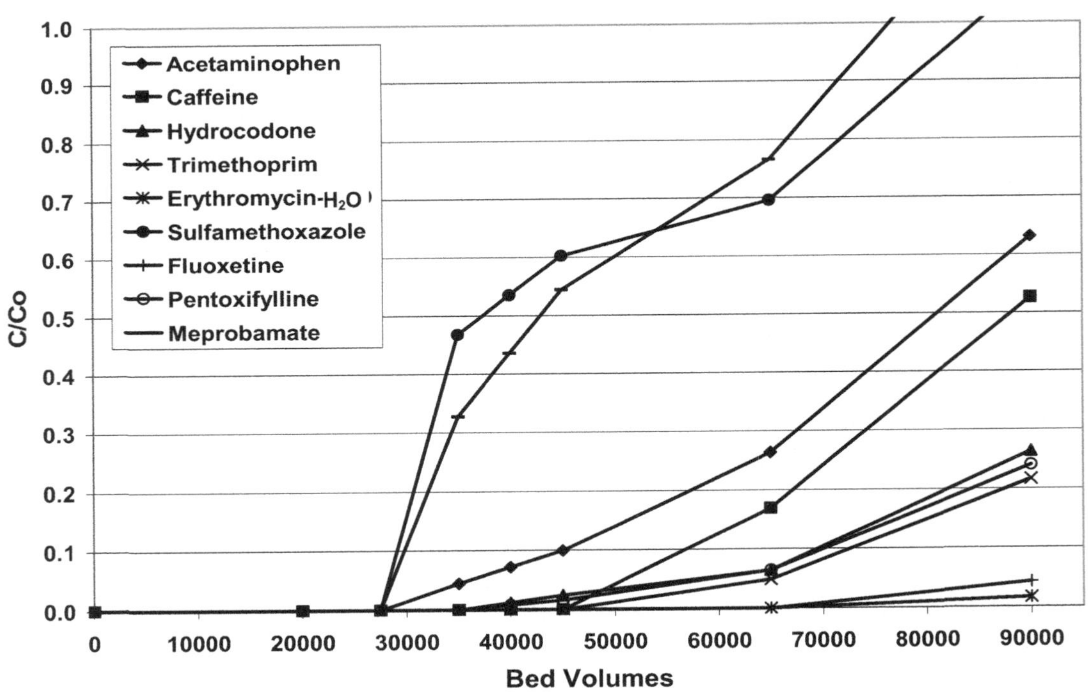

Figure 5.4 Results of select target analytes RSSCT using Superdarco GAC – Part 1 (Data from Snyder et al. 2006a)

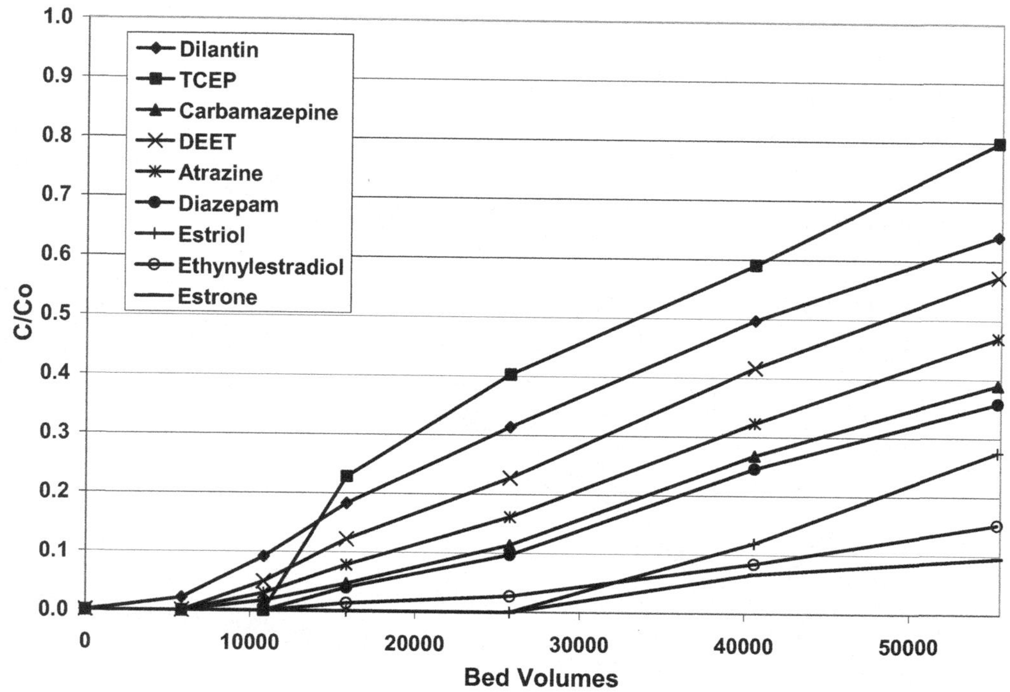

Figure 5.5 Results of select target analytes RSSCT using HD4000 GAC – Part 2 (Data from Snyder et al. 2006a)

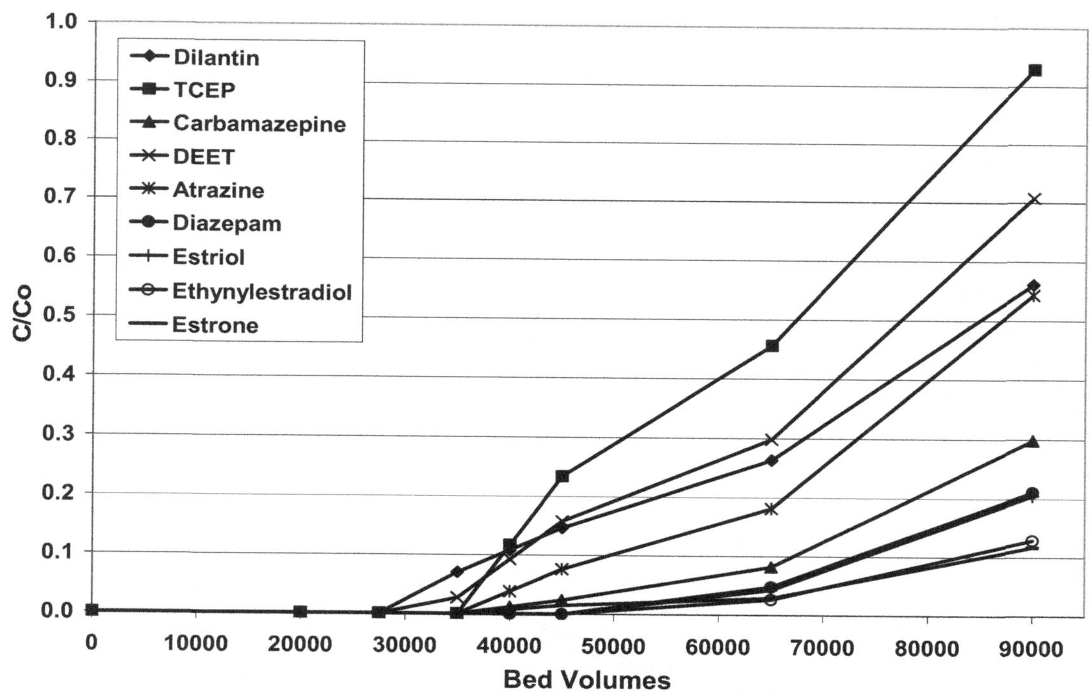

Figure 5.6 Results of select target analytes RSSCT using Superdarco GAC – Part 2 (Data from Snyder et al. 2006a)

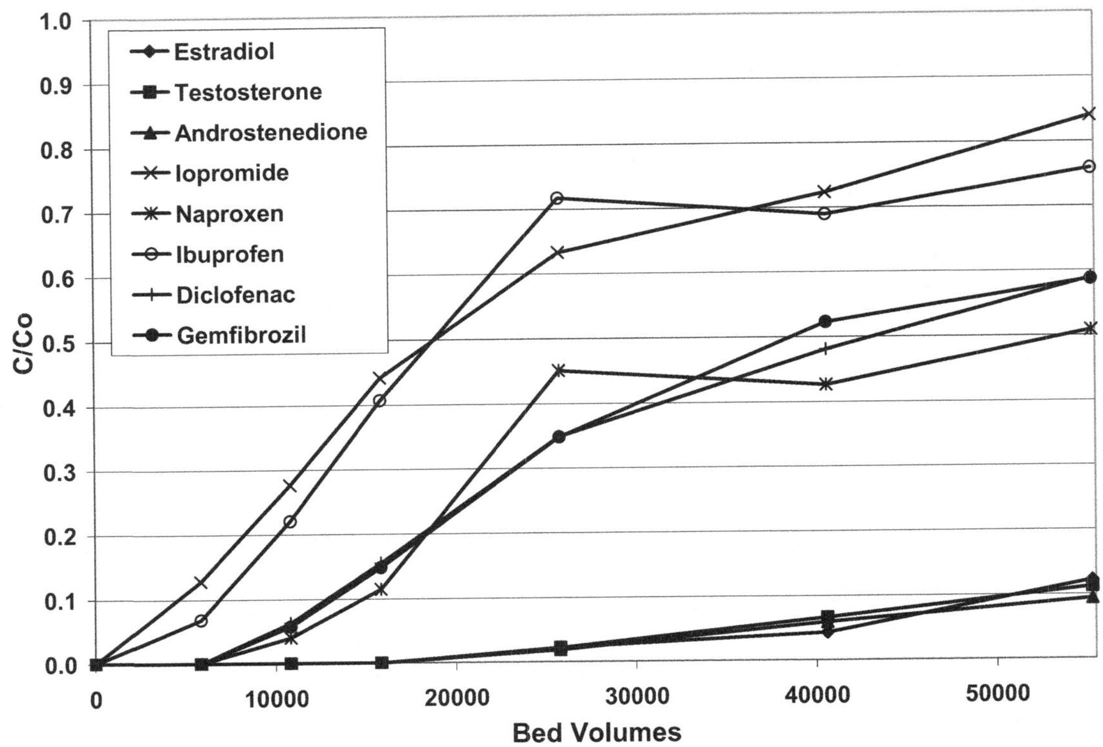

Figure 5.7 Results of select target analytes RSSCT using HD4000 GAC – Part 3 (Data from Snyder et al. 2006a)

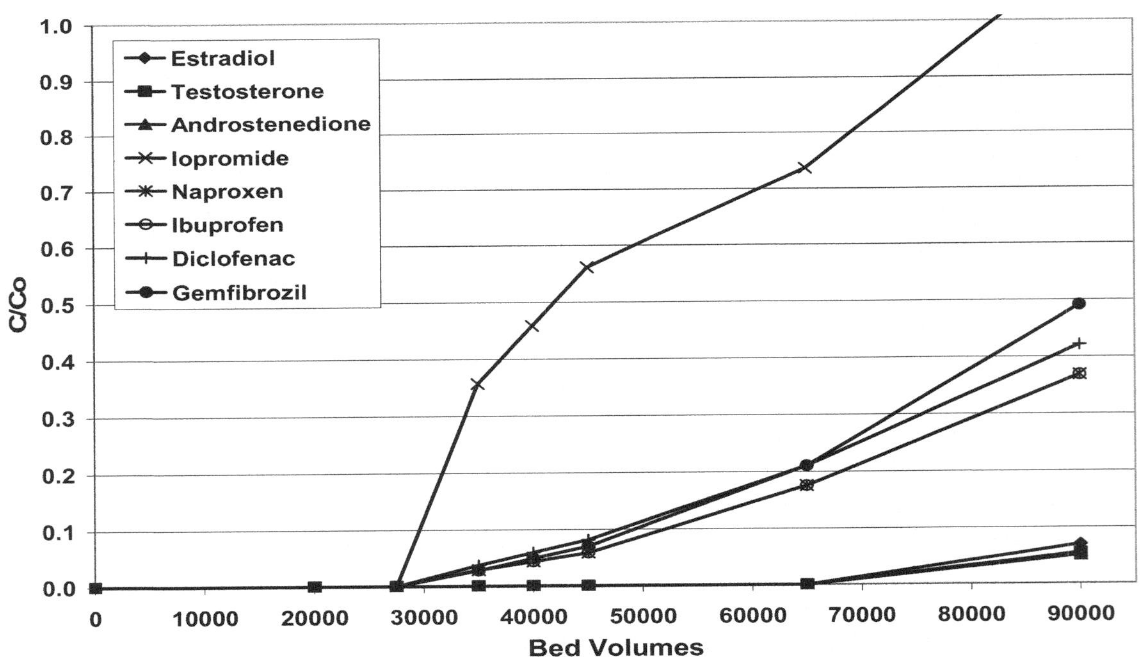

Figure 5.8 Results of select target analytes RSSCT using Superdarco GAC – Part 3 (Data from Snyder et al. 2006a)

Full-Scale Evaluations

Activated carbon was evaluated at three full-scale drinking water utilities and one water reuse facility (Table 5.3). Water reuse facility #6 did not provide information on carbon type. Of these facilities, only drinking water facility #1 regenerates the activated carbon on a regular basis. Drinking water facilities #4 and #19 both operate in biological mode with the carbon essentially exhausted as far as TOC removal is concerned. The GAC at water reuse facility #6 is employed as upflow adsorption beds for organic removal. It is important to consider that these samples were collected as grab samples without consideration of bed volumes passed prior to sample collection. A more detailed study would be required to determine the rate and degree of contaminant break-through over time.

Results of full-scale GAC testing are shown in Tables 5.4-5.5. Only those compounds detectable in either the GAC influent or GAC effluent are shown. These data indicate that fresh GAC is highly effective for the removal of trace organic contaminants, while GAC that is not regularly replaced/regenerated is relatively ineffective. Results for biologically-active GAC can not rule out the possibility of removal by biodegradation in addition to adsorption to the carbon.

Table 5.3
Full-scale GAC facilities investigated

Facility	Carbon type	Carbon source	Biological
Drinking –1	Calgon Filtrasorb-400	Bituminous Coal	No
Drinking –4	Calgon Filtrasorb-300	Bituminous Coal	Yes
Drinking –19	Calgon Filtrasorb-820	Bituminous Coal	Yes
Water reuse -6	Unknown	Unknown	Yes

Table 5.4
Results of drinking water full-scale GAC testing

Compound	Drinking #1			Drinking #4			Drinking #19		
	Influent (ng/L)	Effluent (ng/L)	Percent Removal	Influent (ng/L)	Effluent (ng/L)	Percent Removal	Influent (ng/L)	Effluent (ng/L)	Percent Removal
Atrazine	650	6.1	>99	51	48	5.9	<1.0	<1.0	NA
Caffeine	17	<10	>41	25	16	36	15	22	<1
Carbamazepine	2.2	<1.0	>54	<1.0	<1.0	NA	<1.0	3.4	<1
DEET	1.8	<1.0	>44	7.1	4.4	38	7.1	6.4	10
Dilantin	1.8	<1.0	>44	<1.0	<1.0	NA	2.4	1.7	29
Estradiol	<1.0	<1.0	NA	6.4	<1.0	>84	<1.0	<1.0	NA
Erythromycin	1.8	<1.0	>44	<1.0	<1.0	NA	<1.0	<1.0	NA
Galaxolide	<10	<10	NA	11	<10	>9	<10	<10	NA
Gemfibrozil	1.2	<1.0	>16	<1.0	<1.0	NA	<1.0	<1.0	NA
Ibuprofen	1.1	<1.0	>9	2.4	<1.0	>58	5.0	2.4	52
Iopromide	3.3	<1.0	>69	11	13	<1	37	32	14
Meprobamate	1.2	<1.0	>16	<1.0	<1.0	NA	2.5	2.1	16
Metolachlor	122	<10	>91	<10	<10	NA	<10	<10	NA
Oxybenzone	1.0	<1.0	>1	2.4	<1.0	>58	<1.0	<1.0	NA
Sulfamethoxazole	6.0	<1.0	>83	<1.0	<1.0	NA	<1.0	<1.0	NA
TCEP	<1.0	<1.0	NA	<1.0	<1.0	NA	8.3	5.0	40

Data from Snyder et al. 2006a

Table 5.5
Results of full-scale GAC testing at water reuse utility 6

Compound	GAC Influent (ng/L)	GAC Effluent (ng/L)	Percent Removal (%)
Acetaminophen	10650	47	99
Androstenedione	182.5	72	61
Atrazine	33	32	3.0
Caffeine	42650	35700	16
Carbamazepine	199	168	16
DEET	701	256	63
Diclofenac	3.2	<1.0	69
Dilantin	110	85	23
Erythromycin	139	128	7.9
Estriol	8.3	28	<1
Gemfibrozil	3710	3405	8.2
Hydrocodone	57	<25	56
Ibuprofen	8760	7325	16
Iopromide	5	1.4	72
Meprobamate	226	196	13
Naproxen	3545	3320	6.3
Pentoxifylline	34	<25	26
Sulfamethoxazole	426	69	84
Testosterone	150.5	39	74
Triclosan	3.8	4.2	<1
Trimethoprim	135	48	64

Source: Snyder et al. 2006a. Reprinted from *Desalination*, 202, Snyder, S. A., S. Adham, A. M. Redding, F. S. Cannon, J. DeCarolis, J. Oppenheimer, E. C. Wert, and Y. Yoon. Role of Membranes and Activated Carbon in the Removal of Endocrine Disruptors and Pharmaceuticals. 156-181. Copyright 2006, with permission from Elsevier.

RESULT SUMMARY AND DISCUSSION

Activated carbon can be extremely effective for the removal of EDCs/PPCPs. Both PAC and GAC were capable of removing nearly all compounds investigated. A summary of removal of target compounds using PAC and GAC are shown in Tables 5.6 and 5.7, respectively. Activated carbon efficacy is greatly reduced by the presence of natural organic matter, which competes for binding sites and can block pores within the activated carbon structure. PAC offers the advantage of providing fresh carbon continuously since it is fed as a new product and is not recycled through the treatment process. Since PAC is added dynamically as a feed "chemical", it can be used seasonally when risk of trace organics may be greater (e.g., low-flow events). The ability of PAC to remove organic contaminants depends upon PAC dose and contact time, as well as the molecular structure and behavior of the contaminant of interest. GAC also was found to be highly-effective; however, water soluble contaminants can breakthrough the GAC much more rapidly than strongly bound hydrophobic contaminants. The data shown here also demonstrate that steam-treated GAC has significantly greater absorption capacity as shown by the increase in bed volumes passed prior to contaminant breakthrough. In full-scale GAC applications evaluated, vastly different results were observed. At a drinking water facility with on-site and regular regeneration, removal of trace organics occurring in source water was efficient. In contrast, a facility with relatively high levels of TOC using GAC without regular replacement/regeneration provided very little removal. These data collectively show that activated carbon in both powdered and granular forms has great potential for the removal of trace organic contaminants. PAC dose and GAC regeneration/replacement are important considerations for efficient contaminant removal.

The following specific conclusions were made:

- Addition of 5 mg/L of PAC with a four hour contact time, which is representative of pesticide control applications, removed 50% to >98% of GC-MS/MS compounds and 10% to >95% of LC-MS/MS target compounds. Increasing PAC dosage increased EDC/PPCP removal, with the percentage removal of steroids seemingly independent of initial compound concentration. Octanol-water partition coefficients served as a reasonable indicator of compound removal under controlled PAC test conditions, except for EDC/PPCPs that were protonated or deprotonated at the test pH and some that contained heterocyclic or aromatic nitrogen. For these charged species, D_{ow} that takes protonation/deprotonation into consideration would be more appropriate to represent their hydrophobicity and hence removal tendency.
- Application of PAC with a 5 hour contact time in the presence of a ferric coagulant at the pilot scale resulted in <20% to 80% removal of EDC/PPCPs at a 5 mg/L PAC dose and >70% removal of all detected EDC/PPCPs at a dose of 35 mg/L PAC.
- In continuous flow bench-scale RSSCTs EDC/PPCP removal was dependant upon GAC type. Sulfamethoxazole, meprobamate, dilantin, iopromide, ibuprofen, and diclofenac were removed less effectively (i.e., broke through earlier) than other EDC/PPCPs tested. This is consistent with lower removals of these compounds in PAC tests.
- Full-scale GAC tests are more difficult to evaluate because systems are in different periods of breakthrough and/or have exhausted activated carbon that provides biological degradation as well as adsorption.

Table 5.6
Removal of target compounds by PAC (5 mg/L, 4-5 hour contact time)

> 80% Removal	50-80% Removal	20-50% Removal	< 20% Removal
Benzo(a)pyrene	Acetaminophen	DEET	Ibuprofen
Fluorene	Androstenedione	Diclofenac	Iopromide
Fluoxetine	Atrazine	Dilantin	
Oxybenzone	Caffeine	Erythromycin	
Progesterone	Carbamazepine	Estriol	
Triclosan	DDT	Gemfibrozil	
	Diazepam	Meprobamate	
	Estradiol	Metolachlor	
	Estrone	Naproxen	
	Ethinyl Estradiol	Sulfamethoxazole	
	Galaxolide	TCEP	
	Hydrocodone		
	Lindane		
	Musk Ketone		
	Pentoxifylline		
	Testosterone		
	Trimethoprim		

Table 5.7
Removal of target compounds by GAC

GAC Bed Volumes to 10% Breakthrough (HD4000, EBCT = 7.6 minutes)		
> 50,000 BVs	20,000-50,000 BVs	< 20,000 BVs
Androstenedione	Acetaminophen	Diclofenac
Estradiol	Atrazine	Dilantin
Estrone	Caffeine	Gemfibrozil
Ethynylestradiol	Carbamazepine	Ibuprofen
Fluoxetine	DEET	Iopromide
Oxybenzone	Diazepam	Meprobamate
Progesterone	Erythromycin	Naproxen
Testosterone	Estriol	Sulfamethoxazole
Triclosan	Hydrocodone	TCEP
	Pentoxifylline	
	Trimethoprim	

REFERENCES

Graham, M. R., R. S. Summers, M. R. Simpson, and B. W. MacLeod. 2000. Modeling equilibrium adsorption of 2-methylisoborneol and geosmin in natural waters. *Water Res.*, 34(8): 2291-2300.

Knappe, D. R. U., Y. Matsui, V. L. Snoeyink, P. Roche, M. J. Prados, and M.-M. Bourbigot. 1998. Predicting the capacity of powder activated carbon for trace organic compounds in natural waters. *Environmental Science & Technology* 32(11): 1694-1698.

Li, Q. L., V. L. Snoeyink, B. J. Marinas, and C. Campos. 2003. Pore blockage effect of NOM on atrazine adsorption kinetics of PAC: the roles of PAC pore size distribution and NOM molecular weight. *Water Res.*, 37(20): 4863-4872.

Rangel-Mendez, J. R., and F. S. Cannon. 2005. Improved activated carbon by thermal treatment in methane and steam: Physicochemical influences on MIB sorption capacity. *Carbon* 43(3): 467-479.

Westerhoff, P., Y. Yoon, S. Snyder, and E. Wert. 2005. Fate of endocrine-disruptor, pharmaceutical, and personal care product chemicals during simulated drinking water treatment processes *Environmental Science & Technology* 39(17): 6649-6663.

CHAPTER 6
CHLORINE OXIDATION

BASIC THEORY

Chlorine is an inexpensive disinfectant that is widely used in the water industry. It is generally applied as gaseous chlorine or as concentrated hypochlorite liquid (OCl^-). Both application methods will lead to aqueous chlorine present in the form of either hypochlorite ion or hypochlorous acid (HOCl), the speciation of which is pH dependent. The pK_a for HOCl is 7.5, hence HOCl dominates at pH < 7.5 and OCl^- constitutes a higher percentage at pH > 7.5. HOCl is a stronger oxidizing and substituting agent than OCl^-. Therefore, the oxidative power of aqueous chlorine highly depends on pH. Chlorine reacts selectively with electron rich bonds of organic chemicals (e.g., C=C bonds in aromatic rings). Electron donating substitutes in organic molecules tend to increase reactivity, while electron withdrawing groups decrease reactivity.

Previous research demonstrated that chlorination did not reduce DDT or lindane concentrations in water (Robeck et al. 1965). However, several reports have shown that some EDCs and PPCPs containing reactive functional groups can be oxidized by free chlorine (Adams et al. 2002; Deborde et al. 2004; Lee et al. 2004; Westerhoff et al. 2005). Phenolic steroids (e.g., estradiol, estrone, estriol, and ethynyl estradiol) can be readily oxidized by chlorine, while ketone steroids (e.g., testosterone and progesterone) are not effectively oxidized (Westerhoff et al. 2005). The rapid reactions of phenolic compounds with free chlorine are mainly through the electrophilic attack of the deprotonated phenolate anion (Faust and Hunter 1967; Gallard and von Gunten 2002). This reaction results in sequential chlorine additions to the aromatic ring followed by ring cleavage and subsequent formation of THMs (Greyshock and Vikesland 2006). The selective bond oxidation can be further illustrated by the half-lives of eight aromatic pharmaceuticals, with values ranging from 1.4 to 6,300 minutes (Pinkston and Sedlak 2004). Chlorination (1 mg/L; pH=7.5) transformed >90% of seven amine substituted antibiotics in river water within 40 minutes (Adams et al. 2002). Clearly, chlorination can lead to the apparent removal of several EDCs/PPCPs; however, the formation of oxidation products is a potential concern.

Due to the reactivity of chlorine with organic chemicals, it is necessary to quench chlorine residuals after sample collection. Pinkston and Sedlak (2004) showed that quenching of chlorine using thiosulfate can reverse the chlorination reaction, regenerating the original compound. In the work presented here, the weaker reducing agent ascorbic acid was used to quench chlorine, which did not exert any noticeable impact on spiked controls. Ascorbic acid has been shown to cause only a very minimal reduction of chlorinated by-products when compared with other common chlorine quenching agents (MacCrehan, Bedner, and Helz 2005).

BENCH-SCALE EVALUATIONS

Free chlorine

Free chlorine stock solutions (1200 mg/L) were prepared daily from 5% sodium hypochlorite in Nanopure water. Aliquots of the stock solution were transferred to 1-L silanized amber glass bottles containing the water sample with zero headspace maintained in the bottle.

Free chlorine residuals were measured by the DPD method using a Hach DR4000 spectrophotometer (USEPA-approved Hach Method #8021, Hach Company, Loveland, CO, United States). Free chlorine demand tests were conducted with EDC/PPCP spiked water and a 24-hour contact time before selecting oxidant dosages for experiments. Chlorination experiments were conducted at both ambient and suppressed pH of 5.5 to evaluate the impact of chlorine speciation on contaminant oxidation.

Ascorbic acid (25 mg/L) was added after 24 hours of reaction time to quench residual chlorine (1 ± 0.2 mg Cl_2/L). After chlorination, several EDC/PPCP concentrations were below detection, indicating a high degree of reactivity with chlorine (e.g., acetaminophen, diclofenac, estradiol, estriol, estrone, ethynyl estradiol, naproxen, oxybenzone, sulfamethoxazole, and triclosan). Concentrations of other compounds changed very little after chlorine exposure (e.g., DEET, meprobamate, TCEP, or lindane). Most other compounds exhibited residual concentrations ranging from 25% to 75% of the initial concentrations. The observed "removals" of parent EDC/PPCP compounds were not artifacts of compound recovery changes brought about by reaction of the oxidants with water matrix constituents (e.g., DOC) since consistent recoveries were observed for stable isotopically labeled surrogates spiked prior to solid phase extraction.

The percent removal data from chlorination are summarized in Table 6.1. Figure 6.1 graphically presents removal data for PVW and Figure 6.2 graphically compares percent removal in PVW to removal in CRW. A few compounds (atrazine, DEET, dilantin, fluoxetine, ibuprofen, iopromide, meprobamate) with removals of < 60% of the initial concentration exhibited higher removals in PVW than CRW, ORW and SRW. PVW and CRW experiments at pH 5.5 had different initial EDC/PPCP concentrations, yet chlorination achieved comparable percentage removals for most compounds. The percentage removal of trace-level pollutants is often independent of initial pollutant concentration (Bruce, Westerhoff, and Brawley-Chesworth 2002). Among all four source waters, the following compounds were poorly removed during chlorination: atrazine, DEET, fluoxeine, iopromide, meprobamate, and TCEP. The percentage EDC/PPCP removals after hypochlorite addition at pH 5.5 were mostly equal to or higher than those at ambient pH. It should be noted that several well-oxidized target compounds (>90% removal) have pK_a values between 5.5 and 8.5 (trimethoprim, sulfamethoxazole, dilantin, triclosan, erythromycin); however, the relationship between ionization state and chlorine reactivity was not specifically evaluated.

Table 6.1
Summary of percent removal during bench-scale chlorination

Source Water	Color River Water		Ohio River Water		Passaic River Water		Suwannee River NOM	
Chlorine Type	Free	Free	Free	Free	Free	Free	Free	Free
Chlorine Dose [mg/L]	3.5	3.5	2.8	2.8	3.8	3.8	6.75	6.75
pH	5.5	8.2	5.5	7.9	5.5	6.8	5.5	8.5
Reaction Time [hrs]	24	24	24	24	24	24	24	24
Target Compound	Percent Removal		Percent Removal		Percent Removal		Percent Removal	
Acetaminophen (Tylenol)	>96	>96	>97	>97	97	>98	>90	>90
Androstenedione	40	23	86	25	57	44	22	13
Atrazine	15	2	<1	<1	24	11	25	17
Benzo(a)pyrene	>70	>70	>87	>87	>66	>66	91	91
Caffeine	68	17	>99	11	60	28	10	<1
Carbamazepine	>98	22	>98	20	92	33	40	15
DDT	25	8	11	<1	27	22	37	11
DEET	16	14	2	<1	33	31	15	2
Diazepam (Valium)	81	13	83	25	79	27	>99	13
Diclofenac	>96	>96	>97	>97	>86	>86	>93	>93
Dilantin	33	18	22	<1	53	35	54	22
Erythromycin	>96	>96	82	>97	>96	>96	*	*
Estradiol	>99	>99	*	*	>99	>99	>97	>97
Estriol	>99	>99	*	*	>99	>99	*	*
Estrone	>99	>99	>99	>99	>99	>99	*	*
Ethinyl Estradiol	>99	>99	>99	>99	>99	>99	>98	>98
Fluorene	30	24	41	20	21	14	*	*
Fluoxetine (Prozac)	20	19	13	<1	46	42	<1	<1
Galaxolide	40	15	25	4	56	39	54	40
Gemfibrozil	>99	70	>99	>99	>99	58	>96	56
Hydrocodone	>98	>98	*	*	>99	>99	*	*
Ibuprofen (Advil)	45	37	37	23	78	59	47	47
Iopromide	7	28	5	21	27	27	<1	<1
Lindane (γ-BHC)	21	11	8	1	13	7	19	17
Meprobamate	15	18	18	<1	22	22	4	10
Metolachlor	32	35	36	21	50	41	12	6
Musk Ketone	26	>95	32	73	30	88	55	92
Naproxen	>94	>94	>96	>96	>92	>92	>86	>86
Oxybenzone	>97	>97	>97	>97	>97	>97	>95	92
Pentoxifylline	86	20	>98	30	74	27	12	<1
Progesterone	50	29	93	31	62	51	34	18
Sulfamethoxazole	>97	>97	>96	>96	>79	>79	>78	>78
TCEP	4	<1	<1	<1	22	12	<1	<1
Testosterone	52	26	95	28	61	48	39	19
Triclosan	>98	>98	>97	>97	>98	>98	>96	>96
Trimethoprim	>99	>99	>98	>98	>99	>99	>96	>96

* Target compound not analyzed

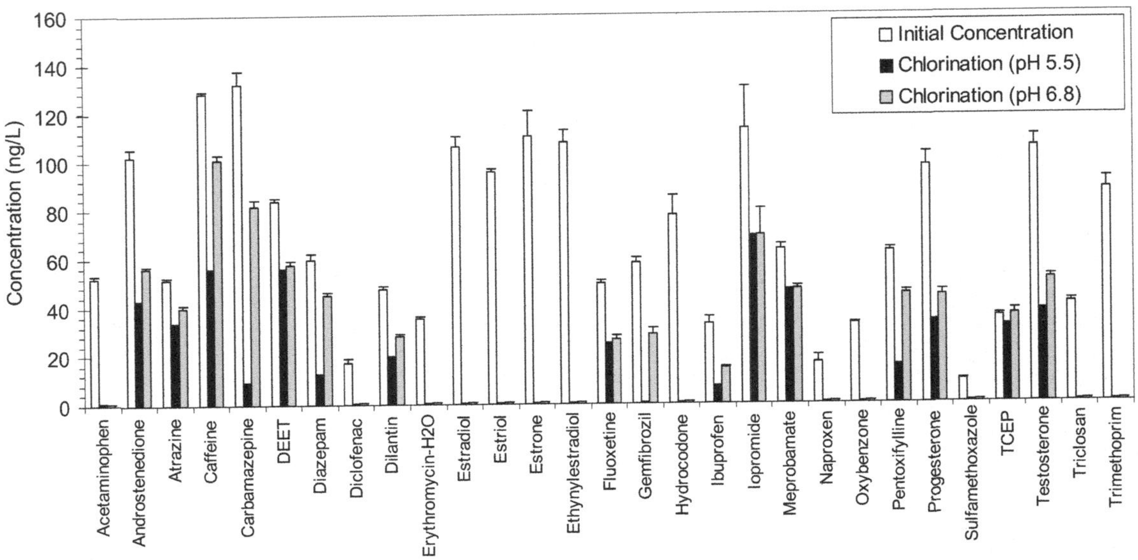

Figure 6.1 Concentrations of spiked EDCs/PPCPs (ng/L) in PVW before chlorination and after hypochlorite (3.8 mg Cl_2/L) addition at pH 5.5 and 6.8

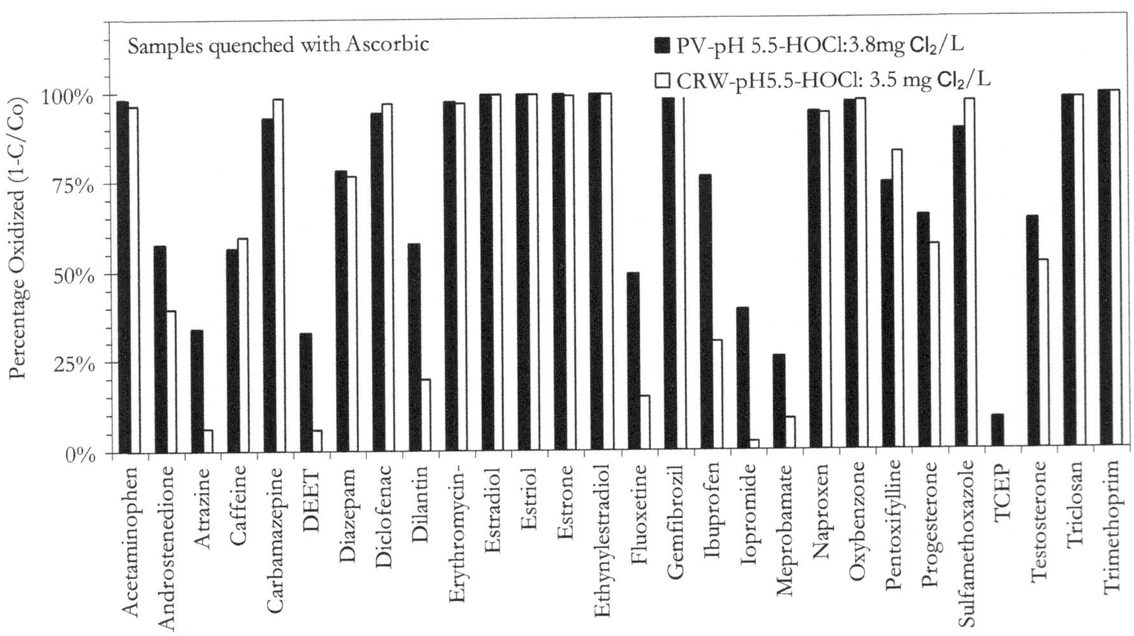

Figure 6.2 Percent removal of spiked EDCs/PPCPs in PVW and CRW after hypochlorite addition (pH 5.5)

Chlorination and Chloramination of CRW

Bench scale chloramine testing was performed with Colorado River water. Chloramine dosages of 2 and 3 mg/L were evaluated. A 4:1 Cl_2:NH_3-N ratio was targeted for the chloramine dose. Application of ammonia followed by subsequent chlorine addition was used to form chloramine residual. Subsequent addition was used instead of preformed chloramine in order to be more representative of full-scale facilities. Simulated distribution system testing was performed with samples collected after 24 and 168 hours of contact time. Chloramine experiments were performed simultaneously with a free chlorine experiment to observe the percent removal differences between the two oxidants under the same water quality conditions. Chlorine residual decay information is provided in Figure 6.3.

Bench scale testing results indicated that 20 of the 36 target compounds were less than 10% removed by 2 or 3 mg/L of chloramine after 24 hours of contact time. Only 4 of the 36 target compounds were removed to below detection under these conditions. Removal of the remaining 12 compounds was clearly related to chloramine dose and contact time (Figure 6.4).

When comparing chloramine to free chlorine with a dose of 3 mg/L after 168 hours of exposure, 24 of the 36 target compounds showed similar removal between both chlorine types. The remaining 12 target compounds are shown in Figure 6.5. These compounds show at least 30% more removal using free chlorine than with chloramines.

Chlorination and Chloramination of CRW using Mixed Oxidants

Mixed oxidants (MIOX) and household bleach were used to generate free and combined chlorine mixtures, designed to quantify any increased target compound removal the mixed oxidant system. Mixed oxidants may contain additional oxidants, such as hydroxyl radicals, thought to be formed during the MIOX generation process. Experiments were performed simultaneously with sodium hypochlorite for comparison. A 4:1 chlorine to ammonia ratio was used during MIOX experiments. During simulated distribution system tests, chlorine was applied to a 1-L sample volume of raw Colorado River water (pH=7.89) in the amounts of 2 mg/L and 3 mg/L for 24 hours. Chlorine residual decay curves are shown in Figures 6.6 and 6.7. Mixed oxidants were generated and used at the time of addition. Based on the product concentration of 230 mg/L, the volume of mixed oxidants was calculated, measured in a 25 mL graduated cylinder, and transferred into the 1-L sample within 10 seconds.

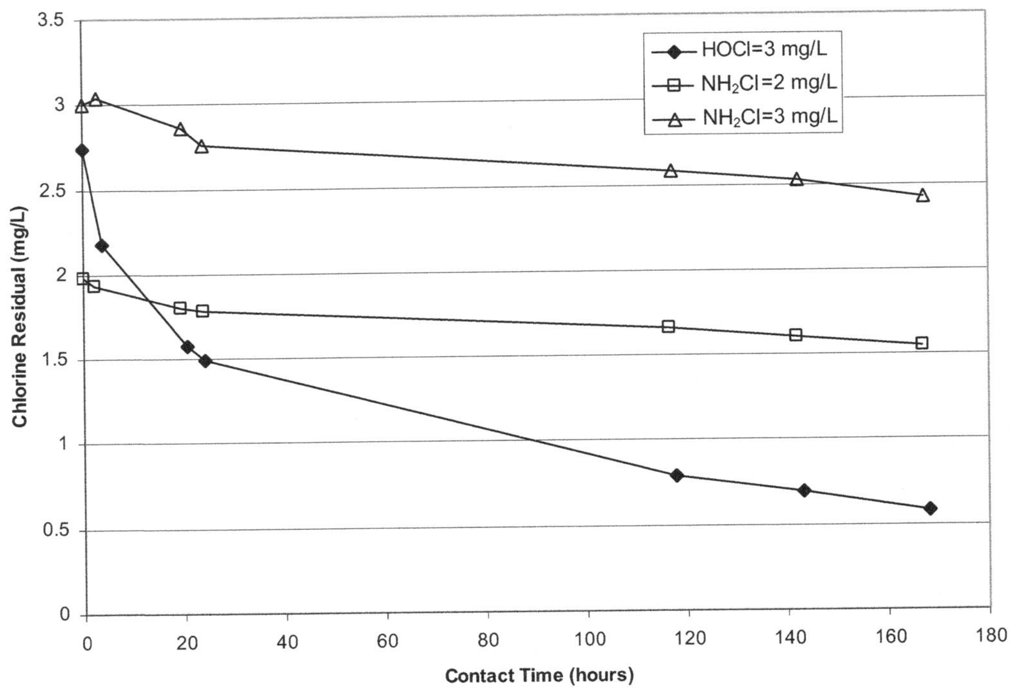

Figure 6.3 Chlorine and chloramine residual decay in Colorado River water

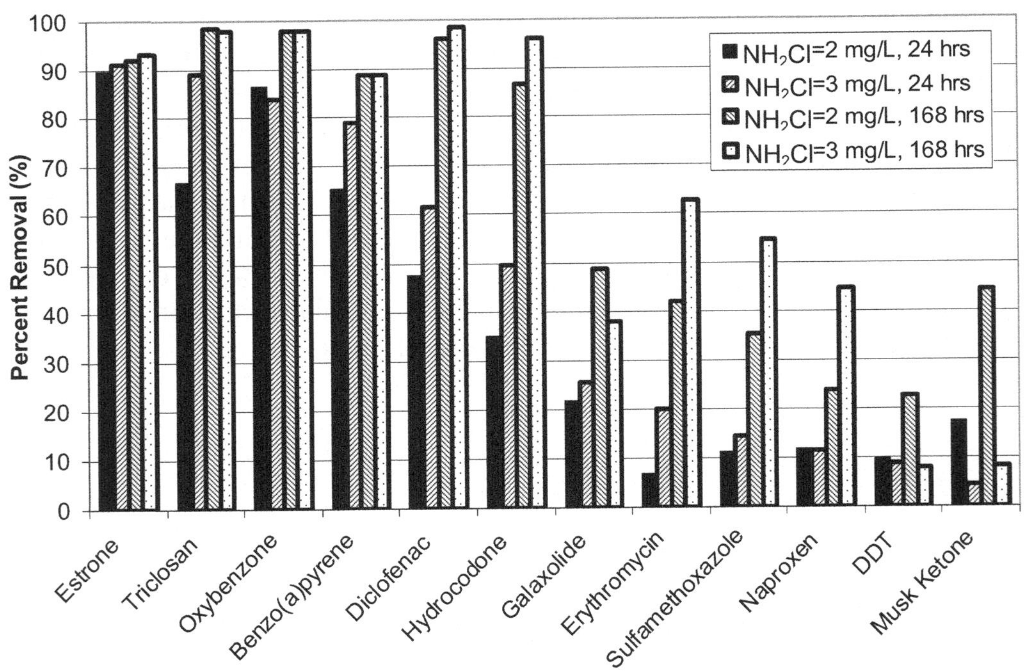

Figure 6.4 Target compound removal by chloramination in Colorado River water

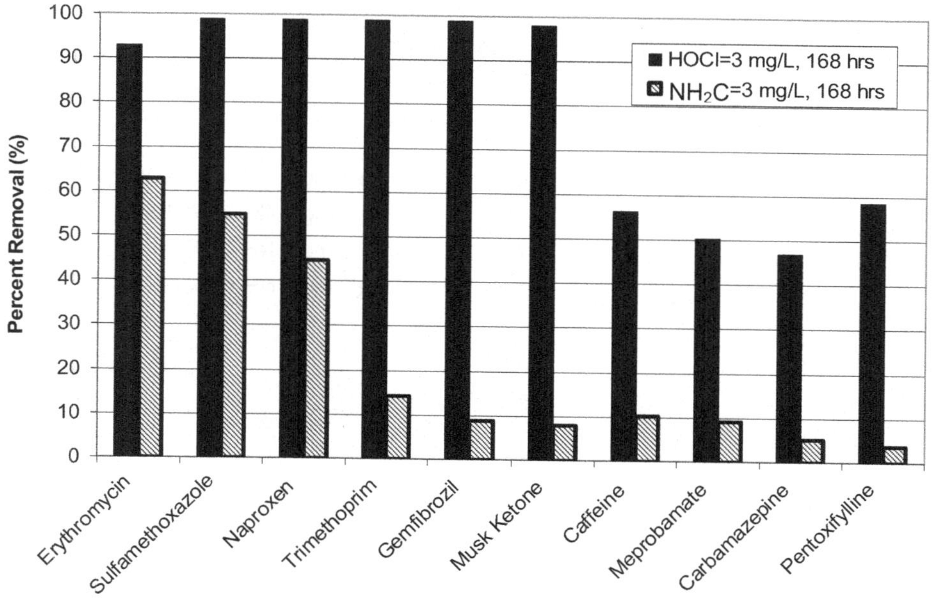

Figure 6.5 Percent removal comparison of target compounds which showed at least 30% greater removal with free chlorine than chloramine (pH 7.9)

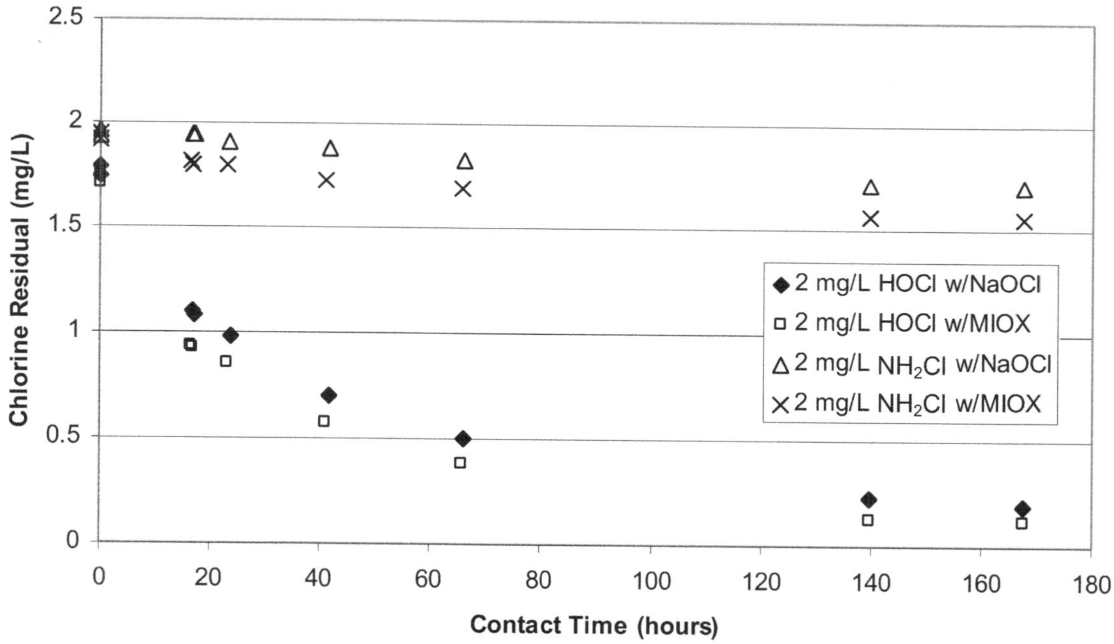

Figure 6.6 Chlorine residual decay information during testing with MIOX and sodium hypochlorite with a chlorine dose of 2 mg/L (pH 7.9)

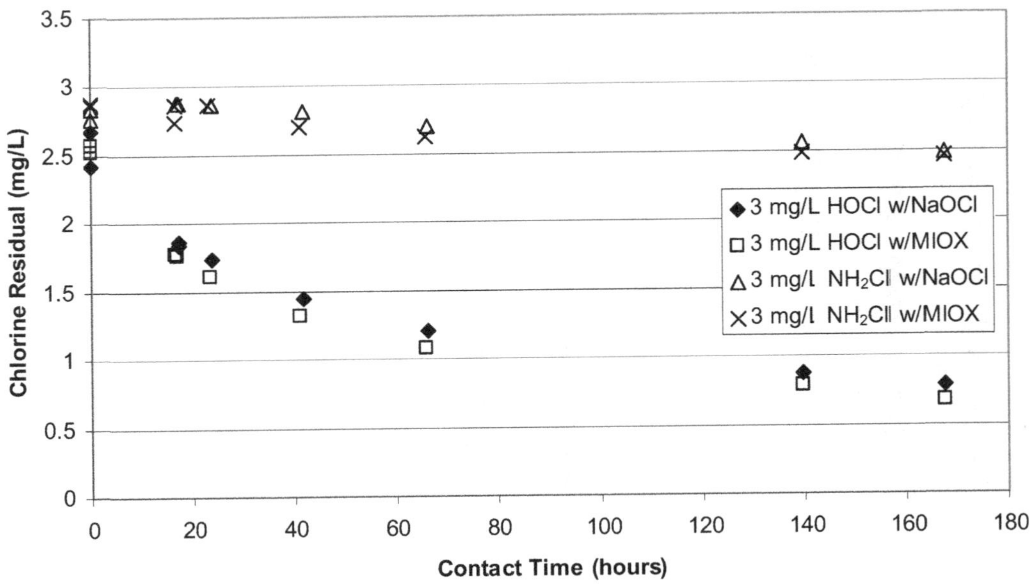

Figure 6.7 Chlorine residual decay information during testing with MIOX and sodium hypochlorite with a chlorine dose of 3 mg/L (pH 7.9)

During the bench scale MIOX testing, 8 of the 36 target compounds previously considered were not analyzed. For the remaining 28 target compounds the percent removal was generally within 10% whether sodium hypochlorite or mixed oxidants was used. Figure 6.8 compares the removal using mixed oxidants versus sodium hypochlorite under the same exposure conditions.

The removal results were similar to previous bench-scale chlorination studies. A 2 mg/L free chlorine dose removed greater than 90% of 11 of the 28 target compounds: acetaminophen, diclofenac, estradiol, estriol, ethynyl estradiol, hydrocodone, naproxen, oxybenzone, sulfamethoxazole, triclosan, and trimethoprim. Once again, chloramines were less effective than free chlorine for the removal of target compounds. The removal rates were very similar when comparing mixed oxidants versus sodium hypochlorite. A 2 mg/L chloramine dose removed only 4 compounds by greater than 90% (acetaminophen, estradiol, estriol, and ethynyl estradiol). These are the same compounds that showed greater than 90% removal in previous bench-scale studies with CRW.

PILOT-SCALE EVALUATIONS

Pilot-scale testing was conducted using both free chlorine and chloramines in Ohio River water. The reaction times are unknown since the chlorine residual was not quenched at the time of collection. Therefore, the data presents what could be the maximum removal attainable using chlorine or chloramines in Ohio River water. Interestingly, the lack of oxidant quenching was apparent analytically due to the addition of ^{13}carbon-labeled estradiol and deuterium-labeled testosterone surrogates to samples prior to extraction. Estradiol is rapidly degraded by chlorine or chloramine while testosterone is not (Table 6.1). Therefore, the absence of the estradiol surrogate and presence of the testosterone surrogate provide excellent evidence of an oxidant residual. In the case of this pilot-plant experiment, the estradiol surrogate was not recovered, while the testosterone surrogate was recovered with minimal loss.

The results showed that 13 of the 29 target compounds detected in raw water were removed to below detection when using either free chlorine or chloramine: acetaminophen, caffeine, diclofenac, estradiol, estriol, estrone, ethynyl estradiol, gemfibrozil, pentoxifyline, trimethoprim, naproxen, oxybenzone, and sulfamethoxazole. Figure 6.9 shows 13 of the remaining 16 target compounds where the difference in percent removal was greater than 10%.

FULL-SCALE EVALUATIONS

Sixteen of the 20 full-scale drinking water facilities included the addition of hypochlorite with free chlorine residual, as a primary and/or secondary disinfectant. Four of the 20 drinking water facilities used chloramination as a secondary disinfectant. Of the processes evaluated at water reuse facilities, one facility utilized free chlorine for disinfection and one facility used chloramine for disinfection.

Table 6.2 provides a removal summary for the drinking water utilities using hypochlorite to achieve a free chlorine residual. Also presented in Table 6.2 is the average removal from bench-scale experiments as a comparison. Chlorine residuals were quenched at the time of sampling using ascorbic acid. Since many of target compounds occur near the MRLs, Table 6.2 shows only compounds occurring at 2x the MRL in order to observe a minimum 50% removal if the contaminant is removed to less than detectable levels. In several cases, compounds were removed to less than the MRL and are reported as >80% removal. This introduces some bias to the data set, as the observed removal can only be calculated as >50% when the compound occurred at just over 2x the MRL and was removed to less than the MRL. Complete results for all utilities investigated are provided in Appendix E.

Complete data are shown for three drinking water utilities using free chlorine disinfection (Table 6.3). Results of three full-scale drinking water utilities using chloramination for disinfection are shown in Table 6.4. Information regarding these utilities can be found in Chapter 3 (Table 3.2). Bench-scale average removal predictions using free chlorine are also shown in Table 6.3 for comparison. Table 6.5 shows the results of the two water reuse facilities evaluated using chlorine and chloramine disinfection. Additional information on water reuse facilities evaluated can be found in Chapter 3 (Table 3.3). Except for water reuse facility #5, oxidant residuals were quenched at the time of sampling. In addition, only 100 mL of sample were collected and subsequently extracted from water reuse facility #5, resulting in a 10 times increase in the MRLs for samples at this facility. Due to the decreased sample volume, GC-MS/MS analyses were not conducted for samples collected reuse facility #5.

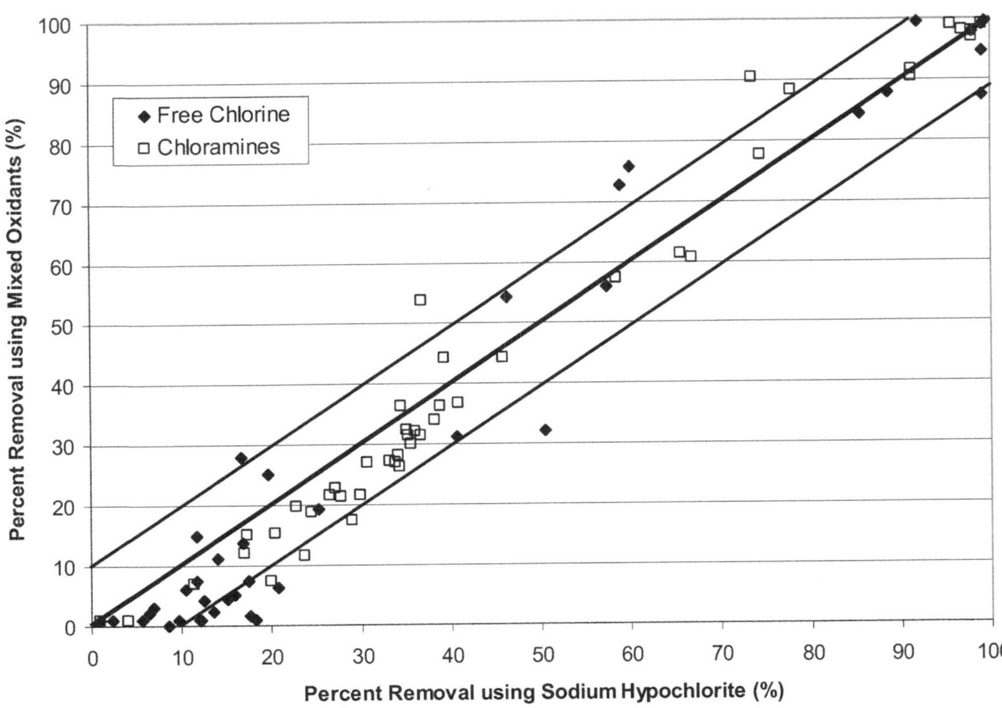

Figure 6.8 Comparison of percent removal using mixed oxidants and sodium hypochlorite. (The line of equality is shown with ±10% error bars)

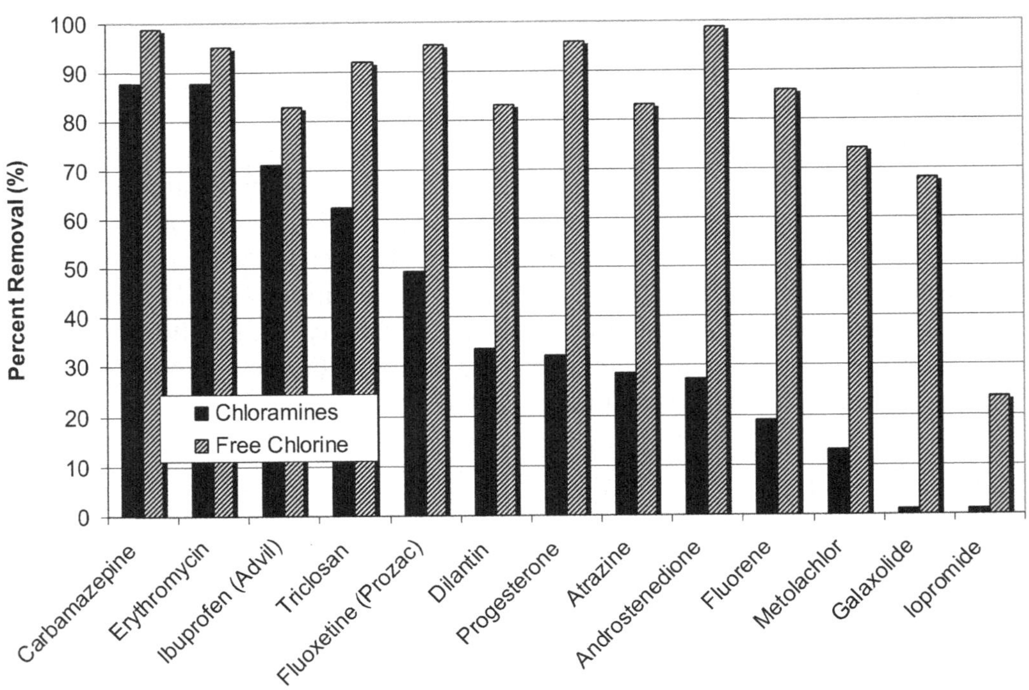

Figure 6.9 Percent removal by free chlorine and chloramine dosed at 4.4 mg/L

Table 6.2
Full-scale removal of detectable compounds (at 2x MRL) by free chlorine

	# Utilities	Observed Percent Removal				Bench-Scale Predicted
		>80%	50-80	20-50	<20	
Acetaminophen	1	1				>80
Atrazine	10			1	9	<20
Caffeine	8		2		6	<20
Carbamazepine	8	1	1	3	3	<20
DEET	12			1	11	<20
Dilantin	8			3	5	<20
Erythromycin-H_2O	3	2	1			>80
Estrone	1				1	>80
Galaxolide	2				2	20-50
Gemfibrozil	5	2	1	1	1	50-80
Ibuprofen	8			2	6	<20
Iopromide	8	1		5	2	<20
Meprobamate	8				8	<20
Metolachlor	4				4	<20
Naproxen	4	4				>80
Oxybenzone	2	2				>80
Sulfamethoxazole	7	6	1			>80
TCEP	5		1		4	<20
Triclosan	1	1				>80
Trimethoprim	2	2				>80

Table 6.3
Chlorination at three full-scale utilities

	Utility #3			Utility #7			Utility #12			Bench-Scale Predicted Percent Removal (%)
	Pre Chlorine (ng/L)	Post Chlorine (ng/L)	Percent Removal (%)	Pre Chlorine (ng/L)	Post Chlorine (ng/L)	Percent Removal (%)	Pre Chlorine (ng/L)	Post Chlorine (ng/L)	Percent Removal (%)	
Acetaminophen	2	<1	>50	<1	<1	NA	1.5	<1	>33	>80
Atrazine	427	432	<1	8.5	8.2	4	25	26	<1	<20
Caffeine	87	29	67	27	25	7	74	83	<1	<20
Carbamazepine	2.3	1	>57	2.8	3.7	<1	20	3.5	83	<20
DEET	18	18	<1	4.6	6.2	<1	28	30	<1	<20
Dilantin	3.2	2.2	31	2.1	2.7	<1	13	6.7	48	<20
Erythromycin	<1	<1	NA	2.6	<1	>62	3.2	<1	>69	>80
Galaxolide	<25	<25	NA	<25	<25	NA	28	<25	>11	50-20
Gemfibrozil	2	<1	>50	10	6.5	35	5.8	<1	>83	80-50
Hydrocodone	<1	<1	NA	<1	<1	NA	1.9	<1	>47	
Ibuprofen	4.4	4.7	<1	8.1	9.1	<1	10	18	<1	<20
Iopromide	7.9	9	<1	7.8	5.9	24	46	16	65	<20
Meprobamate	2.8	2.9	<1	12	13	<1	10	7.6	24	<20
Metolachlor	176	156	11	<10	<10	NA	14	15	<1	<20
Naproxen	2.4	<1	>58	<1	<1	NA	16	<1	>94	>80
Oxybenzone	4.2	<1	>76	<1	<1	NA	<1	<1	NA	>80
Sulfamethoxazole	8.8	<1	>89	2.1	<1	>52	35	<1	>97	>80
TCEP	15	15	<1	22	18	18	19	19	0	<20
Triclosan	2	<1	>50	<1	<1	NA	2.7	<1	>62	>80
Trimethoprim	<1	<1	NA	<1	<1	NA	2.2	<1	>55	>80

Table 6.4
Chloramination at full-scale drinking water utilities

Utility Number	Utility #8			Utility #17			Utility #18			Bench-Scale
	Pre Chloramine (ng/L)	Post Chloramine (ng/L)	Percent Removal (%)	Pre Chloramine (ng/L)	Post Chloramine (ng/L)	Percent Removal (%)	Pre Chloramine (ng/L)	Post Chloramine (ng/L)	Percent Removal (%)	Percent Removal (%)
Acetominophen	1.7	<1	<1	<1	<1	NA	<1	<1	NA	>80
Atrazine	42	38	10	33	36	<1	42	40	5	<20
Caffeine	48	35	27	26	31	<1	16	14	13	<20
Carbamazepine	3.8	3.7	3	3.2	3.6	<1	<1	<1	NA	<20
DEET	3.9	3	23	12	14	<1	7.6	7.4	3	<20
Diclofenac	<1	<1	NA	<1	<1	<1	<1	<1	NA	80-50
Dilantin	3.7	4	0	2.7	3.1	<1	<1	<1	NA	<20
Erythromycin	1.2	1.3	<1	<1	<1	NA	<1	<1	NA	<20
Estrone	<1	3.7	<1	<1	<1	NA	<1	<1	NA	>80
Fluorene	<10	<10	NA	82	<10	>88	<10	<10	NA	<20
Fluoxetine	<1	<1	NA	<1	<1	NA	<1	<1	NA	<20
Galaxolide	<25	<25	NA	<10	<10	NA	<10	<10	NA	50-20
Gemfibrozil	5.2	5.1	2	1.5	1.3	13	<1	<1	NA	<20
Hydrocodone	<1	<1	NA	<1	<1	NA	<1	<1	NA	50-20
Ibuprofen	4	3.8	5	4.5	4.5	<1	3.5	3.1	11	<20
Iopromide	12	11	8	5.9	6.5	<1	2.4	2.1	13	<20
Meprobamate	1.4	1.6	<1	28	32	<1	3.3	3.4	<1	<20
Metolachlor	<10	<10	NA	14	15	<1	14	14	<1	<20
Musk Ketone	14	<10	>29	16	17	<1	<10	<10	NA	<20
Naproxen	10	8	20	<1	<1	NA	<1	<1	NA	<20
Oxybenzone	<1	<1	NA	<1	<1	NA	<1	<1	NA	<20
Progesterone	<1	<1	NA	<1	<1	NA	<1	<1	NA	<20
Sulfamethoxazole	40	20	50	<1	<1	NA	2.4	2.1	13	<20
TCEP	<10	<10	NA	<10	<10	NA	<10	<10	NA	<20
Triclosan	30	43	<1	<1	<1	NA	<1	<1	NA	>80
Trimethoprim	1.1	1.3	<1	<1	<1	NA	<1	<1	NA	<20

NA=Not applicable

Table 6.5
Elimination of target analytes at full-scale water reuse facilities

	Reuse Facility #4			Reuse Facility #5		
	Pre Chlorine (ng/L)	Post Chlorine (ng/L)	Percent Removal (%)	Pre Chloramine (ng/L)	Post Chloramine (ng/L)	Percent Removal (%)
Acetaminophen	3.4	1.2	65	<10	<10	<1
Androstenedione	2.8	3.0	<1	37	38	<1
Caffeine	123	64	48	260	155	40
Carbamazepine	143	86	40	63	67	<1
DEET	323	269	17	39	<10	>74
Diazepam	2.8	1.9	32	238	244	<1
Diclofenac	5.5	<1.0	>81	855	507	41
Dilantin	88	78	11	49	<10	>80
Erythromycin	131	<1.0	>99	<10	<10	<1
Estriol	4.0	2.1	48	71	52	27
Estrone	25	2.5	90	<10	<10	<1
Fluoxetine	8.2	5.3	35	289	<10	>97
Gemfibrozil	100	8.4	92	107	44	59
Hydrocodone	62	<1.0	>98	116	110	5
Ibuprofen	14	7.0	50	107	101	6
Iopromide	65	29	55	534	571	<1
Meprobamate	665	594	11	<10	<10	<1
Naproxen	45	<1.0	>97	<10	<10	<1
Oxybenzone	13	4.3	67	133	<10	>92
Pentoxifylline	20	16	20	36	<10	>72
Sulfamethoxazole	206	15	93	14	<10	>29
TCEP	200	187	7	<10	<10	<1
Testosterone	7.8	6.8	13	<10	<10	<1
Triclosan	18	19	<1	47	<10	>79
Trimethoprim	44	<1.0	>97	190	<10	>95

RESULT SUMMARY AND DISCUSSION

The bench, pilot, and full-scale results show that removal of target compounds by chlorine and chloramine oxidation was quite consistent. Using free chlorine, most compounds exhibited equal or higher removal at pH 5.5 in comparison to ambient pH due to the equilibrium between HOCl and OCl^-. Tables 6.6 and 6.7 summarize the average percent removal of target compounds during bench-scale experiments at ambient pH using free chlorine and chloramine, respectively. The more reactive compounds (>80% removal) generally have aromatic rings with either hydroxyl, amine, or methoxy groups, which react rapidly and selectively with chlorine (Pinkston and Sedlak 2004). In contrast, the least reactive compounds (<20%) lack C=C bonds (i.e., aliphatic unsaturated bonds), have electron withdrawing functional groups (e.g., carboxyl groups, Cl), or contain stable heterocyclic nitrogen. Examination of compound structure can lead to reasonable estimation of potential removal by chlorine oxidation.

The results show that free chlorine removes more target compounds to a higher degree than chloramine. Increasing the reaction time and dose of chloramine resulted in improved removal for a number of compounds likely due to the formation of $NHCl_2$ and HOCl over long term from the autodecomposition of NH_2Cl. Greyshock and Vikesland point out that $NHCl_2$ and HOCl are responsible for a significant percentage of triclosan decay at pH below 9.5 (Greyshock and Vikesland 2006). Other contaminants are expected to be affected the same way by these autodecomposition species at pH 5.5 and ambient pH. Mixed oxidants did not appear to have any advantage over sodium hypochlorite. The removal with both free chlorine and chloramine produce by conventional bleach or by mixed oxidations were within ±10%, indicating no clear advantage in contaminant removal by mixed oxidations.

Removal of target compounds at full-scale was generally in agreement with bench-scale observations. Full-scale results are somewhat confounded by the limited and trace occurrence of the target compounds and by analytical variability. Additionally, precise oxidant dose and contact times were often not provided by the utilities. Considering these variables, when the data are binned into removal categories based on poor (<20%), moderate (>20 and <50%), good (>50 and <80%), and excellent (>80%), agreement among scales was remarkable.

By grouping the selected compounds according their functional groups, some general patterns in the removal of these compounds are discovered. Compounds with phenolic groups tend to react quickly with chlorine. Phenolic target compounds include acetaminophen, estradiol, estriol, estrone, ethynyl estradiol, oxybenzone and triclosan. These phenolic compounds all exhibited greater than 95% removal by chlorination under the conditions tested. The reaction mechanisms of phenolic compounds with free chlorine proceeds via an electrophilic substitution pathway, with a mixture of substituted products at the otho and para positions (Burttschell et al. 1959; Lee and Morris 1962). Further chlorine addition results in cleavage of the aromatic ring. It is important to note that some reaction products (e.g., compounds with chlorinated ring structure) may pose taste and odor problems and may pose equivocal health risks.

The effect of chlorine on compounds with amine or amide groups was variable, from nearly no removal to over 95% removal. By examining these compounds, it is noted that compounds with a primary amine attached to a conjugated ring structure, such as trimethoprim and sulfamethoxazole, exhibited great than 90% removal. Conversely, a decreased reactivity with chlorine was observed for compounds with amide structures. For instance, less than 30% of iopromide and meprobamate was removed by chlorination.

Table 6.6
Summary of percent removal by free chlorine at drinking water dosages

Chlorine Dose = 3 mg/L, Contact Time = 24 hours, pH=7.9-8.5			
> 80% Removal	50-80% Removal	20-50% Removal	< 20% Removal
Acetaminophen	Gemfibrozil	Diazepam	Androstenedione
Benzo(a)pyrene		Galaxolide	Atrazine
Diclofenac		Pentoxifylline	Caffeine
Erythromycin-H_2O			Carbamazepine
Estradiol			DDT
Estriol			DEET
Estrone			Dilantin
Ethynylestradiol			Fluorene
Hydrocodone			Fluoxetine
Musk Ketone			g-BHC
Naproxen			Ibuprofen
Oxybenzone			Iopromide
Sulfamethoxazole			Meprobamate
Triclosan			Metolachlor
Trimethoprim			Progesterone
			TCEP
			Testosterone

Table 6.7
Summary of percent removal by chloramines at drinking water dosages

Chloramine Dose = 3 mg/L, Contact Time = 24 hours			
> 80% Removal	50-80% Removal	20-50% Removal	< 20% Removal
Acetaminophen	Benzo(a)pyrene	Hydrocodone	Androstenedione
Estradiol	Diclofenac	Galaxolide	Atrazine
Estriol	Oxybenzone		Caffeine
Estrone			Carbamazepine
Ethynylestradiol			DDT
Triclosan			DEET
			Diazepam
			Dilantin
			Erythromycin-H_2O
			Fluorene
			Fluoxetine
			g-BHC
			Gemfibrozil
			Ibuprofen
			Iopromide
			Meprobamate
			Metolachlor
			Musk Ketone
			Naproxen
			Pentoxifylline
			Progesterone
			Sulfamethoxazole
			TCEP
			Testosterone
			Trimethoprim

Unsubstituted aromatic rings showed limited reactivity to chlorine. For example, biphenyl is not highly reactive with chlorine. Highly alkylated benzenes are more reactive with chlorine due to the increased electron density from the substituents. Examples of this case are gemfibrozil and hydrocodone, both of which are subject to electrophilic reactions with free chlorine that lead to increased apparent removal. Polycyclic aromatic hydrocarbons (PAHs) with electron donating substituents or compounds with highly conjugated systems tend to exhibit high reactivity with chlorine, which explains the relatively high removal of carbamazepine, benzo(a)pyrene, diclofenac and naproxen. Chlorine mostly reacts via electrophilic substitution and addition. Under some conditions, chlorine can also act as an oxidant. The oxidation of acetone to lactic acid and phenyl ketones in the α position to the corresponding acidic products have been reported, with the latter presumably due to the benzilic acid rearrangements (Guthrie and Cossar 1986; Guthrie, Cossar, and Lu 1991). For the selected compounds in this research, the ring cleavage is actually an oxidation process, even though at much slower kinetics.

In addition to functional groups and the overall molecular structures, many operational factors also affect the destruction of EDCs and PPCPs by free chorine, among which pH, water matrices and contact time are important. Based on these findings, it should be feasible to estimate the removal of organic contaminants by free chlorine under conditions evaluated in this study.

REFERENCES

Adams, C., Y. Wang, K. Loftin, and M. Meyer. 2002. Removal of antibiotics from surface and distilled water in conventional water treatment processes. *Journal of Environmental Engineering* 128(3): 253-260.

Bruce, D., P. Westerhoff, and A. Brawley-Chesworth. 2002. Removal of 2-methylisoborneol and geosmin in surface water treatment plants in Arizona. *Journal of Water Supply: Research and Technology - AQUA*, 51(4): 183-198.

Burttschell, R. H., A. A. Rosen, F. M. Middleton, and M. B. Ettinger. 1959. Chlorine derivatives of phenol causing taste and odor. *Journal American Water Works Association*, 51: 205-213.

Deborde, M., S. Rabouan, H. Gallad, and B. Legube. 2004. Aqueous chlorination kinetics of some endocrine disruptors. *Environmental Science & Technology*, 38: 5577-5583.

Faust, S. D., and J. V. Hunter. 1967. Principles and Applications of Water Chemistry. John Wiley and Sons: New York.

Gallard, H., and U. von Gunten. 2002. Chlorination of phenols: kinetics and formation of chloroform. *Environ. Sci. Technol.*, 36(5): 884-890.

Greyshock, A. E., and P. J. Vikesland. 2006. Triclosan reactivity chloraminated waters. *Environ. Sci. Technol.*, 40(8): 2615-2622.

Guthrie, J. P., and J. Cossar. 1986. The chlorination of acetone - a complete kinetic-analysis. *Canadian Journal of Chemistry-Revue Canadienne De Chimie*, 64(6): 1250-1266.

Guthrie, J. P., J. Cossar, and J. Q. Lu. 1991. Dihydroxyacids from the chlorination of ketones - an unexpected process. *Canadian Journal of Chemistry-Revue Canadienne De Chimie*, 69(12): 1904-1908.

Lee, B.-C., M. Kamata, Y. Akatsuka, M. Takeda, K. Ohno, T. Kamei, and Y. Magara. 2004. Effects of chlorine on the decrease of estrogenic chemicals. *Water Research* 38(3): 733-739.

Lee, G. F., and J. C. Morris. 1962. Kinetics of chlorination of phenol: chlorophenolic tastes and odors. *Internat. J. Air Water Pollut.*, 6(419-431).

MacCrehan, W. A., M. Bedner, and G. R. Helz. 2005. Making chlorine greener: Performance of alternative dechlorination agents in wastewater. *Chemosphere*, 60(3): 381-388.

Pinkston, K. E., and D. L. Sedlak. 2004. Transformation of aromatic ether- and amine-containing pharmaceuticals during chlorine disinfection. *Environmental Science & Technology* 38(4019-4025).

Robeck, G. G., K. A. Dostal, J. M. Cohen, and J. F. Kreissl. 1965. Effectiveness of water treatment processes in pesticide removal. *Journal American Water Works Association* 57(2): 181-200.

Westerhoff, P., P. Chao, and H. Mash. 2004. Reactivity of natural organic matter with aqueous chlorine and bromine. *Water Res.*, 38(6): 1502-1513.

Westerhoff, P., Y. Yoon, S. Snyder, and E. Wert. 2005. Fate of endocrine-disruptor, pharmaceutical, and personal care product chemicals during simulated drinking water treatment processes. *Environ. Sci. Technol.*, 39(17): 6649-6663.

CHAPTER 7
OZONE AND OZONE/HYDROGEN PEROXIDE OXIDATION

BACKGROUND

Ozone is a strong oxidant and disinfectant. Unlike free chlorine or chloramine, ozone decays rapidly within minutes after its addition to water and results in the formation of fewer halogenated organic disinfection by-products. Ozone reacts with organic contaminants through either the direct reaction with molecular ozone or through the formation of free radicals, including the hydroxyl radical ($HO^{\bullet}$). The formation of $HO^{\bullet}$ occurs through a series of promotional steps, which involve initiators such as hydrogen peroxide, ultraviolet radiation, natural organic matter, or hydroxide ion. Both molecular ozone (O_3) and hydroxyl radical pathways can lead to transformation of organic compounds (Hoigne and Bader 1983a; Hoigne and Bader 1983b). In the direct pathway, ozone is a selective electrophile that reacts quickly with amines, phenols, and double bonds in aliphatic compounds. The $HO^{\bullet}$ reacts less selectively than ozone and with faster reaction rates, with second-order rate constants (k_{OH}) in the order of 10^8 to 10^{10} $M^{-1}s^{-1}$ (Buxton et al. 1988; Haag and Yao 1992).

In most water treatment applications, the concentration of hydroxyl radicals is extremely low, usually in the order of 10^{-12} M. In order to increase the transformation of recalcitrant micropollutants, advanced oxidation processes (AOPs), such as UV/hydrogen peroxide (H_2O_2), O_3/H_2O_2, and UV/O_3, are applied to increase $HO^{\bullet}$ concentrations (Hoigne 1998; Acero, Stemmler, and von Gunten 2000; Acero and von Gunten 2001). In most ozone AOP applications, H_2O_2 is added to promote the initiation step in ozone decomposition that will lead to higher radical production. The overall kinetics of the ozone decomposition is dependent on water quality parameters. Given information on water quality (e.g. alkalinity, pH, and NOM), it may be possible to estimate the concentrations of O_3 and $HO^{\bullet}$ present during ozonation. Such data can be used with second order rate constants to predict the rate of transformation of EDCs/PPCPs during an ozonation process.

EDC/PPCP compounds exhibit a wide range of reactivity toward molecular ozone (Huber, Canonica, and von Gunten 2002), largely due to the difference in their 2nd order reaction rate constants with ozone (Huber et al. 2003). Most k_{OH} second-order rate constant values were close to the diffusion controlled limit and on the order of 5×10^9 $M^{-1}s^{-1}$. However, under conditions encountered in water treatment systems ($[HO^{\bullet}]/[O_3] \approx 10^{-8}$), the molecular ozone reaction pathway is dominant and hence only those compounds with ozone rate constants (k_{O3}) greater than ~50 $M^{-1}s^{-1}$ will be transformed to an appreciable degree (>50%) through direct reactions with ozone.

Many EDCs/PPCPs react directly with ozone resulting in significant decreases in concentrations under conditions typically used for disinfection (Carlile et al. 1996; West 2000). In one example, a linear relationship between ozone dose and fraction of 17β-estradiol (E2) remaining in model waters was observed during O_3 and O_3/H_2O_2 processes (Kosaka et al. 2000). The use of an AOP will increase the destruction of target contaminants. For example, clofibric acid could not be as efficiently removed as diclofenac, carbamazepine, and bezafibrate by direct ozonation (Sacher et al. 2000). The addition of hydrogen peroxide (0.4 to 0.7 mg H_2O_2 /mg ozone) improved removal of clofibric acid, ibuprofen, and diclofenac as compared to ozone alone (Carlson, Sanchez, and Muylwyk 2000).

The ozone and ozone-AOP experiments conducted in this study were designed to encompass a large range of chemical diversity, while offering a comparison of scale. Colorado River water was evaluated at bench, pilot, and full scale. Spiked experiments at bench and pilot scale agreed well with observations from full scale ozonation systems.

BENCH-SCALE EVALUATIONS

Bench-scale testing was conducted using one synthetic water prepared from Suwannee River NOM isolate (SRW) and three natural waters (CRW, PVW, and ORW). Due to the differences in NOM content and concentration, ozone demand tests were performed on each of the waters to determine the ozone dosage required to achieve an ozone residual of 0.2 to 0.3 mg/L after 3 minutes of contact time, and zero residual within 10 minutes. This ozone residual and contact time yield a USEPA regulatory-based CT value of approximately 0.8 min-mg/L, which is appropriate for primary disinfection when used for *Giardia* or virus inactivation in most cases. Ozone exposure (OE) values were calculated as the area beneath the preliminary ozone decay curve. OE values at the dosages illustrated in Figure 7.1 were 12, 22, 2.8, and 8.2 min-mg/L for SRW, CRW, ORW, and PVW, respectively. Ozone demand and decay rates vary due to differences in water quality; therefore, each water required a different ozone dose to meet the target CT value. A low dosage of hydrogen peroxide was added to select experiments to promote hydroxyl radical formation as an AOP.

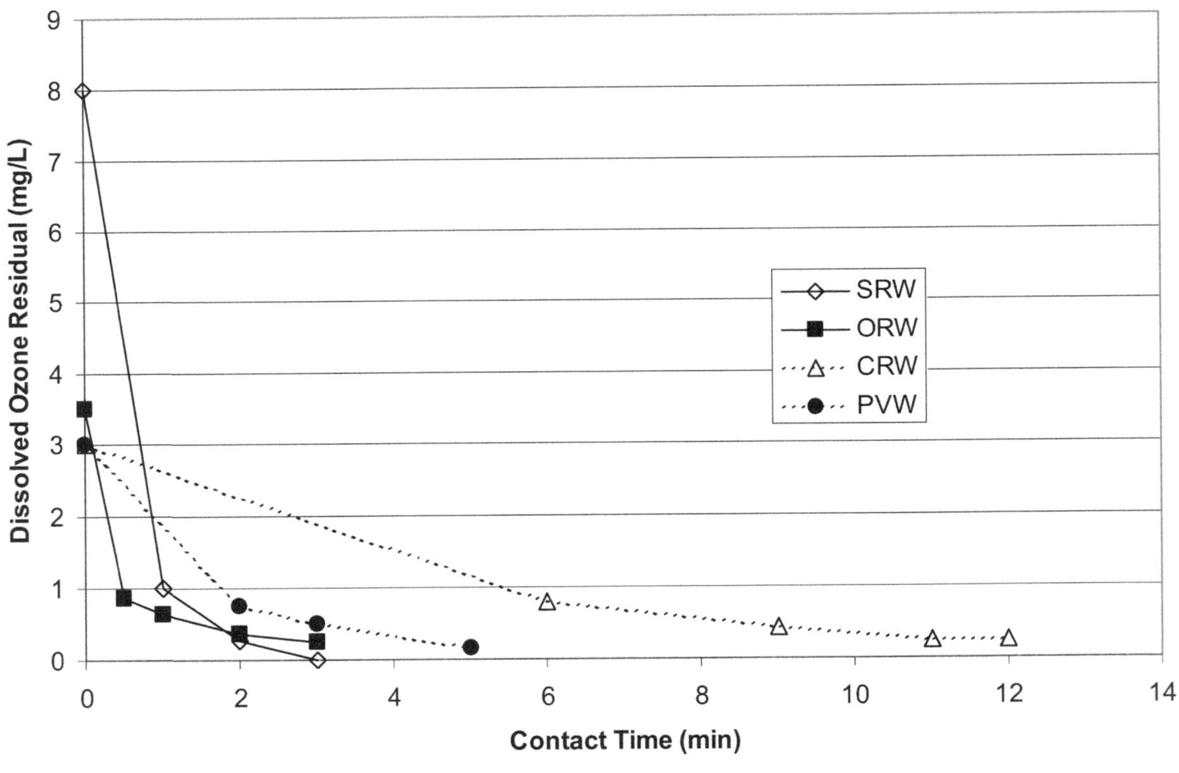

Figure 7.1 Summary of ozone decay for the waters examined

Bench-Scale Results

Table 3.4 in Chapter 3 indicates that SRW has the highest DOC level. As expected, this water exhibited most rapid ozone decay rate. The ozone decay rate was related to the level of DOC in each water (Figure 7.1). Although kept to a minimum, solvents in the EDC/PPCP spiking solution could have affected the rate of ozone decay by scavenging hydroxyl radicals and initiating ozone decay. However, as shown in the results section, predictions from spiked bench-scale testing were in good agreement with full-scale sampling where solvents were not used.

The ozone dose, hydrogen peroxide dose, and contact time were determined from ozone decay information as shown in Table 7.1 along with target compound removal information for each of the waters after five minutes of reaction time. The SRW results are not shown as 17 of 25 target compounds had raw water concentrations less than 10 ng/L, likely due to loss through adsorption on particles or surfaces. Most target compounds investigated exhibited >50% removal, except for the herbicides, atrazine, the pesticides, DDT and Lindane, the flame-retardant TCEP, the pharmaceutical, meprobamate, and. The PVW results showed that 15 of 29 target compounds were removed to below detection. The CRW and ORW results showed 20 of 29 target compounds were removed to below detection. Increasing ozone dose increased the percentage of the target compound removed. The addition of hydrogen peroxide generally led to only a slight increase in the removal of target compounds as compared to ozone alone.

PILOT-SCALE EVALUATIONS

Pilot-scale ozone testing was conducted at the Southern Nevada Water Authority (SNWA) using two dynamic pilot testing systems with flow rates of 0.26 gpm and 6 gpm. A bench-top pilot plant (BTPP) with a flow rate of 0.26 gpm was used to conduct multiple ozonation and advanced oxidation experiments. The larger pilot plant with a flow rate of 6 gpm was limited in application due to the large quantities of target compounds required for spiking and the waste volume generated.

Bench-Top Pilot Plant Evaluation Using Colorado River Water

The BTPP testing was performed using two 45-gallon batches of CRW spiked with target compounds, with each batch used to perform two tests. Ozone dosages of 1.25 and 2.50 mg/L were selected to replicate conditions used during bench-scale studies. During AOP experiments, hydrogen peroxide was added approximately 30 seconds prior to the ozone contactor. The first set of AOP experiments were designed to model the bench-scale conditions by using a 0.0625 mg/L hydrogen peroxide dose. The second set of AOP experiments were performed targeting a 0.2 hydrogen peroxide to ozone mass ratio. The BTTP system was operated for an hour to assure that steady state conditions were achieved before sampling. During all experiments, samples were collected after 2, 6, 14, and 24 minutes of contact time to determine the relative rates of reaction.

Table 7.1
Summary of percent removal during bench-scale ozonation testing of surface waters

Source Water	CRW		ORW			PRW		SR-NOM		
Ozone Dose [mg/L]	2.5	2.5	3.5	7	3.5	3	3	4	8	8
Peroxide Dose [mg/L]		0.0625			0.0875		0.075			0.2
Reaction Time [min]	5	5	5	5	5	5	5	5	5	5
Target Compound	Percent Removal		Percent Removal			Percent Removal		Percent Removal		
Acetaminophen (Tylenol)	>96	>96	>97	>97	>97	96	96	>90	>90	>90
Androstenedione	>99	>99	>98	>98	>98	78	95	85	>98	>98
Atrazine	45	52	71	68	80	17	24	60	96	96
Benzo(a)pyrene	>71	>71	>91	>91	>91	>67	>67	>91	>91	>91
Caffeine	>99	>99	>99	>99	>99	83	87	87	>95	>95
Carbamazepine	>98	>98	>98	>98	>98	>99	>99	>99	>99	>99
DDT	57	61	58	65	71	35	27	77	>85	>85
DEET	76	84	95	94	96	63	68	90	99	>99
Diazepam (Valium)	79	86	96	94	96	66	69	90	96	>99
Diclofenac	>96	>96	>97	>97	>97	>86	>86	>93	>93	>93
Dilantin	82	89	>98	>98	>98	72	73	92	>93	>93
Erythromycin	>96	>96	>98	>98	>98	>96	>96	*	*	*
Estradiol	>99	>99	99	99	>99	>99	>99	>97	>97	>97
Estriol	>99	>99	*	*	*	>99	>99	*	*	*
Estrone	>99	>99	*	*	*	>99	>99	*	*	*
Ethinyl Estradiol	>99	>99	>99	>99	>99	>99	>99	>98	>98	>98
Fluorene	94	>95	>94	>94	>94	83	83	*	*	*
Fluoxetine (Prozac)	>98	>98	92	>97	>97	82	81	89	>93	>93
Galaxolide	87	89	91	90	91	76	77	79	>89	>89
Gemfibrozil	>99	>99	>98	>98	>98	>99	>99	>96	>96	>96
Hydrocodone	>98	>98	*	*	*	>99	>99	*	*	*
Ibuprofen (Advil)	82	88	94	94	96	56	60	91	99	>99
Iopromide	46	60	75	77	86	52	57	69	>92	>92
Lindane (γ-BHC)	4	14	24	19	26	1	3	<1	27	32
Meprobamate	50	61	84	80	85	40	43	72	99	99
Metolachlor	80	86	92	88	94	73	82	81	86	86
Musk Ketone	30	34	50	41	68	20	3	47	76	79
Naproxen	>94	>94	>95	>95	>95	>92	>92	>86	>86	>86
Oxybenzone	>97	>97	>97	>97	>97	>97	>97	>95	>95	>95
Pentoxifylline	>98	>98	>98	>98	>98	86	90	89	>93	>93
Progesterone	>98	>99	>97	>97	>97	83	96	90	>97	>97
Sulfamethoxazole	>97	>97	>95	>95	>95	>79	>79	>78	>78	>78
TCEP	13	16	5	<1	16	<1	<1	<1	<1	<1
Testosterone	>99	>99	>98	>98	>98	83	96	95	>99	>99
Triclosan	79	82	>97	>97	>97	>98	>98	>96	91	>96
Trimethoprim	>99	>99	>99	>99	>99	>99	>99	>96	>96	>96

* Target compound not analyzed

BTPP CRW Results

During BTTP experiments, the raw water temperature increased from 15.3°C to 20.3°C as ambient air warmed the raw water while EDC spiking and mixing were performed. The raw water pH also increased from 7.7 to 7.9 during the mixing process. The TOC concentration was 3.2 mg/L and remained unchanged after the addition of target compounds. Results from dissolved ozone residual sampling are provided in Table 7.2. CT was provided based integration on an ozone decay curve. During the AOP experiments, ozone demand and CT could not be accurately calculated since there was no measurable dissolved ozone residual in cell #3; a complete ozone curve could not be obtained.

Table 7.2
Summary of ozone operating conditions during BTPP experiments

Ozone Parameter	Test #1	Test #2	Test #3	Test #4	Test #5	Test #6
Applied Ozone Dose (mg/L)	1.31	1.20	2.69	2.64	1.30	2.57
Hydrogen Peroxide Dose (mg/L)	-	0.0625	-	0.0625	0.25	0.50
Dissolved Ozone Residual Cell#1 (mg/L)	0.42	0.40	1.54	1.03	0.42	0.62
Dissolved Ozone Residual Cell#3 (mg/L)	0.15	0.05	1.04	0.42	-	-
CT (mg-min/L)	0.80	0.48	9.62	2.90	-	-
Ozone Demand (mg/L)	0.85	0.76	1.28	1.70	-	-

During the BTPP ozonation experiments, 16 of 29 target compounds were removed by greater than 95% within the first two minutes of contact time at an ozone dose of 1.25 mg/L. Removal of the remaining compounds corresponded to the dissolved ozone residual and the contact time (Figure 7.2). Raising the ozone dose to 2.50 mg/L resulted in an increased removal of all compounds, although the flame-retardant TCEP remained difficult to oxidize (Figure 7.3).

During the AOP experiments, the majority of contaminant removal occurred within the first 2 minutes of contact time. The same 16 of 29 target compounds removed by ozone only also were removed by the AOP within the first two minutes of contact time. Figure 7.4 shows the percent removal comparison between ozone only and the AOP (0.0625 mg/L hydrogen peroxide). The AOP showed approximately 10% greater removal for certain compounds as compared to ozone alone (ozone doses of 1.2 – 1.3 mg/L). For tests five and six, efficient removal was achieved within the first six minutes of contact time for both AOP experiments. Figure 7.5 compares the 24-minute removal from ozone only to the 6-minute removal from the AOP using similar ozone dosages. The comparison shows that the net removals were similar among all of the compounds with a constant ozone dose. As expected, the AOP accelerated the reaction rates, experiencing similar compound removal in six minutes versus 24 minutes for ozone only. Both molecular ozone and AOP showed that at higher ozone dosages, 21 of 29 of the target compounds could be removed to greater than 95%. Again, TCEP remained difficult to remove at all conditions investigated.

Bench-Top Pilot Plant Evaluation Using Tertiary Treated Wastewater

The BTPP testing was performed using one 45-gallon batch of tertiary treated wastewater. The water was collected and transported to SNWA within one hour. The wastewater had a TOC of 7.0 mg/L. Since the wastewater already contained detectable concentrations of several target compounds, spiking of the wastewater was not performed. The concentrations of ozone dissolved into water, also defined as transferred ozone dosages (4.9, 7.3, and 8.7 mg/L), were evaluated after 18 minutes of contact time in order to observe the dose-response relationship. Once the operating conditions were set, the system was operated for an hour to assure that steady-state conditions were achieved before sampling.

Additional testing was conducted with tertiary treated wastewater to quantify the production of $HO^{\bullet}$ radicals during both ozone and ozone AOP. $HO^{\bullet}$ radical exposures were estimated using a probe compound, para-chlorobenzoic acid (pCBA), which was added to the evaluated waters in separate bench scale tests. For this investigation, three ozone doses were used (2, 7, and 11 mg/L) with and without the addition of H_2O_2.

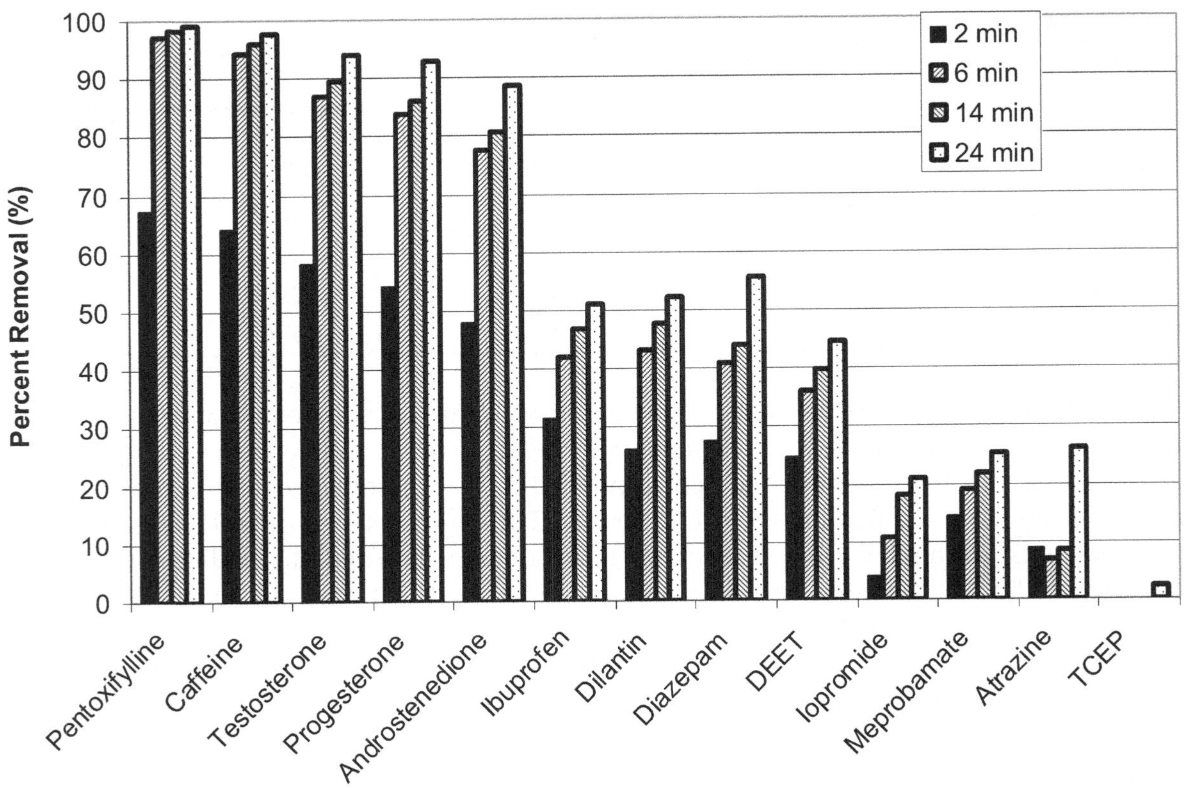

Figure 7.2 BTPP removal of target compounds using an ozone dose of 1.31 mg/L

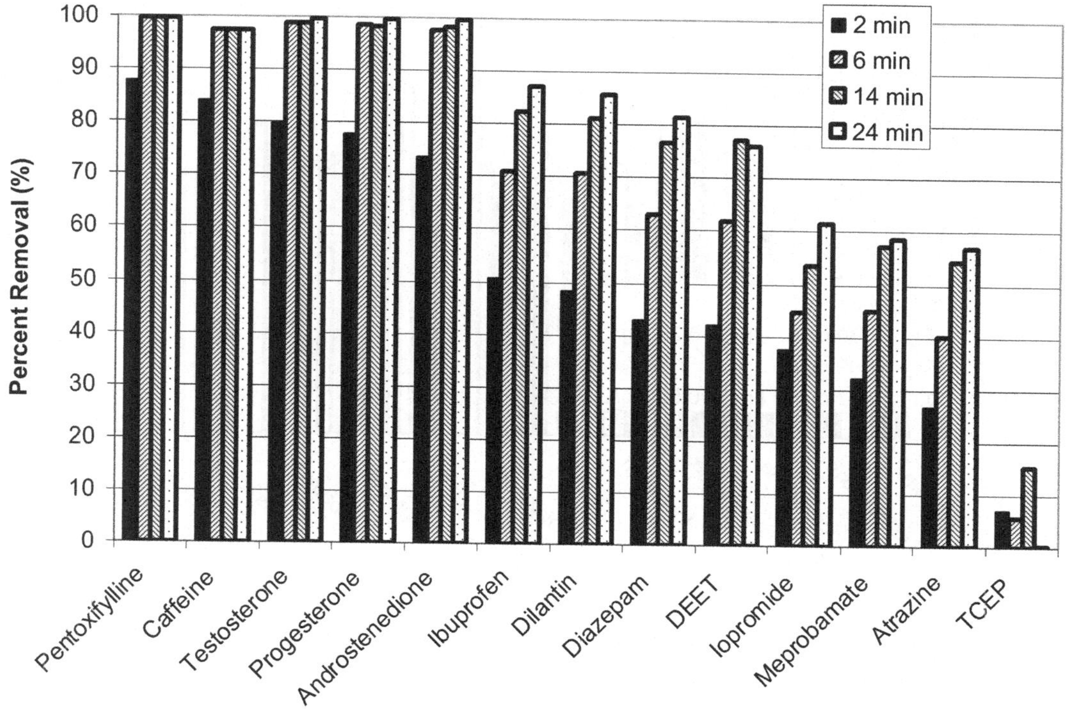

Figure 7.3 BTPP removal of target compounds using an ozone dose of 2.69 mg/L

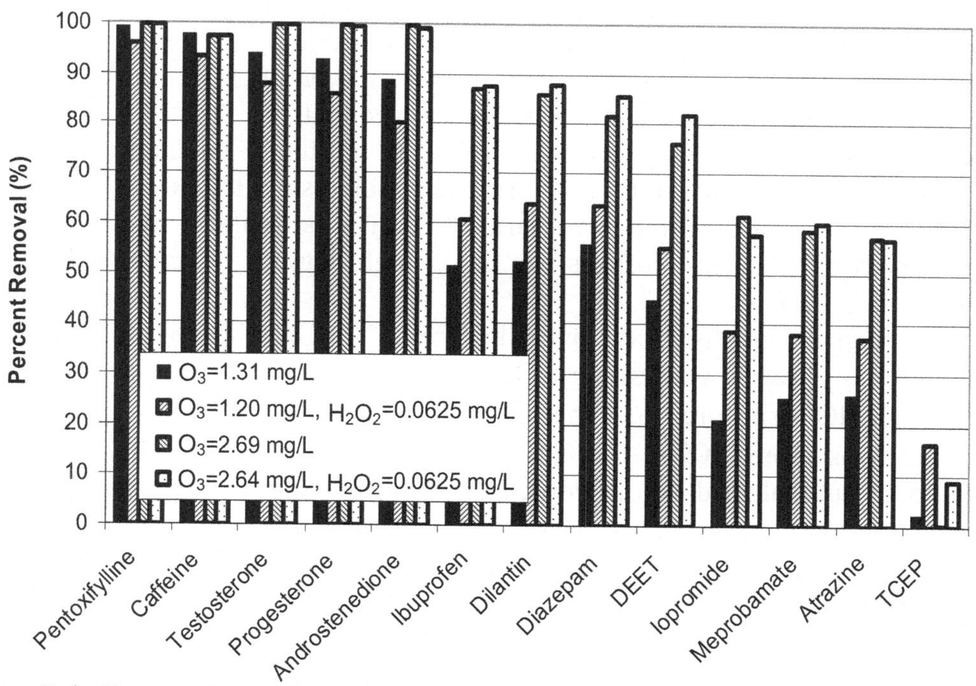

Figure 7.4 Comparison of percent removal when using a 24- minute contact time

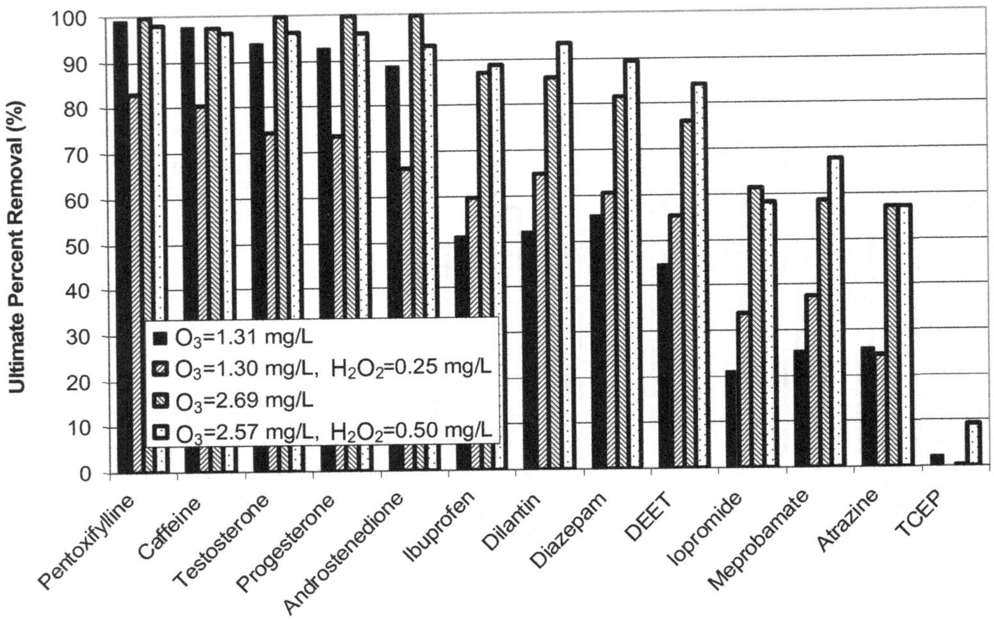

Figure 7.5 Comparison of percent removal when using a 24-minute contact time for ozonation and a 6-minute contact time for the AOP

BTPP Tertiary Wastewater Results

Ozone decay rates were much faster in tertiary treated wastewater than in CRW. The tertiary water had almost twice as much TOC as the CRW, contributing to the faster decay rate. As observed in Figure 7.6, the dissolved ozone residual had decayed after 12 minutes of contact time under three dose conditions. Analysis of the water showed that 15 of the 29 analyzed target compounds were detectable in the tertiary effluent (Table 7.3). Of the 15 contaminants detected, eight target compounds were removed to less than detection using the lowest ozone dose evaluated of 4.9 mg/L. Once again, the flame-retardant TCEP was removed by less than 10%. These results indicate that low ozone dosages can attain high degrees of removal for a large number of organic contaminant.

Figure 7.7 provides the results for the overall decrease of pCBA during ozone and ozone peroxide addition to tertiary treated wastewater effluent. The decomposition of pCBA serves as a measure of the overall $HO^{\bullet}$ exposure. As expected, the concentration of pCBA is reduced with increasing ozone doses as more $HO^{\bullet}$ is formed. A more rapid loss of pCBA occurs when peroxide is added to an equivalent dose of ozone. This suggests that the addition of peroxide increases $HO^{\bullet}$ radical formation as expected. The net amount of pCBA decomposition is similar between ozone alone and ozone with peroxide addition after six minutes of contact time. For example, at an ozone dose of approximately 2.3 mg/L, the pCBA concentration was reduced to approximately 62% of the initial pCBA concentration. When 2.3 mg/L of ozone is combined with 2.0 mg/L of hydrogen peroxide, pCBA was approximately 60% of its initial concentration after six minutes. The rate at which pCBA decomposes is clearly faster with the addition of peroxide, while the total pCBA decomposition is only marginally greater after only six minutes of contact time. These data are comparable to the removal of spiked target compounds with ozone and ozone/peroxide as shown in Figures 7.4 and 7.5. These data show that the addition of peroxide generally yielded only a meager increase in removal at similar ozone doses. The pCBA decomposition data offers additional evidence of the reduced advantage of peroxide addition as compared to ozone alone during water reuse applications.

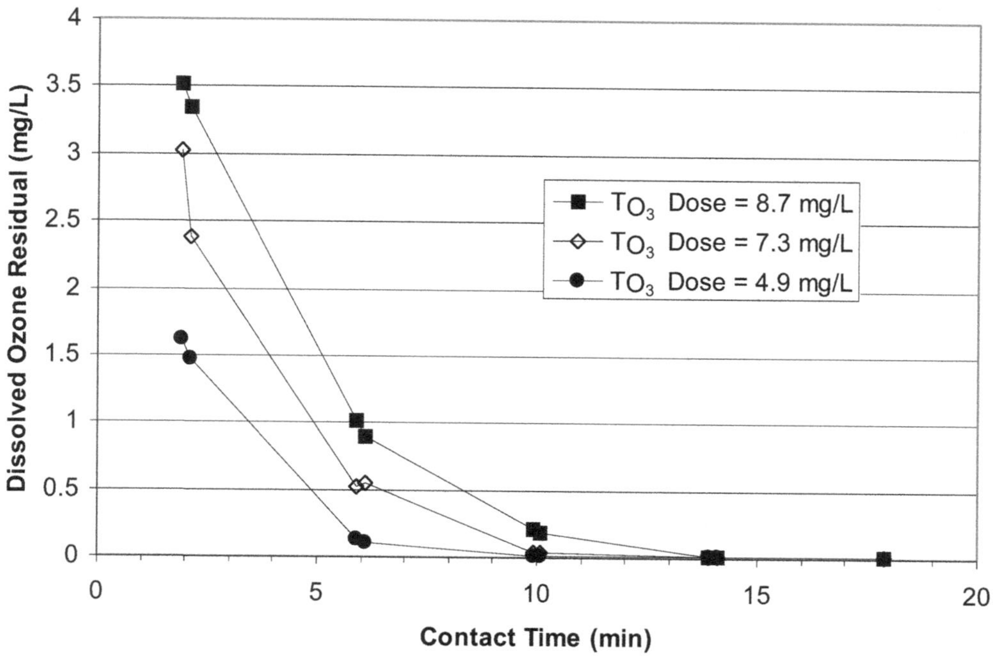

Figure 7.6 BTPP ozone decay in tertiary treated wastewater

Table 7.3
BTPP ozone removal of target compounds from reuse water

Source Water	Raw Sewage	Tertiary Effluent	Ozonated Tertiary Effluent		
Transferred Ozone Dose*			4.9 mg/L	7.3 mg/L	8.7 mg/L
Reaction Time			18 min	18 min	18 min
Target Compound	ppt	ppt	Percent Removal		
Caffeine	97800	51	>80	>80	>80
Carbamazepine	99	210	>99	>99	>99
DEET	413	188	79	95	98
Diclofenac	28	54	>98	>98	>98
Dilantin	94	154	89	98	>99
Erythromycin	285	133	>99	>99	>99
Hydrocodone	218	240	>99	>99	>99
Ibuprofen (Advil)	11950	19	94	>95	>95
Iopromide	37	22	72	91	>95
Meprobamate	739	332	58	81	87
Naproxen	13200	13	>92	>92	>92
Oxybenzone	2925	6	<1	>83	75
Sulfamethoxazole	590	841	>99	>99	>99
TCEP	453	373	<1	6	10
Trimethoprim	319	35	>97	>97	>97

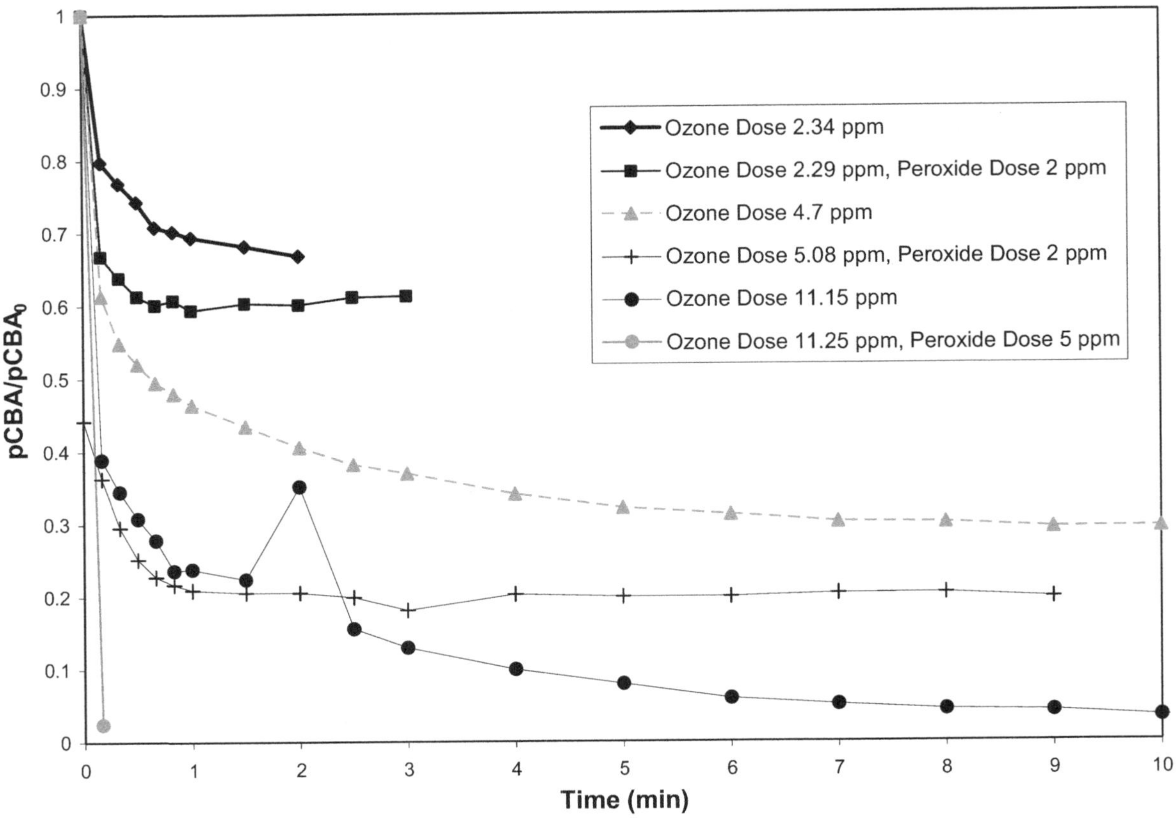

Figure 7.7 Concentration profiles of pCBA under various ozone and H_2O_2 conditions

Conventional Pilot Plant Evaluation

Conventional pilot plant testing was performed at SNWA using CRW targeting a 1.0-log *Cryptosporidium* inactivation according to the proposed Long Term 2 Enhanced Surface Water Treatment Rule (LT2ESWTR). Prior to testing with spiked target compounds, a solvent-only test was performed to investigate the ozone demand resulting from the solvents contained in the EDC/PPCP spiking solution. The ozone dosage was adjusted accordingly to account for the additional ozone demand while achieving the disinfection goal.

Solvent Test Results

Bench scale testing showed that natural waters spiked with the target compounds exerted greater ozone demands than the natural water alone due to solvents contained in the EDC spiking solution. Organic solvents can act as hydroxyl radical scavengers or react directly with ozone, both of which cause faster ozone decay and higher ozone demand. Pilot plant testing was performed in three stages to investigate and account for the additional demand. During Stage 1 of pilot plant testing, a solvent mixture without the target compounds was infused ahead of the ozone contactor. The solvent mixture was designed to match the quantities of methanol, dichloromethane, hexane, toluene, and acetone found in the actual spiking solution. Once the disinfection goal was achieved, Stage 2 began by turning off the solvent injection and allowing raw water to pass through the ozone contactor at the same ozone dose. The dissolved ozone residual and corresponding disinfection level increased as steady state was achieved, illustrating the demand from the solvent solution (Figure 7.8). In Stage 3, spiking solution containing the target compounds was infused. After steady state was reached, the ozone dose was adjusted to meet the desired disinfection goal observed with the solvent addition. The changes in ozone operating conditions are summarized in Table 7.4.

Conventional Pilot Plant Results

During the conventional pilot plant experiments, the raw water pH was 8.14 and the temperature was 17.5°C. Ozone operating conditions are provided in Table 7.4. During pilot plant testing, 16 of the 29 target compounds analyzed were greater than 95% removed within the first two minutes of contact time. The compounds removed were the same as those removed during BTPP experiments (Figures 7.2, 7.3), with the exception of acetaminophen and Pentoxifylline. The pilot testing also provided some kinetic data from samples collected at 2, 6, and 24 minutes (Figure 7.9). Removal of the remaining 13 compounds once again showed a response to dissolved ozone residual and contact time. Percent removals were remarkably similar to those observed during the BTPP experiments. Once again, TCEP was poorly removed by ozone.

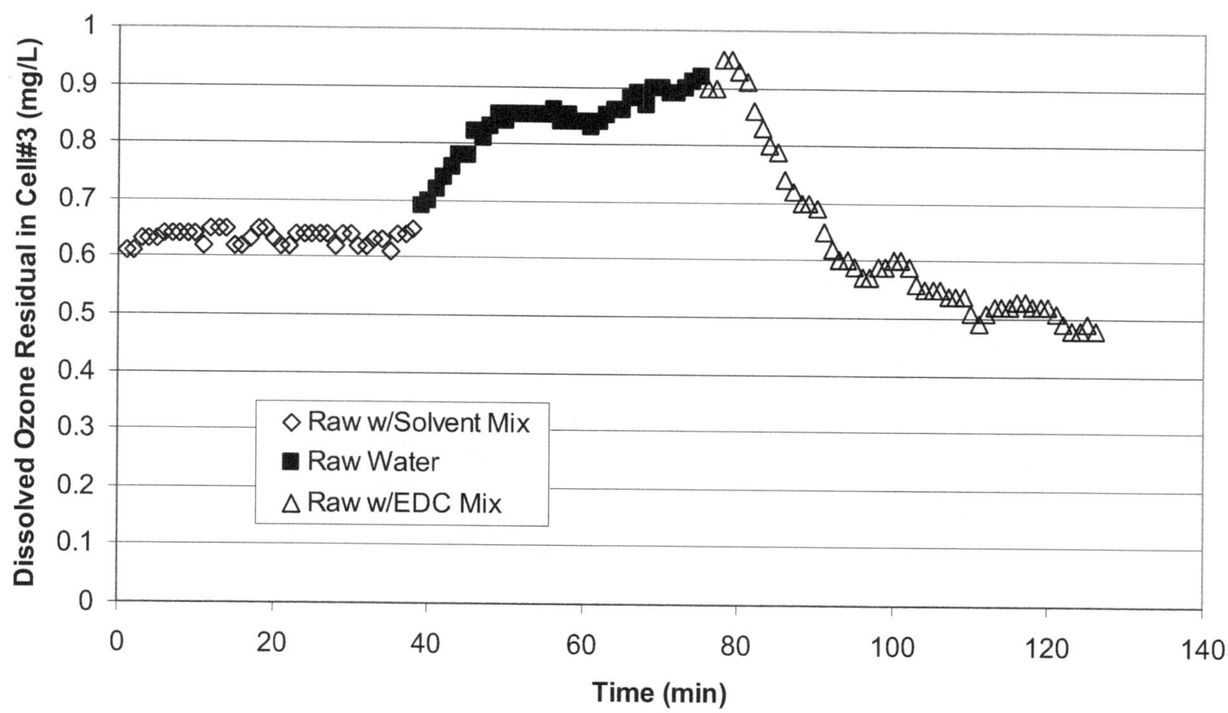

Figure 7.8 Ozone demand of the solvents added during the spiking of target compounds

Table 7.4
Summary of ozone operating parameters

	Raw CRW w/Solvent	Raw CRW	Raw CRW w/EDC Spike
Transferred Ozone Dose (mg/L)	2.29	2.29	2.39
Ozone Demand (mg/L)	0.91	0.66	1.33
Half Life (min)	3.74	4.49	4.30
Ozone Decay Rate (1/min)	-0.186	-0.154	-0.161
Ozone Residual @ 6 min (mg/L)	0.63	0.85	0.54
Ozone Residual @ 10 min (mg/L)	0.33	0.40	0.24
Ozone Residual @ 18 min (mg/L)	0.06	0.12	0.07
CT (mg-min/L)	4.77	6.68	4.20
Cryptosporidium Inactivation (log)*	1.33	1.87	1.17

Estimated based on inactivation credit from EPA Long Term 2 ESWTR

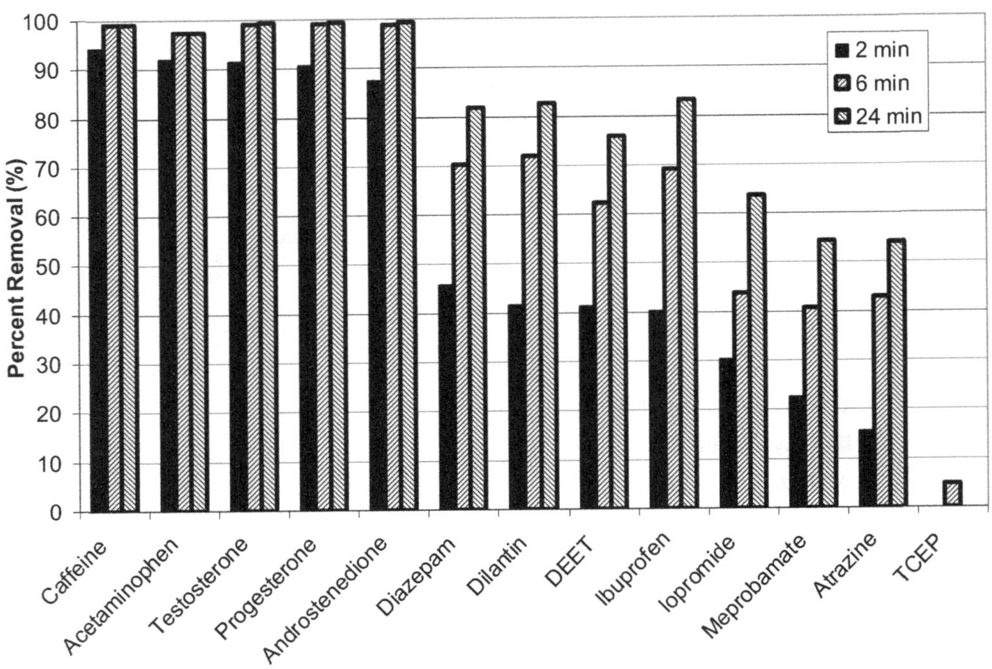

Figure 7.9 Conventional-pilot plant removal of target compounds while using an ozone dose of 2.4 mg/L

FULL-SCALE EVALUATIONS

Full-scale observations were limited since the majority of the target compounds were either not detected or detected at concentrations near the analytical method report limit. The raw and finished water at three full-scale utilities (facilities #13, 14, and 19) was sampled before and after ozone disinfection.

Full-Scale Results

The results from the three utilities evaluated are presented in Table 7.5 along with removal ranges observed during bench-scale testing. At drinking water utilities 13 and 14, the ozone dose ranged from 1.0-1.2 mg/L, each with an ozone contact time of 24 minutes. Only 15 of the target compounds were detectable in at least one of the three utilities investigated (Table 7.5). The concentrations reported here generally fall with the concentration ranges reported previously in US surface waters. For instance, Lindsey et al. reported the occurrence of sulfamethoxazole from less than detection to 1020 ng/L in 138 surface water samples across the US (Lindsey, Meyer, and Thurman 2001). The concentration of TCEP found in this investigation is much lower than that reported by Stakelberg et al., who observed 70 to 150 ng/L of TCEP (Stackelberg et al. 2004). Excellent agreement was observed between average removal from bench-scale experiments and removal observed at full-scale. These results indicate that contaminant removal studies performed at bench or pilot-scale are accurate indicators of full-scale performance.

Table 7.5
Full-scale removal by ozone

Utitilty #		13			14			19		Bench
	Pre	*Post*	*% Rem.*	*Pre*	*Post*	*% Rem.*	*Pre*	*Post*	*% Rem.*	*% Rem.*
Atrazine	1.9	<1.0	>47	1.4	<1.0	>28	<1.0	<1.0	NA	20-50
Caffeine	<1.0	<1.0	NA	4.1	<1.0	>75	36	1.9	94	>80
Carbamazepine	4.8	<1.0	>79	3.5	<1.0	>71	16	<1.0	>93	>80
DEET	7.9	<1.0	>87	4.0	2.1	47.5	13	6.8	48	50-80
Dilantin	4.2	1.8	57	3.1	1.5	52	4.8	2.1	56	20-50
Erythromycin	<1.0	<1.0	NA	<1.0	<1.0	NA	2.5	<1.0	>60	>80
Estrone	<1.0	<1.0	NA	1.4	<1.0	>28	<1.0	<1.0	NA	>80
Galaxolide	<1.0	<1.0	NA	<1.0	<1.0	NA	29	13	55	
Gemfibrozil	2	<1.0	>50	<1.0	<1.0	NA	4.8	<1.0	>79	>80
Ibuprofen	1.7	<1.0	>41	<1.0	<1.0	NA	11	2.6	76	50-80
Iopromide	2.2	1.1	50	<1.0	<1.0	NA	48	36	25	20-50
Meprobamate	16	12	25	13	9.4	28	4.0	2.9	28	20-50
Naproxen	<1.0	<1.0	NA	<1.0	<1.0	NA	12	<1.0	>91	>80
Sulfamethoxazole	12	<1.0	>91	11	<1.0	>90	10	<1.0	>90	>80
TCEP	<1.0	<1.0	NA	<1.0	<1.0	NA	5.4	11	-104	<20
Triclosan	<1.0	<1.0	NA	<1.0	<1.0	NA	3.2	<1.0	>68	>80

RESULT SUMMARY AND DISCUSSION

During the ozonation tests, most of the target compounds investigated reacted quickly with ozone at typical drinking water dosages and were removed within the first two minutes of contact time. Ozone is a selective oxidant, with the highest reactivity toward olefins and slower rates with unsubstituted aromatic compounds. As expected, electron-donating groups enhance the reactivity of aromatic compounds toward ozone, while electron-withdrawing groups inhibit the reactivity. As a result, phenols react 600 times faster with ozone than with benzene (Hoigné 1982). All target compounds with phenolic structures, such as acetaminophen, oxybenzone, and several estrogens, have confirmed this point showing removal to below detection limits. A number of compounds also have exhibited apparent complete removals (carbamazepine, gemibrozil, hydrocodone, diclofenac, erythromycin, naproxen, oxybenzone, sulfamethoxazole, and trimethoprim) due to their increased affinity to electrophiles (ozone in this case) from highly conjugated systems and/or multiple electron donating substituents.

Ozone oxidized most remaining compounds by >50%, except atrazine, iopromide, meprobamate, and TCEP (Table 7.7). Androstenedione, progesterone, and testosterone were oxidized less efficiently than any of the estrogen hormones due to the ketone functional group, which decreases the reactivity of ozone with the adjacent carbons. Oxidation of ethynyl estradiol and other estrogens occurs in the phenolic ring, while ozone likely attacks the double bonds of the other ketone steroids.

For a subset of EDCs/PPCPs, molecular ozone second order rate constants are available and varied between <1 to $7x10^9$ $M^{-1}s^{-1}$ while hydroxyl radical rate constants were generally orders of magnitude higher (Table 7.6). There is good agreement between observed removal trends and order of magnitude molecular ozone second order rate constants (Figure 7.10) (Buxton et al. 1988; Huber et al. 2003). This indicates that despite the presence of hydroxyl radicals during the ozonation process, oxidation by molecular ozone is still the major pathway because the non-selective hydroxyl radicals can be readily quenched by a large number of other solutes in the water (e.g. humic substances, bicarbonate).

The overall percent removal during AOP testing was similar to ozone alone. The addition of hydrogen peroxide prior to ozonation generally improved removal by only 5 to 15%. As expected, the AOP accelerated the reaction rates, experiencing similar compound removal in 6 minutes versus 24 minutes for ozone only. An AOP should be considered when contact time is unavailable for destruction by ozone alone. Once ozone disinfection requirements have been met, immediate dissolved ozone residual quenching with hydrogen peroxide provides another source for additional contaminant removal. Hydrogen peroxide addition increased the rate of molecular ozone decay (i.e., lower molecular ozone concentrations) but increased hydroxyl radical concentration. Clofibric acid and ibuprofen were poorly oxidized (~10%) by ozone alone, but the use of 3.7 mg/L ozone and 1.4 mg/L hydrogen peroxide increased removal to >90% (Zwiener and Frimmel 2000). For a subset of 8 EDC/PPCPs, Huber et al. previously suggested that advanced oxidation systems are promising for efficient removal of some pharmaceuticals (Huber et al. 2003).

Removals observed at pilot- and full-scale were consistent with bench-scale results and confirmed that bench-scale tests are valid and efficient ways to predict the removal of EDCs/PPCPs at full-scale plants. Although ozone is highly efficient for contaminant oxidation, it should noted that oxidation by ozone generally does not result in mineralization. Therefore, oxidation products should be expected. The organic oxidation products from ozone are general biodegradable and thus may be removed during subsequent biological filtration. In the case of estrogencity in wastewater effluent, it was determined that ozone readily eliminated *in vitro* estrogenicity without formation of estrogenic oxidation products. A more thorough discussion regarding oxidation by ozone can be found in Snyder et al. 2006.

Table 7.6
Rate constants for the reaction of O_3 and ·OH with selected pharmaceuticals

	k_{O3} ($M^{-1}s^{-1}$)	k_{OH} ($M^{-1}s^{-1}$)
Carbamazepine	3.0×10^5	8.8×10^9
Diazepam	0.75	7.2×10^9
Diclofenac	1.0×10^6	7.5×10^9
Ethynyl estradiol	7.0×10^9	9.8×10^9
Ibuprofen	9.6	7.4×10^9
Iopormide	0.80	3.3×10^9
Sulfamethoxazole	2.5×10^6	5.5×10^9

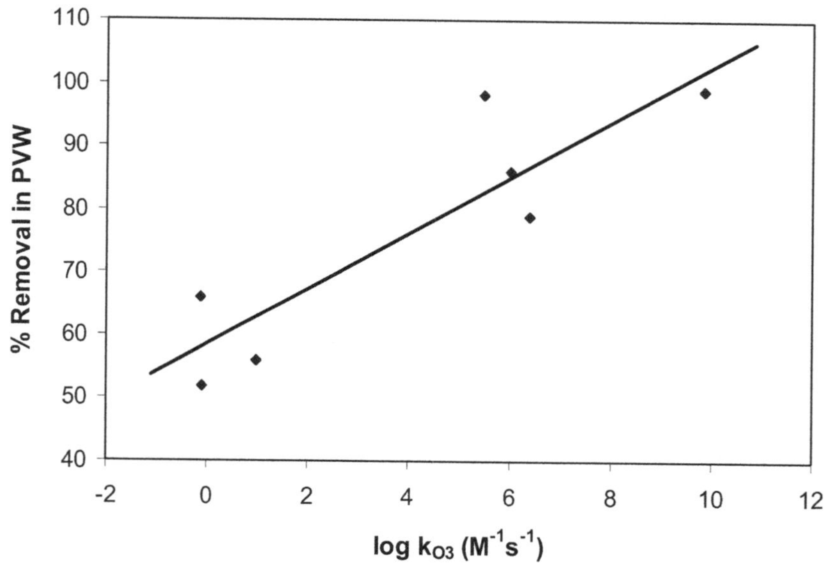

Figure 7.10 Correlation between 2nd order reaction rate constants with ozone and percent removal in PVW

Table 7.7
Summary of minimum percent removal by ozonation at drinking water dosages

2 Minutes Contact Time	24 Minutes Contact Time			
> 95% Removal	> 80% Removal	50-80% Removal	20-50% Removal	< 20% Removal
Acetaminophen	Androstenedione	DEET	Atrazine	TCEP
Carbamazepine	Caffeine	Diazepam	Iopromide	
Diclofenac	Pentoxifylline	Dilantin	Meprobamate	
Erythromycin-H_2O	Progesterone	Ibuprofen		
Estradiol	Testosterone			
Estriol				
Estrone				
Ethynylestradiol				
Fluoxetine				
Gemfibrozil				
Hydrocodone				
Naproxen				
Oxybenzone				
Sulfamethoxazole				
Triclosan				
Trimethoprim				

REFERENCES

Acero, J. L., K. Stemmler, and U. von Gunten. 2000. Degradation kinetics of atrazine and its degradation products with ozone and OH radicals: A predictive tool for drinking water treatment. *Environ. Sci. Technol.*, 34(4): 591-597.

Acero, J. L., and U. von Gunten. 2001. Characterization of oxidation processes: ozonation and the AOP O_3/H_2O_2. *Journal American Water Works Association* 93(10): 90-100.

Buxton, G. V., C. L. Greenstock, W. P. Helman, and A. B. Ross. 1988. Critical review of rate constants for reactions of hydrated electrons, hydrogen atoms and hydroxyl radicals (×OH/×O-) in aqueous solution. *Journal of Physical Reference Data* 17(2): 513-851.

Carlile, P., M. Fielding, L. Harding, J. Hart, J. Hutchison, and R. Kanda.1996. Effect of water treatment processes on oestrogenic chemicals. UK WIR Report 96/DW/05/01

Carlson, M., D. Sanchez, and Q. Muylwyk. 2000. Treatment for Endocrine Disrupters. In *AWWA 2000 Annual Conference - Sunday Workshop*. Denver,CO.

Haag, W. R., and C. C. D. Yao. 1992. Rate constants for reaction of hydroxyl radicals with several drinking-water contaminants. *Environ. Sci. Technol.*, 26(5): 1005-1013.

Hoigne, J. 1998. Chemistry of aqueous ozone and transformation of pollutants by ozonation and advanced oxidation processes. In *Handbook of Environmental Chemistry*. Edited by J. Hrubec. Berlin, Germany: Springer-Verlag.

Hoigné, J. 1982. Mechanism, rates, and selectivities of oxidations of organic compounds initiated by ozonation of water. In *Handbook of Ozone Technology and Application*. Edited by R. G. Rice and A. Netzer. Ann Arbor, MI: Ann Arbor Sci. Pub.

Hoigne, J., and H. Bader. 1983a. Rate constants of reactions of ozone with organic and inorganic-compounds in water .1. Non-dissociating organic- compounds. *Water Res.*, 17(2): 173-183.

Hoigne, J., and H. Bader. 1983b. Rate constants of reactions of ozone with organic and inorganic-compounds in water .2. Dissociating organic-compounds. *Water Res.*, 17(2): 185-194.

Huber, M., S. Canonica, and U. von Gunten. 2002. Oxidative treatment of pharmaceuticals in drinking waters. In *AWWA EDC Workshop*. Cincinnati, OH.

Huber, M. M., S. Canonica, G.-Y. Park, and U. von Gunten. 2003. Oxidation of pharmaceuticals during ozonation and advanced oxidation processes. *Environ. Sci. Technol.*, 37(5): 1016-1024.

Kosaka, K., H. Yamada, S. Matsui, and K. Shishida. 2000. The effects of the co-existing compounds on the decomposition of micropollutants using the ozone/hydrogen peroxide process. *Water Sci. Technol.*, 42(7-8): 353-361.

Lindsey, M. E., M. Meyer, and E. M. Thurman. 2001. Analysis of trace levels of sulfonamide and tetracycline antimicrobials in groundwater and surface water using solid-phase extraction and liquid chromatography/mass spectrometry. *Anal. Chem.*, 73(19): 4640-4646.

Sacher, F., B. Haist-Gulde, H.-J. Brauch, G. Preuß, U. Wilme, N. Zullei-Seibert, M. Meisenheimer, H. Welsch, and T. A. Ternes. 2000. Behavior of selected pharmaceuticals during drinking water treatment. In *219th ACS National Meeting*. San Francisco, CA: American Chemical Society.

Snyder, S. A., E. C. Wert, D. J. Rexing, R. E. Zegers, and D. D. Drury. 2006. Ozone oxidation of endocrine disruptors and pharmaceuticals in surface water and wastewater. *ozone Science & Engineering*, 28(6): 445-460.

Stackelberg, P. E., E. T. Furlong, M. T. Meyer, S. D. Zaugg, A. K. Henderson, and D. B. Reissman. 2004. Persistence of pharmaceutical compounds and other organic wastewater contaminants in a conventional drinking-water-treatment plant. *Sci. Total Environ.*, 329(1-3): 99-113.

West, P. 2000. State of endocrine disruptors and the water industry. In *AWWA 2000 Annual Conference - Sunday Workshop*. Denver,CO: AWWA.

Zwiener, C., and F. H. Frimmel. 2000. Oxidative treatment of pharmaceuticals in water. *Water Res.*, 34(6): 1881-1885.

CHAPTER 8
ULTRAVIOLET IRRADIATION AND ULTRAVIOLET/HYDROGEN PEROXIDE OXIDATION

BASIC THEORY

Since early 1970's, ultraviolet (UV) light-based processes were identified as promising alternatives to conventional treatment technologies for organic pollutant removal from contaminated waters. Ultraviolet light oxidizes organics in water via two mechanisms. First, UV can directly cleave bonds in organic molecules by direct photolysis. Second, UV reacts with water or inorganic constituents in water to form highly reactive intermediates, with the formation of hydroxyl radicals ($HO^\bullet$) being the most important in oxidation of organic compounds. $HO^\bullet$ radicals are highly reactive, non-selective oxidants that react with organic compounds. The presence of hydrogen peroxide (H_2O_2) can increase the concentration of $HO^\bullet$ in solution (H_2O_2 + UV => 2 $HO^\bullet$). Quantum yield is a measure of light absorption efficiency and hence photo reactivity of a species. By definition, it is the number of molecules participating in a given photo physical process or reaction divided by the number of photons absorbed. The quantum yield ranges between zero and one for photoreactions induced by a single photon; values larger than one are possible for photon induced chain reactions.

UV is an effective disinfectant, with the US EPA providing 3-log *Cryptosporidium* inactivation credit using a 12 mJ/cm^2 germicidal dose (U.S. Environmental Protection Agency 2006). The energy carried by the UV-C radiations (200-280 nm) is often sufficient to break chemical bonds in organic compounds capable of absorbing those wavelengths. This process is considered direct UV photolysis, and its efficiency is strongly dependent on both contaminant and water quality. However, at a typical disinfection dose of 40 mJ/cm^2, oxidation of contaminants is extremely small.

UV advanced oxidation processes (AOPs) processes rely on the ability of certain chemicals dissolved in the water (i.e., hydrogen peroxide) that absorb UV light and generate highly reactive species, such as $HO^\bullet$ which react indiscriminately with most organic pollutants ($k_{OH,P} = 10^8 - 10^{10}$ $M^{-1}s^{-1}$). UV/H_2O_2 processes are currently applied in full-scale UV systems to remove recalcitrant contaminants, such as: 1,4-dioxane, methyl *tert*-butyl ether (MTBE), trichloroethylene (TCE), N-nitrosodimethylamine (NDMA) and other organic compounds.

UV and UV/H_2O_2 AOPs are viable options for treatment of trace organic compounds in water. Ahel *et al.* (1994) demonstrated that 4-nonylphenol and its polyethoxylates can be photolyzed in aqueous solution with medium pressure lamp radiations. Mazellier *et al.* (2003) studied the photodegradation of 4-*tert*-octylphenol at 253.7 nm in pure aqueous solution and in simulated lake water. This study determined the photolysis quantum yield as 0.058±0.004 and the $HO^\bullet$ rate constant $k = (6.4\pm0.5)\times10^9$ $M^{-1}s^{-1}$. The UV/H_2O_2 process was found very effective for removal of 5-methyl-1,3,4-thiadiazole-2-methylthiol (MMTD-Me), a pharmaceutical intermediate as well as a reagent in the synthesis of cefazolin, a cephalosporin antibiotic (Bozzi et al. 2002). Huber *et al.* (2003) published a thorough study on the fundamentals of advanced oxidation of nine pharmaceuticals (bezafibrate, carbamazepine, diazepam, diclofenac, 17α-ethynyl estradiol, ibuprofen, iopromide, sulfamethoxazole, roxithromycin) and determined their rate constants for the reaction with hydroxyl radical, which ranged from 3.3×10^9 $M^{-1}s^{-1}$ (iopromide) to 9.8×10^9 $M^{-1}s^{-1}$ (17α-ethynyl estradiol). The anti-inflammatory drug diclofenac was completely degraded and 39% mineralized after exposure to a 17W low-pressure mercury arc lamp for 90 min in the presence of 5 mM H_2O_2 (Vogna et al. 2004).

Rosenfeldt and Linden (2004) examined the degradation of bisphenol A, 17α-ethynyl estradiol, and 17β-estradiol via UV and UV/H_2O_2 processes using either low- or medium-pressure arc lamp collimated beam reactors. Low degradation yields were reported for all compounds, ranging from 2 to 5% and from 15 to 22% at a UV dose of 1000 mJ/cm^2 delivered by a low or a medium-pressure lamp, respectively. The addition of 15 mg/L H_2O_2 increased the removal to more than 90% with either low- or medium pressure irradiation of 1000 mJ/cm^2 under the same conditions. Large rate constants for the reaction of these three EDCs were measured (1.0-1.4×10^{10} M^{-1}s^{-1}) and applied in a model to predict the degradation of these compounds by UV and UV AOP processes.

Pereira *et al.* (2005) compared the degradation kinetics of six pharmaceuticals (ciprofloxacin, carbamazepine, iohexol, clofibric acid, naproxen, ketoprofen) in drinking water and wastewater, using low- and medium-pressure collimated beam reactors. The HO$^•$ radical rate constants were large for all pharmaceuticals, ranging from 3×10^9 (ciprofloxacin) to 1.2×10^{10} (ketoprofen) M^{-1}s^{-1}. High degradation yields (>80%) of selected pharmaceuticals (diclofenac, ibuprofen, carbamazepine) and bisphenol A were reported by Kruithof *et al.* (Kruithof et al. 2005) in a pilot study using a Swift 4L12 Trojan UV reactor at a water treatment plant in Andijk, The Netherlands, when an electrical energy dose of 0.56 kWh/m^3 and 6 mg/L H_2O_2 were applied.

In a recently published study, Dodd *et al* (2006) examined the reactivity of 14 antibacterial compounds with molecular ozone and OH radical. These pharmaceuticals represent nine important structural classes, such as macrolides, sulfonamides, fluoroquinolones, trimethoprim, and *β*-lactames. The HO$^•$ radical rate constants were found to be very large (>10^9 M^{-1} s^{-1}), approaching those characteristics to the diffusion-controlled processes.

In the current study, UV and UV AOP were evaluated at bench, pilot, and full-scale. UV at typical disinfection doses (≈40 mJ/cm^2) and at higher energy doses (>400 mJ/cm^2) are compared.

BENCH-SCALE EVALUATIONS

Metropolitan Water District of Southern California

Bench-scale UV experiments conducted at the Metropolitan Water District of Southern California (MWD) used a collimated beam apparatus (Calgon Carbon, Pittsburgh, PA, United States) to conduct the experiments using a medium-pressure lamp. The spectrum for this lamp is shown in Figure 8.1. Prior to testing, an IL 1700 radiometer (International Light, Newberry Port, MA, United States) was used to determine the petri factor, which accounts for the variation in irradiance across the surface of the sample. Exposure times were determined based on the desired UV dose and the average irradiance through the water (E_{avg}) estimated from water absorbance, lamp spectral distribution, and geometry factors, such as Petri factor, divergence factor, sensitivity factor, and reflection factor. A 500 mL sample volume mixed using a stir plate during irradiation. UV germicidal doses of 40 and 1000 mJ/cm^2 were evaluated for DI water, Colorado River water (CRW), Ohio River water (ORW), Passaic River water (PVW), and synthetic Suwannee River NOM (SRW). Water qualities for the waters investigated are presented in Chapter 3 Table 3.4. Advanced oxidation using 5 mg/L of hydrogen peroxide was evaluated with a germicidal UV dose of 1000 mJ/cm^2. Hydrogen peroxide was added 1 minute prior to UV exposure. After UV exposure, the peroxide was quenched by bovine catalase (0.2 mg/L final concentration). The samples were then acidified to pH 2.0 for preservation.

MWD Results

Results are presented as percent removals in the Table 8.1. Minimal removal was observed for most target analytes using a typical drinking water disinfection dose of 40 mJ/cm^2. Additional removal was observed when using higher UV exposures and AOP conditions. The use of hydrogen peroxide did not appear to increase compound removal above that observed with UV alone. The AOP did not significantly increase removal, and in some cases decreased removal. The apparent inefficiency of the AOP is likely due to the relatively large amount of solvent carrier that was used in this experiment conducted early in the project. The methanol carrier added to the samples (around 100 μL) likely acted as a scavenger for hydroxyl radicals and thus limited contaminant oxidation.

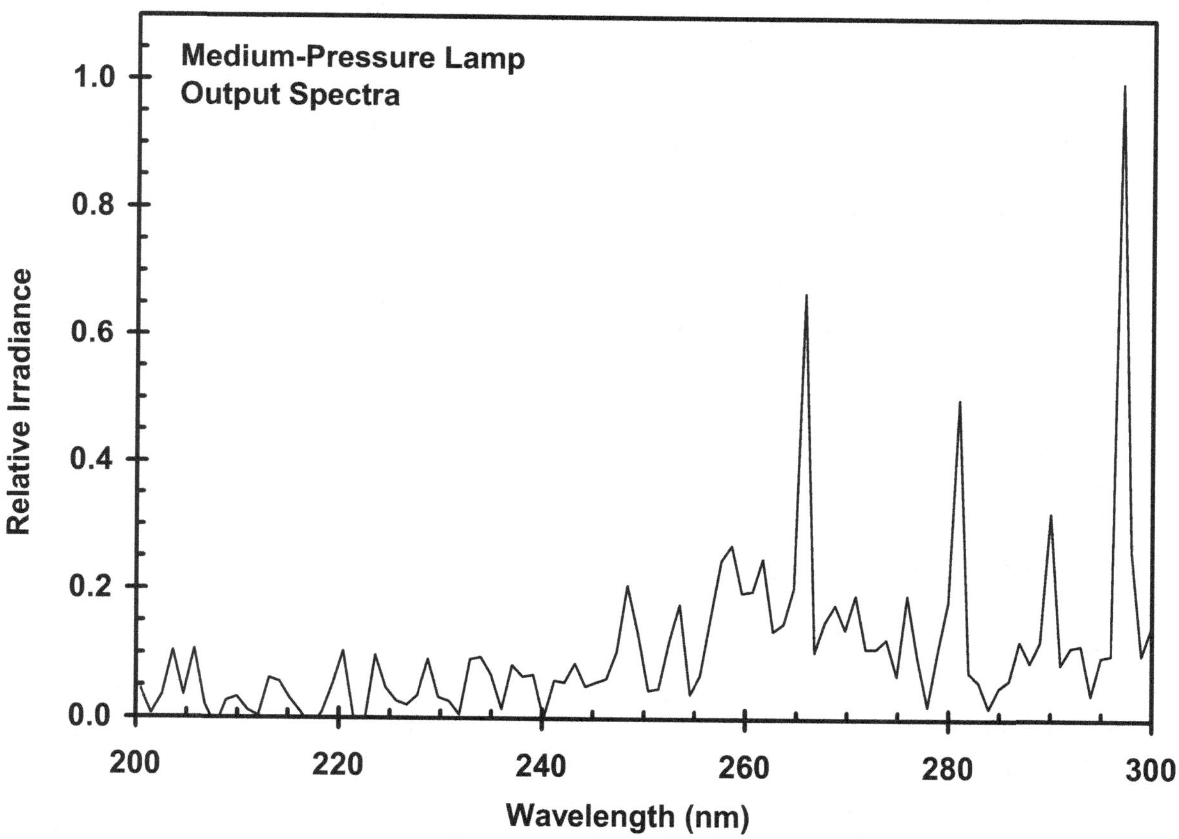

Figure 8.1 Spectrum of the medium-pressure lamp in MWD bench-scale experiments

Table 8.1
Summary of percent removal during bench-scale UV testing of surface waters

Source Water	Colorado River			Ohio River			Passaic River		
UV Dose [mJ/cm^2]	40	1000	1000	40	1000	1000	40	1000	1000
H_2O_2 Dose (mg/L)			5			5			5
Bovine Catalase (mg/L)			0.2			0.2			0.2
Target Compound	Percent Removal			Percent Removal			Percent Removal		
Acetaminophen (Tylenol)	37	>99	68	13	99	92	16	>98	71
Androstenedione	<1	51	53	<1	41	51	<1	41	48
Atrazine	5	70	65	<1	62	67	2	56	58
Benzo(a)pyrene	29	63	77	<1	74	82	<1	71	64
Caffeine	<1	9	4	<1	<1	8	<1	<1	8
Carbamazepine	<1	9	8	<1	<1	22	<1	<1	13
DDT	19	77	79	<1	83	85	<1	77	73
DEET	1	6	3	<1	<1	14	<1	<1	11
Diazepam (Valium)	<1	5	4	<1	<1	15	<1	<1	7
Diclofenac	55	>99	>99	36	>99	>99	34	>99	>99
Dilantin	7	72	67	<1	61	62	<1	63	64
Erythromycin	<1	15	6	<1	12	19	<1	<1	<1
Estradiol	31	>99	88	8	99	89	12	97	83
Estriol	26	>99	86	8	>99	88	20	98	82
Estrone	30	>99	92	8	>99	96	14	>99	96
Ethinyl Estradiol	30	>99	87	9	99	88	7	97	82
Fluorene	4	8	31	13	17	54	<1	<1	12
Fluoxetine (Prozac)	15	97	95	<1	92	93	5	92	91
Galaxolide	<1	<1	23	7	19	42	<1	11	17
Gemfibrozil	<1	<1	2	<1	<1	12	<1	<1	<1
Hydrocodone	6	79	74	<1	62	42	<1	44	36
Ibuprofen (Advil)	5	35	41	<1	27	34	14	32	43
Iopromide	21	77	56	<1	76	82	<1	70	72
Lindane (γ-BHC)	33	<1	<1	24	15	3	<1	<1	<1
Meprobamate	<1	4	3	<1	<1	5	<1	<1	<1
Metolachlor	10	77	74	9	76	68	7	65	64
Musk Ketone	<1	<1	<1	<1	33	30	<1	32	33
Naproxen	25	99	94	15	97	96	<1	96	92
Oxybenzone	5	63	23	<1	25	27	<1	6	7
Pentoxifylline	<1	9	20	<1	<1	22	<1	<1	9
Progesterone	3	55	57	<1	50	60	<1	41	49
Sulfamethoxazole	54	>98	>98	54	>96	>96	44	>85	79
TCEP	<1	5	<1	<1	<1	15	<1	<1	5
Testosterone	<1	50	53	<1	41	52	<1	42	47
Triclosan	67	>99	>99	40	>99	>99	21	>99	>99
Trimethoprim	<1	<1	56	<1	<1	4	<1	<1	<1

Trojan UV Technologies

Bench-scale UV experiments were conducted at the Trojan UV Technologies research facility in London, Ontario, Canada. The source water for these experiments was chlorinated drinking water originating from Lake Huron. The source water was dechlorinated using UV irradiation at >1000 mJ/cm^2. Dechlorination was verified by lack of a detectable chlorine residual post UV exposure. UV irradiations were performed under controlled irradiance using a Trojan Technologies collimated beam apparatus equipped with a medium pressure mercury arc lamp with a known spectral power distribution. The irradiance measurements were taken with a 1700 International Light Research Radiometer detector, calibrated at each wavelength within 200-340 nm range. Water samples were collected from water spiked during UV pilot testing to correlate bench-scale data with pilot-scale data. Therefore, the water quality parameters in the bench- and pilot-scale experiments were identical. The contaminant and hydrogen peroxide initial concentrations as well as the water quality parameters were as described in the Bench-Scale section in Chapter 3 and also as provided in Table 8.2. The spectra of water spiked with the target contaminants and hydrogen peroxide along with the emission spectrum of the medium pressure Hg lamp are provided in Figure 8.2.

Table 8.2
Trojan UV water quality parameters

Chlorine residual/ mg/L	0.7
Raw temperature/ ºC	14
Test temperature/ ºC	18-19
pH	7.4
DOC/ mg/L	1.3
Hardness/ mg/L $CaCO_3$	100
Alkalinity/ mg/L $CaCO_3$	100
%*T* (254nm, 1 cm)	98.5

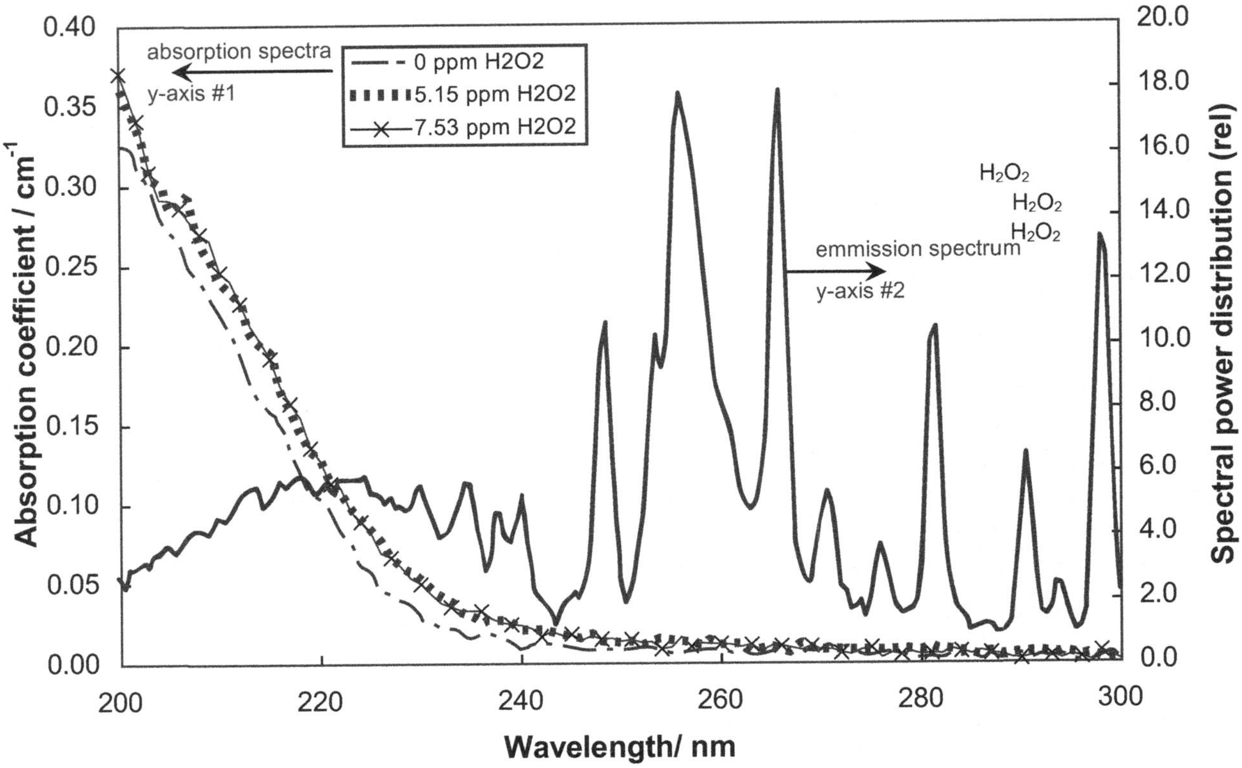

Figure 8.2 Water absorption spectra and relative emission spectrum of MP light source

Trojan UV Technologies Results

All contaminants exhibited *pseudo*-first order kinetics when the natural logarithm of their concentration at various exposure times was plotted against the UV fluence (UV dose) applied. Such behavior is expected at very low concentrations of chemical pollutants when the main absorber of light, as well as the OH radical scavenger, is the water background.

The Trojan UV bench scale results again showed poor removal at drinking water disinfection dosages consistent with results found at MWD. Removal increased with increasing UV energy, also consistent with those findings at MWD with three natural waters. However, the AOP proved to be more effective for compound oxidation during the Trojan UV testing, as the solvent carrier was eliminated from the spiking mixture by evaporation with nitrogen. As shown in Table 8.3, direct UV photolysis results in low degradation levels at 40 mJ/cm^2, which represents a common UV dose used for disinfection. For most target analytes, greater than 80% removal was achieved at an applied UV dose of approximately 1000 mJ/cm^2.

Unexpected high removal rates were calculated for the phenolic hormones ethynyl estradiol, estriol, estrone, and estradiol at 40 mJ/cm^2. The removal of estradiol by low pressure UV lamp was reported previously, while the photolysis of estrone required high pressure mercury lamp with emission wavelength above 365 nm and 10 times higher light intensity (Liu and Liu 2004). Interestingly, compounds with high reactivity to chlorine showed the greatest removal at the 40 mJ/cm^2 dose; however, chlorine was removed to less than detection from the source water.

Control samples collected post-peroxide addition but pre-UV irradiation resulted in significant removal of several target compounds (Table 8.4). Control samples for both hydrogen peroxide and bovine catalase were conducted with the MWD experiments with no significant degradation of target compounds. In previous experiments, antimicrobial preservation was performed by pH suppression using sulfuric acid; this method was abandoned for sodium azide in later experiments. Several compounds not impacted in the controls, but showed significant removal by UV-AOP conditions evaluated. Compounds degraded in the controls from the azide/peroxide dark reaction appear to be those highly sensitive to oxidation. These same compounds also were degraded by chlorination and ozonation (Chapters 6 and 7, respectively). The competition kinetics between bovine catalase and azide/peroxide are not well understood and may account for observed losses. Percent removal of all compounds was based upon contaminant concentrations in the spiked raw water without peroxide addition, which may result in an apparent enhanced removal of those compounds sensitive to the peroxide/azide dark reaction.

Table 8.3
Summary of percent removal during bench-scale UV testing of Lake Huron water

Source Water	Lake Huron Water									
UV Dose [mJ/cm^2]	39.5	439	787	1318	372	671	969	371	668	966
Peroxide Dose [mg/L]					5	5	5	7.5	7.5	7.5
	Percent Removal (%)				Percent Removal (%)			Percent Removal (%)		
Acetaminophen	81	>98	>98	97	97	>98	>98	>98	>98	>98
Androstenedione	<1	21	11	88	77	89	89	88	95	95
Atrazine	3	57	88	95	67	90	96	77	92	97
Caffeine	<1	16	51	61	64	88	91	79	93	95
Carbamazepine	<1	28	57	77	90	97	99	97	99	>99
DEET	<1	12	51	55	68	89	95	82	94	97
Diazepam	<1	15	32	63	73	90	94	85	95	97
Diclofenac	88	>99	97	>99	>99	>99	>99	>99	>99	>99
Dilantin	3	77	90	>99	93	99	>99	97	>99	>99
Erythromycin	16	52	66	68	69	87	94	93	95	96
Estradiol	74	99	96	>99	>99	>99	>99	>99	>99	>99
Estriol	71	>97	>97	>97	>97	>97	>97	>97	>97	>97
Estrone	70	92	87	>99	>99	>99	>99	>99	>99	>99
Ethinyl Estradiol	73	98	94	>99	>99	>99	>99	>99	>99	>99
Fluoxetine	17	93	92	>99	99	>99	99	>99	>99	>99
Gemfibrozil	<1	21	43	72	82	94	96	90	96	98
Hydrocodone	25	95	>99	>99	98	>99	>99	>99	>99	>99
Ibuprofen	5	49	80	89	88	97	>99	93	99	>99
Iopromide	15	69	94	98	77	94	98	86	95	99
Meprobamate	<1	<1	22	26	42	69	81	56	78	86
Naproxen	29	>99	>99	>99	>99	>99	>99	>99	>99	>99
Oxybenzone	55	>92	>92	>92	91	>92	>92	>92	>92	>92
Pentoxifylline	<1	20	51	64	70	89	95	81	94	98
Progesterone	<1	34	37	93	84	90	87	91	97	94
Sulfamethoxazole	78	>99	>99	>99	>99	>99	>99	>99	>99	>99
TCEP	<1	<1	<1	<1	<1	8	5	<1	3	<1
Testosterone	<1	44	58	92	85	95	96	93	97	98
Triclosan	83	97	78	>99	>99	>99	>99	>99	>99	>99
Trimethoprim	<1	35	66	82	83	95	98	90	98	>99

Table 8.4
Bench-scale spiked concentrations

H_2O_2 Dose	Spiked Concentration (ng/L)		
	None	5 mg/L	7.5 mg/L
Acetaminophen	57	<1.0	<1.0
Androstenedione	125	154	151
Atrazine	155	184	175
Caffeine	156	183	186
Carbamazepine	163	76	46
DEET	148	174	158
Diazepam	96	114	111
Diclofenac	107	8.8	9.0
Dilantin	126	156	145
Erythromycin	219	198	84
Estradiol	121	<1.0	<1.0
Estriol	144	<5.0	<5.0
Estrone	100	<1.0	<1.0
Ethynylestradiol	113	<1.0	<1.0
Fluoxetine	211	263	184
Gemfibrozil	87	121	120
Hydrocodone	136	15	<1.0
Ibuprofen	130	137	139
Iopromide	185	219	186
Meprobamate	156	180	193
Naproxen	112	102	51
Oxybenzone	12	14	10
Pentoxifylline	157	186	187
Progesterone	441	572	532
Sulfamethoxazole	130	150	133
TCEP	144	166	177
Testosterone	100	119	117
Triclosan	168	7.8	<1.0
Trimethoprim	153	183	174

PILOT-SCALE EVALUATIONS

Pilot testing was performed in order to evaluate the electrical energy per order parameter, which describes the electrical energy efficiency of a treatment process. By definition, for a first order kinetic decay of a chemical pollutant, the Electrical Energy per Order (E_{EO}) is the electrical energy (kWh) required to reduce the concentration of a pollutant by one order of magnitude in 1000 L of water (Bolton et al. 2001). This figure-of-merit can be expressed in units of kWh/m^3/order or kWh/kgal/order. The E_{EO} expression for a flow-through system is given as:

$$E_{EO} = \frac{P}{F \times \log(c_i/c_f)} \quad (8.1)$$

where P is the rated power (kW), F is the flow rate (m^3/h or kgal/h), and c_i and c_f are the influent and the effluent concentration of chemical pollutant. This metric is dependent on several parameters, including pollutant kinetic characteristics, water quality, oxidant concentration, reactor design (including electrical, hydraulic, optical efficiencies), lamp type, etc. Therefore, E_{EO} is a site-specific metric. It can be used to compare various optimized AOPs tested at the same site. E_{EO} values were calculated as the inverse of the slope of the graph of decimal logarithm of the concentration ratio versus electrical energy dose (EED). The electrical energy dose is the ratio of power to the flow rate, and is expressed in kWh/m^3 or kWh/kgal. The germicidal UV doses in the pilot tests were estimated through modeling, accounting for the water absorption spectrum, flow rate and lamp conditions (power and spectral distribution) in each particular run. The germicidal UV doses were based on the MS2 action spectrum and varied from 56 to 806 mJ/cm^2, depending on the experimental conditions used. The parameters used for the pilot investigation here are provided in Table 8.5.

Table 8.5
UV pilot-scale evaluation conditions

Run No.	Flow rate (gpm)	Reactor power (kW)	EED (kWh/kgal)	Germicidal UV dose (mJ/cm^2)	Influent H_2O_2 (mg/L)
0	50	0	0	0	0
1	330	7.02	0.35	290	0
2	166	7.02	0.7	548	0
3	110	7.02	1.06	806	0
4	370	7.02	0.32	216	4.6
5	207	7.02	0.57	366	5.8
6	142	7.02	0.82	518	5.2
7	353	7.02	0.33	234	6.7
8	207	7.02	0.57	379	7.6
9	142	7.02	0.82	537	7.5
10	710	2.87	0.07	56	0

Hydrogen peroxide and TOC were also monitored before and after UV exposure. As expected, less than 10% of the peroxide was destroyed by UV light and virtually no change in TOC was observed. The residual peroxide should be considered in full-scale applications, as peroxide is well known as a chlorine quenching agent and will greatly increase chlorine demand if a secondary disinfectant is required.

Results from the pilot experiments are shown as percent removal in Tables 8.6 and 8.7, and as E_{EO} values in Figure 8.3. The addition of H_2O_2 led to increased removal rates of most target compounds, which translated to significantly lower E_{EO} values as compared with the direct photolysis. The increase of the H_2O_2 concentration correlates with the decrease of the E_{EO}. The effluent concentrations of hydrocodone, acetaminophen, oxybenzone, estriol, ethynyl estradiol, estrone, estradiol, triclosan, and naproxen were below the analytical reporting limits, which indicates high removal through the processes investigated. The values for TCEP were larger than 20 kWh/kgal/order even in the presence of H_2O_2 and are not shown in Figure 8.3. In some cases, the E_{EO} decreased slightly as the EED decreased (i.e., flow rate increased), which is explained by a higher reactor hydraulic efficiency at higher flow rates. Such an effect is more prominent over a wider range of flow rates than used in this pilot testing. The large E_{EO} data in the absence of H_2O_2 as compared with that in the presence of H_2O_2 indicates that direct UV photolysis is not a viable treatment option for most of the studied target contaminants. Relatively low E_{EO} data observed for some contaminants, such as sulfamethoxazole, fluoxetine, dilantin, naproxen, diclofenac, testosterone, and progesterone suggests that those compounds may exhibit significant quantum yields. With a few exceptions (i.e., TCEP and meprobamate), the molecular structures of almost all tested compounds contain chromophores (conjugated or isolated double bonds, aromatic rings, carboxylate or amide functional groups) that absorb the UV light, which is the first condition for photolysis.

The pilot results indicate that the increase of H_2O_2 concentration leads to lower E_{EO} values, even though the H_2O_2 concentration range is not sufficiently wide to illustrate this trend. The chemical structures of the investigated compounds are prone to the OH radical attack at high rates, which explains the low E_{EO} values for the UV/H_2O_2 process. One can conclude that the UV/H_2O_2 process in combination with medium pressure lamps is an effective means of treatment of the EDCs/PPCPs investigated in this study, at practical concentrations of oxidant.

Table 8.6
Summary of percent removal during pilot-scale UV photolysis of Lake Huron water

UV Dose [mJ/cm^2]		56	290	548	806
	Spiked Concentration (n=3) (ng/L)	Percent Removal (%)			
Acetaminophen	35	34	31	45	73
Androstenedione	73	32	39	58	89
Atrazine	165	19	35	55	92
Caffeine	169	18	11	19	44
Carbamazepine	164	17	18	27	60
DEET	166	16	15	31	52
Diazepam	84	18	17	25	52
Diclofenac	62	58	>98	>98	>98
Dilantin	142	26	48	69	96
Erythromycin-H_2O	99	14	<1	11	39
Estradiol	84	38	37	62	93
Estriol	113	39	38	62	90
Estrone	72	42	64	>99	94
Ethynylestradiol	78	38	38	63	93
Fluoxetine	52	40	69	90	>98
Gemfibrozil	111	<1	15	35	57
Hydrocodone	72	17	25	35	64
Ibuprofen	158	5	24	42	70
Iopromide	198	<1	49	75	99
Meprobamate	177	17	9	14	29
Naproxen	135	15	65	84	99
Oxybenzone	3	<1	<1	3	50
Pentoxifylline	172	19	10	22	50
Progesterone	51	31	41	61	92
Sulfamethoxazole	120	36	78	93	>99
TCEP	150	14	<1	4	10
Testosterone	106	32	38	61	91
Triclosan	32	44	95	>97	>97
Trimethoprim	118	23	18	29	59

Table 8.7

Summary of percent removal during pilot-scale UV AOP of Lake Huron water

UV Dose (mJ/cm^2)		216	366	518	234	379	537
Peroxide Dose (mg/L)		4.6	5.8	5.2	6.7	7.6	7.5
	Spiked Concentration (n=3) (ng/L)	Percent Removal (%)					
Acetaminophen	35	>97	>97	>97	>97	>97	>97
Androstenedione	73	81	83	93	73	89	96
Atrazine	165	61	61	74	45	67	80
Caffeine	169	66	68	82	57	76	89
Carbamazepine	164	16	49	59	53	67	>88
DEET	166	64	67	83	58	78	89
Diazepam	84	73	74	87	64	81	93
Diclofenac	62	>98	>98	>98	>98	>98	>98
Dilantin	142	84	86	94	76	91	97
Erythromycin-H_2O	99	-20	35	8	36	19	64
Estradiol	84	>98	>98	>98	>98	>98	>98
Estriol	113	>99	>99	>99	>99	>99	>99
Estrone	72	>99	>99	>99	>99	>99	>99
Ethynylestradiol	78	>98	>98	>98	>98	>98	>98
Fluoxetine	52	92	93	>98	83	96	>98
Gemfibrozil	111	76	78	91	65	85	95
Hydrocodone	72	>98	>98	>98	>98	>98	>98
Ibuprofen	158	73	74	89	63	83	94
Iopromide	198	76	80	90	59	79	91
Meprobamate	177	48	45	65	39	58	75
Naproxen	135	>99	>99	>99	>99	>99	>99
Oxybenzone	3	>66	>66	>66	>66	>66	>66
Pentoxifylline	172	68	68	82	59	78	90
Progesterone	51	84	86	95	75	91	98
Sulfamethoxazole	120	95	97	99	88	97	>99
TCEP	150	10	-3	9	4	8	16
Testosterone	106	83	85	94	74	90	97
Triclosan	32	>97	>97	>97	>97	>97	>97
Trimethoprim	118	76	77	88	69	85	94

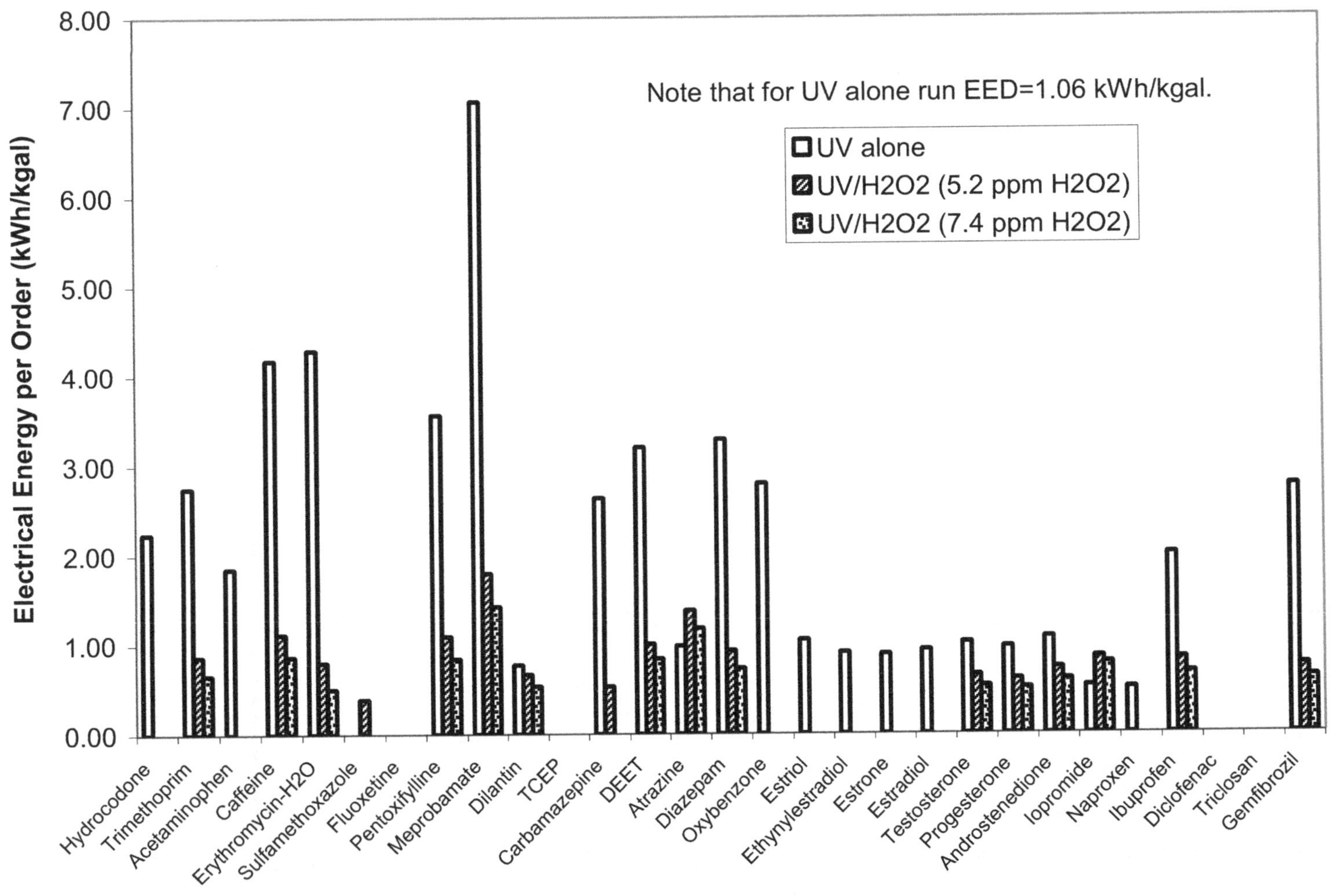

Figure 8.3 E_{EO} **as a function of the influent** H_2O_2 **concentration at 0.82 kWh/kgal.**

FULL-SCALE EVALUATIONS

UV processes were evaluated at three drinking water and three water reuse facilities at full-scale (Tables 8.8 and 8.9). Samples for the evaluation of UV processes were collected immediately before and immediately after the UV reactor.

Three drinking water facilities provided samples from UV disinfection systems (Facilities 5, 6, & 7). These facilities have similar treatment trains consisting of preoxidation, coagulation (each with aluminum sulfate), sedimentation, filtration, UV primary disinfection, chlorine secondary disinfection, and fluoridation. Facility 5 used permanganate for preoxidation, while facilities 6 and 7 used chlorine dioxide. TOC was similar in all three plants (ranging from 3 – 4 mg/L), while turbidity ranged from 5 to >25 NTU. Samples were collected post-filtration (pre-UV); however, the post-filtration sample from utility 5 was destroyed during shipping and removal reported is based upon raw water at this facility (Table 8.8). Although residual chlorine was quenched using ascorbic acid, it cannot be ruled out that some observed removal might have occurred through chemical oxidation in addition to, or instead of, UV photolysis. Results from the three drinking water facilities are difficult to interpret relative to bench-scale testing. Table 8.8 shows results from the three drinking water facilities and the results from the 40 mJ/cm^2 UV irradiation of Colorado River water. Interpretation of the full-scale results is complicated by the presence of a preoxidant (especially at facility 5), the detection of few target compounds, and the minute concentrations at which the detectable compounds occurred. Despite these complications, the removal observed at full-scale UV disinfection agreed well with data from collimated beam irradiation of the Colorado River at a 40 mJ/cm^2 dose. At facility 5, removal of sulfamethoxazole and gemfibrozil were greater than removal predicted at bench-scale. The enhanced removal is likely due to preoxidation with permanganate, since the pre-UV sample was raw water as opposed to post-filtration.

Full-scale UV applications were also evaluated at three water reuse facilities. Water reuse facility #2 processes secondary wastewater effluent using submerged MF membranes followed by RO and UV AOP (low-pressure high-output of ~400 mJ/cm^2 combined with approximately 3 mg/L hydrogen peroxide), while facilities #3 and #6 used UV for disinfection only with doses of approximately 40 mJ/cm^2. Table 8.9 shows the results from the three reuse facilities along with the collimated beam results from the Colorado River. Removal at facilities 3 and 6 were generally lower than expected from collimated beam experiments, which is likely due to the lower transmittance of the reuse water.

Table 8.8
UV process at full-scale drinking water facilities

Utility Number	Drinking #5			Drinking #6			Drinking #7			Bench-Scale Predicted Percent Removal (%)
	Pre-UV (ng/L)	Post-UV (ng/L)	Percent Removal (%)	Pre-UV (ng/L)	Post-UV (ng/L)	Percent Removal (%)	Pre-UV (ng/L)	Post-UV (ng/L)	Percent Removal (%)	
Acetaminophen	1.8	<1.0	>44	<1.0	<1.0	-	<1.0	<1.0	-	50-20
Androstenedione	<1.0	<1.0	-	<1.0	<1.0	-	<1.0	<1.0	-	<20
Atrazine	28	27	4	10	11	<1	8	8.5	<1	<20
Benzo[a]pyrene	<10	<10	-	<10	<10	-	<10	<10	-	50-20
Caffeine	24	14	42	12	15	<1	25	27	<1	<20
Carbamazepine	1.2	<1.0	>17	<1.0	<1.0	-	2.2	2.8	<1	<20
DDT	<10	<10	-	<10	<10	-	<10	<10	-	<20
DEET	5	3.9	22	2.1	1.7	19	3.9	4.6	<1	<20
Diazepam	<1.0	<1.0	-	<1.0	<1.0	-	<1.0	<1.0	-	<20
Diclofenac	<1.0	<1.0	-	<1.0	<1.0	-	<1.0	<1.0	-	80-50
Dilantin	1.3	1.1	15	<1.0	<1.0	-	1.5	2.1	<1	<20
Erythromycin	1	<1.0	>1	<1.0	<1.0	-	2.5	2.6	<1	<20
Estradiol	<1.0	<1.0	-	<1.0	<1.0	-	<1.0	<1.0	-	50-20
Estriol	<5.0	<5.0	-	<5.0	<5.0	-	<5.0	<5.0	-	50-20
Estrone	<1.0	<1.0	-	<1.0	<1.0	-	<1.0	<1.0	-	50-20
Ethynylestradiol	<1.0	<1.0	-	<1.0	<1.0	-	<1.0	<1.0	-	50-20
Fluorene	<10	<10	-	<10	<10	-	<10	<10	-	<20
Fluoxetine	<1.0	<1.0	-	<1.0	<1.0	-	<1.0	<1.0	-	<20
Galaxolide	11	10	9	13	<10	>23	12	11	8	<20
Gemfibrozil	7.4	2.3	69	<1.0	<1.0	-	9.5	10	<1	<20
Hydrocodone	<1.0	<1.0	-	<1.0	<1.0	-	<1.0	<1.0	-	<20
Ibuprofen	2.4	2.5	<1	5.1	8.3	<1	6.2	8.1	<1	<20
Iopromide	4.1	4.1	<1	<1.0	<1.0	-	7.3	7.8	<1	50-20
Lindane	<10	<10	-	<10	<10	-	<10	<10	-	50-20
Meprobamate	8.0	8.0	<1	1.5	2	<1	10	12	<1	<20
Metalochlor	<10	<10	-	<10	<10	-	<10	<10	-	<20
Musk Ketone	<10	<10	-	<10	<10	-	<10	<10	-	<20
Naproxen	<1.0	<1.0	-	<1.0	<1.0	-	<1.0	<1.0	-	50-20
Oxybenzone	<1.0	<1.0	-	<1.0	<1.0	-	<1.0	<1.0	-	<20
Pentoxifylline	<1.0	<1.0	-	<1.0	<1.0	-	<1.0	<1.0	-	<20
Progesterone	<1.0	<1.0	-	<1.0	<1.0	-	<1.0	<1.0	-	<20
Sulfamethoxazole	5.8	<1.0	>83	<1.0	<1.0	-	1.7	2.1	<1	80-50
TCEP	19	18	5	5.5	8.3	<1	21	22	<1	<20
Testosterone	<1.0	<1.0	-	<1.0	<1.0	-	<1.0	<1.0	-	<20
Triclosan	<1.0	<1.0	-	<1.0	<1.0	-	<1.0	<1.0	-	80-50
Trimethoprim	<1.0	<1.0	-	<1.0	<1.0	-	<1.0	<1.0	-	<20

Table 8.9
UV process at full-scale water reuse facilities

Utility Number	Reuse #2			Reuse #3			Reuse #6			Bench-Scale Predicted Percent Removal (%)
	Pre-UV (ng/L)	Post-UV (ng/L)	Percent Removal (%)	Pre-UV (ng/L)	Post-UV (ng/L)	Percent Removal (%)	Pre-UV (ng/L)	Post-UV (ng/L)	Percent Removal (%)	
Acetaminophen	<1.0	<1.0	-	<1.0	<1.0	-	10700	8640	19	50-20
Androstenedione	<1.0	<1.0	-	<1.0	<1.0	-	183	181	1	<20
Atrazine	<1.0	<1.0	-	<1.0	<1.0	-	33	33	<1	<20
Benzo[a]pyrene	<10	<10	-	NM	NM	-	<10	<10	-	50-20
Caffeine	5	3.2	36	38	37	3	42700	40300	6	<20
Carbamazepine	<1.0	<1.0	-	85	82	4	199	197	1	<20
DDT	<10	<10	-	NM	NM	-	<10	<10	-	<20
DEET	4.6	2.3	50	222	246	<1	701	734	<1	<20
Diazepam	<1.0	<1.0	-	1.8	1.8	<1	<1.0	<1.0	-	<20
Diclofenac	<1.0	<1.0	-	8	6.6	18	3.2	1.6	50	80-50
Dilantin	<1.0	<1.0	-	49	56	<1	110	105	5	<20
Erythromycin	<1.0	<1.0	-	238	241	<1	139	148	<1	<20
Estradiol	<1.0	<1.0	-	<1.0	<1.0	-	<1.0	<1.0	-	50-20
Estriol	<5.0	<5.0	-	2	1.4	30	8.3	7.7	7	50-20
Estrone	<1.0	<1.0	-	2.4	<1.0	>58	<1.0	<1.0	-	50-20
Ethynylestradiol	<1.0	<1.0	-	<1.0	<1.0	-	1.0	<1.0	>1	50-20
Fluorene	<10	<10	-	NM	NM	-	<10	<10	-	<20
Fluoxetine	<1.0	<1.0	-	9.5	9.0	5	<1.0	<1.0	-	<20
Galaxolide	<10	<10	-	NM	NM	-	421	491	<1	<20
Gemfibrozil	1.9	<1.0	>47	7.9	13	<1	3710	3560	4	<20
Hydrocodone	<1.0	<1.0	-	65	58	11	57	49	14	<20
Ibuprofen	<1.0	<1.0	-	2.7	5.9	<1	8760	9210	<1	<20
Iopromide	<1.0	<1.0	-	48	66	<1	5.0	4.1	18	50-20
Lindane	<10	<10	-	NM	NM	-	<10	<10	-	50-20
Meprobamate	<1.0	<1.0	-	164	181	<1	226	212	6	<20
Metalochlor	<10	<10	-	NM	NM	-	34	<10	>71	<20
Musk Ketone	<10	<10	-	NM	NM	-	20	20	<1	<20
Naproxen	<1.0	<1.0	-	3.1	4.15	<1	3550	3520	1	50-20
Oxybenzone	6.4	2.4	63	3.1	4.0	<1	<1.0	<1.0	-	<20
Pentoxifylline	<1.0	<1.0	-	<1.0	<1.0	-	34	30	12	<20
Progesterone	<1.0	<1.0	-	<1.0	<1.0	-	1.4	1.9	<1	<20
Sulfamethoxazole	1.8	<1.0	-	184	138	25	426	364	15	80-50
TCEP	2	<1.0	>50	104	113	<1	<1.0	<1.0	-	<20
Testosterone	<1.0	<1.0	-	1.2	1.6	<1	151	137	9	<20
Triclosan	<1.0	<1.0	-	9	13.5	<1	3.8	2.0	47	80-50
Trimethoprim	<1.0	<1.0	-	25	22	12	135	128.5	5	<20

RESULT SUMMARY AND DISCUSSION

UV technology at practical dosages used for disinfection only removed a small number of target analytes (Table 8.10). The chemical structure of the target analyte controlled whether or not it was oxidized by UV light. For example, the pesticide lindane was not well removed by UV, while diclofenac and triclosan were readily removed. Lindane is an aliphatic compound and lacks conjugated bonds, thus having very little ability to absorb UV light, while diclofenac and triclosan are conjugated aromatic compounds and would be expected to have strong UV absorbance. In order for a compound to be oxidized by UV light, it must absorb light at the same wavelength as produced by the UV process. For instance, the UV/VIS absorption spectra for six compounds are illustrated in Figure 8.4 and the light intensity spectrum for a medium-pressure UV lamp was presented in Figure 8.1. Diclofenac, sulfamethoxazole, and triclosan have significant UV absorbance peaks at wavelengths above 240 nm, which overlap with emission spectra from the medium-pressure UV lamps (Figures 8.1 & 8.2). These compounds exhibited the highest removal during UV treatment. The three analytes with the lowest percent removal show minimal UV absorbance at wavelengths above 220 nm. UV spectra provide a means to explain the efficacy of UV irradiation in removing organic compounds.

The addition of H_2O_2 for advanced oxidation with UV increased the removal of some compounds with efficacy highly dependent on H_2O_2 dosage and UV fluence. Table 8.11 summarizes UV removal efficiency based a UV dosage of 372 mJ/cm^2 with 5 mg/L of H_2O_2. The advantage of peroxide addition is clear when comparing Table 8.11 to Table 8.12 where a similar UV dose was applied without peroxide The results for this study have demonstrated the benefit of AOP with a diversity of contaminants representing a broad range of physicochemical properties.

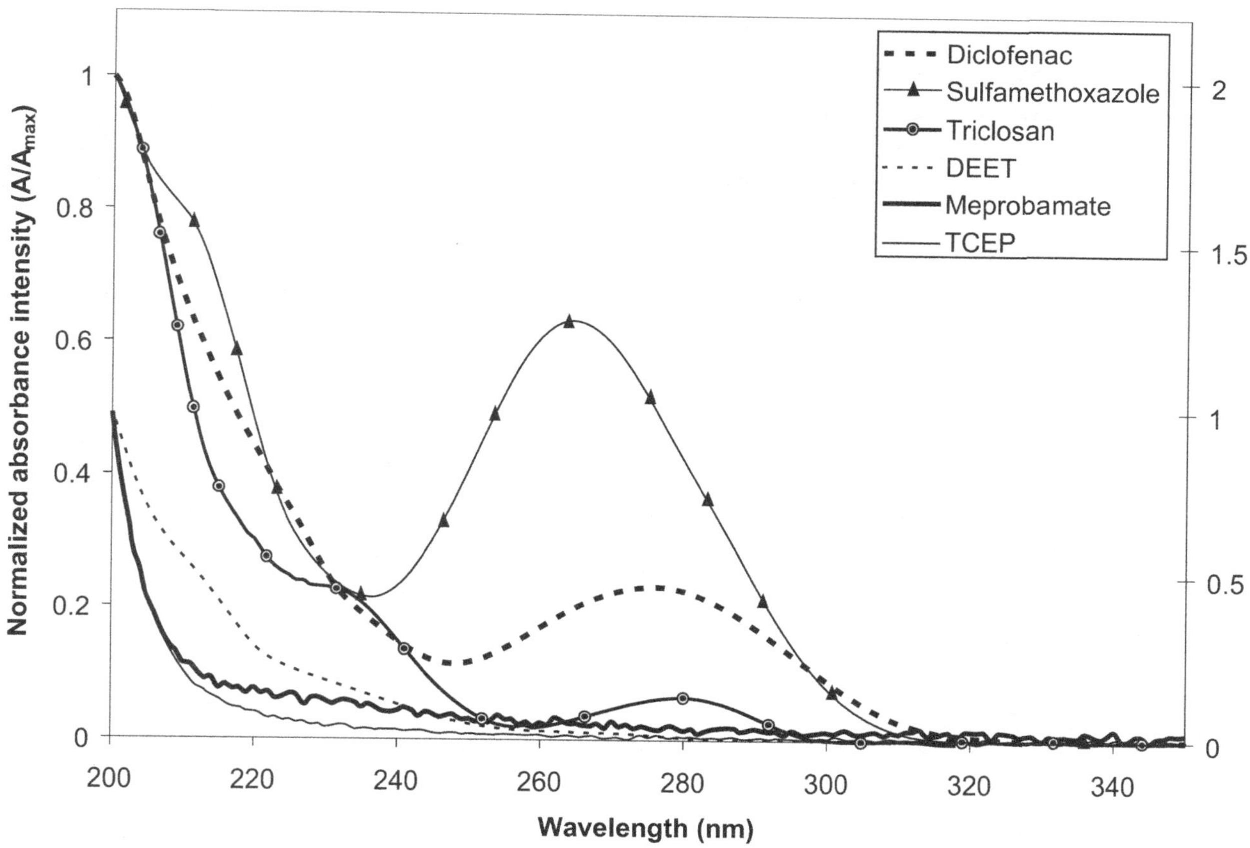

Figure 8.4 Absorbance spectra of six analytes with low and high removals by a medium-pressure UV lamp.
For illustration purpose, DEET, meprobamate and TCEP are plotted on 2nd y-scale shown on the right of the figure.

Table 8.10
Summary of percent removal by UV disinfection (40 mJ/cm^2)

> 80% Removal	50-80% Removal	20-50% Removal	< 20% Removal
	Diclofenac	Acetaminophen	Androstenedione
	Sulfamethoxazole		Atrazine
	Triclosan		Caffeine
			Carbamazepine
			DEET
			Diazepam
			Dilantin
			Erythromycin-H_2O
			Estradiol
			Estriol
			Estrone
			Ethynylestradiol
			Fluoxetine
			Gemfibrozil
			Hydrocodone
			Ibuprofen
			Iopromide
			Meprobamate
			Naproxen
			Oxybenzone
			Pentoxifylline
			Progesterone
			TCEP
			Testosterone
			Trimethoprim

Table 8.11
Summary of percent removal by UV/H_2O_2 process
UV dosage of 372 mJ/cm^2, H_2O_2 at 5 mg/L

> 80% Removal	50 –80% Removal	20-50% Removal	< 20% Removal
Acetaminophen	Androstenedione	Meprobamate	TECP
Carbamazepine	Atrazine		
Diclofenac	Caffeine		
Dilantin	DEET		
Estradiol	Diazepam		
Estriol	Erythromycin		
Estrone	Iopromide		
Ethynyl estradiol	Pentoxifylline		
Fluoxetine			
Gemfibrozil			
Hydrocodone			
Ibuprofen			
Naproxen			
Oxybenzone			
Progesterone			
Sulfamethoxazole			
Testosterone			
Triclosan			
Trimethoprim			

Table 8.12
Summary of percent removal by UV process
UV dosage of 439 mJ/cm^2

> 80% Removal	50 –80% Removal	20-50% Removal	< 20% Removal
Acetaminophen	Atrazine	Carbamazepine	Androstenedione
Diclofenac	Dilantin	Gemfibrozil	Caffeine
Estradiol	Erythromycin	Ibuprofen	DEET
Estriol	Iopromide	Pentoxifylline	Diazepam
Estrone		Progesterone	Meprobamate
Ethynyl estradiol		Testosterone	TECP
Fluoxetine		Trimethoprim	
Hydrocodone			
Naproxen			
Oxybenzone			
Sulfamethoxazole			
Triclosan			

REFERENCES

Ahel, M., F. E. Scully Jr., J. Hoigne, and W. Giger. 1994. Photochemical degradation of nonylphenol and nonylphenol polyethoxylates in natural waters. *Chemosphere* 28: 1361-1368.

Bolton, J. R., K. G. Bircher, W. Tumas, and C. A. Tolman. 2001. Figures-of-merit for the technical development and application of advanced oxidation technologies for both electric- and solar-driven systems - (IUPAC Technical Report). *Pure and Applied Chemistry*, 73(4): 627-637.

Bozzi, A., A. Lopez, G. Mascolo, and G. Tiravanti. 2002. Pharmaceuticals degradation by UV and UV/H_2O_2 treatments. *Water Science and Technology: Water Supply*, 2(2): 19-26.

Dodd, M. C., M. O. Buffle, and U. von Gunten. 2006. Oxidation of antibacterial molecules by aqueous ozone: Moiety-specific reaction kinetics and application to ozone-based wastewater treatment. *Environ. Sci. Technol.*, 40(6): 1969-1977.

Huber, M. M., S. Canonica, G.-Y. Park, and U. von Gunten. 2003. Oxidation of pharmaceuticals during ozonation and advanced oxidation processes. *Environ. Sci. Technol.*, 37(5): 1016-1024.

Kruithof, J. C., P. C. Kamp, M. Belosevic, and M. I. Stefan. 2005. UV/H_2O_2 retrofit of PWN's water treatment plant Andijk for primary disinfection and organic contaminant control. pp. 397-404. In *Proceedings of the Conference on "Photocatalytic and Advanced Oxidation Processes for Treatment of Air, Water, Soil, and Surfaces", San Diego, CA, 2004*: Redox Technologies Inc., London, ON, Canada.

Liu, B., and X. L. Liu. 2004. Direct photolysis of estrogens in aqueous solutions. *Sci. Total Environ.*, 320(2-3): 269-274.

Mazellier, P., J. Leverd, and J. De Laat. 2003. Elimination of 4-*tert*-octylphenol by UV and H_2O_2/UV in water. In *Proceedings of the 2nd International Congress on Ultraviolet Technologies, IUVA*. Vienna, Austria.

Pereira, V. J., K. G. Linden, and H. S. Weinberg. 2005. Direct photolysis and UV advanced oxidation processes of pharmaceuticals in surface water. In *Proceedings of the 3rd International Congress on Ultraviolet Technologies, IUVA*. May 24-27, Whistler, BC, Canada

Rosenfeldt, E. J., and K. G. Linden. 2004. Degradation of endocrine disrupting chemicals bisphenol A, ethynyl estradiol, and estradiol during UV photolysis and advanced oxidation processes. *Environ. Sci. Technol.*, 38(20): 5476-5483.

U.S. Environmental Protection Agency. 2006. 40 CFR Parts 9, 141 and 142. National Primary Drinking Water Regulations: Long Term 2 Surface Water Treatment Rule (LT2 rule); Final Rule. *Fed. Reg. (Part II)*, 71(3): 653-702.

Vogna, D., R. Marotta, R. Andreozzi, A. Napolitano, and M. d'Ischia. 2004. Kinetic and chemical assessment of the UV/H_2O_2 treatment of antiepileptic drug carbamazepine. *Chemosphere* 54(4): 497-505.

CHAPTER 9
MEMBRANES

BASIC THEORY

Membrane filtration separates contaminants from water based upon molecular size and/or electrostatic interactions on the membrane surface. In some cases, organic contaminants adsorb onto membrane surfaces and/or particulates in the water, which impacts the removal of the contaminant by membrane filters. Size, charge, and hydrophobicity of EDCs and PPCPs emerge as critical characteristics influencing the removal effectiveness of filtration membranes. Microfiltration and ultrafiltration, both considered low pressure membranes, generally remove only those EDCs/PPCPs that adsorb onto colloids or particulates in the water. They were found to be ineffective at removing two steroid hormones, six different acidic drugs and beta blockers at two different full-scale water recycling systems treating tertiary municipal wastewater effluent (Huang and Sedlak 2001). EDCs/PPCPs can be removed by high-pressure membranes, such as reverse osmosis (RO) and tight nanofiltration (NF), commonly used for rejection of inorganic and organic contaminants due to their smaller pore sizes and corresponding lower molecular weight cuff-offs. For example, greater than 90% removal of steroid hormones has been reported in one RO study (Huang and Sedlak 2001). The removal of organic compounds by RO and NF can be qualitatively predicted based on compound structure and membrane properties, which was discussed in a literature review (Bellona et al. 2004). In this chapter, conventional membrane applications that do not involve bioactivity, membrane bioreactors (MBRs), and MBR/RO are evaluated.

BENCH-SCALE EVALUATIONS

NF and UF membranes were studied using dead-end stirred-cell membrane units (see Chapter 3 for details). For this setup, the retentate concentration close to membrane surface in the stirred cell is much higher than the bulk aqueous concentration due to concentration polarization caused by the build up of retained materials. There are two possible impacts from concentration polarization. Higher rejection caused by the increased repulsion between retained species and lower rejection due to the increased membrane surface concentration. In UF, the rejection tends to increase due to concentration polarization, while in NF both increase and decrease in rejection can be observed depending on specific compounds and dominant rejection mechanisms. While dead-end membrane units do not provide an ideal simulation of full-scale systems, they are effective in screening membranes for relative compound rejection.

The rejection of EDCs/PPCPs by UF and NF membranes using the stirred-cell membrane testing units are summarized in Table 9.1. Percentage removals were calculated based upon the EDC/PPCP concentration in the permeate divided by the initial measured concentrations in the spiked raw water. Control tests on the apparatus without a membrane indicated minimal EDC/PPCP compound loss from the testing equipment. As a result, the percent removal reflects membrane rejection along with membrane surface adsorption. Because water was passed through the membranes until 1 L permeate was obtained, adsorption may not have reached equilibrium, which may impact a compound's long-term rejection. For instance, a lower rejection of four steroid hormones was reported when membrane filtration was operated over one day and equilibrium was approached (Nghiem et al. 2004). However, adsorption kinetics was not the focus of this study.

Of the compounds listed in Table 9.1, diclofenac, dilantin, erythromycin, gemfibrozil, ibuprofen, naproxen, and sulfamoxazole are negatively charged at environmental relevant pH values due to deprotonation of a carboxylic acid or protonation of an amine group. Only two compounds, fluoxetine and trimethoprim, may carry a positive charge based on the predicted pKa. Despite these differences, molecular charge does not seem to impact the rejection of these compounds by the membranes investigated here as a result of electrostatic repulsion. For instance, dilantin with a negative charge, a molecular weight (MW) of 252 Daltons and a log K_{ow} of 2.47 was removed at only a marginally higher degree than carbamazepine and estriol, which are neutral compounds with similar MW and log K_{ow} values.

As expected, EDC/PPCP concentrations were lower in the permeate for both UF and NF membranes than in the initial feedwater. This is illustrated by results obtained with CRW as one example presented in Figure 9.1. Error bars in Figure 9.1 concentrations represent triplicate measurements. NF permeate concentrations ranging from 2 to 35 ng/L were significantly lower than UF values in most cases. Rejection by the NF membrane ranged from 44% to 93% (except naproxen) was greater than that by the UF membrane that typically had less than 40% rejection (except triclosan, oxybenzone, and progesterone). For some compounds, UF permeate concentrations were slightly higher than feed concentrations, which could possibly be due to analytical variance. For instance, abnormal higher removal of fluoxetine by UF than NF in ORW is clearly related to the fact that the feed concentration was very close to the analytical detection limit, which can cause erroneous results when calculating percent removal. For ORW, the results follow the similar trends as observed with CRW: retention by the NF membrane was greater than that by the UF membrane for most test compounds. However, the trend was not observed for PVW, where a relatively high removal by the UF membrane was obtained for most compounds.

Table 9.1
Summary of percent removal during bench-scale membrane filtration

Source Water	CRW		ORW		PRW	
Membrane	UF	NF	UF	NF	UF	NF
Target Compounds	Percent Removal					
Acetaminophen (Tylenol)	64	83	12	14	29	<1
Androstenedione	<1	64	4	53	93	46
Atrazine	6	65	6	79	70	11
Benzo(a)pyrene	>94	>94	>88	>90	>75	>75
Caffeine	<1	31	1	56	66	9
Carbamazepine	<1	60	6	77	66	6
DDT	>93	>92	>89	>88	>78	>78
DEET	<1	57	6	75	77	10
Diazepam (Valium)	7	76	14	79	78	14
Diclofenac	50	75	2	74	59	<1
Dilantin	<1	18	5	69	67	14
Erythromycin	61	78	48	96	93	36
Estradiol	<1	<1	<1	71	75	<1
Estriol	<1	63	*	*	74	<1
Estrone	14	66	*	*	71	<1
Ethinyl Estradiol	12	77	18	77	80	9
Fluorene	>95	>95	89	>93	>94	>94
Fluoxetine (Prozac)	98	92	>97	<1	93	>97
Galaxolide	78	82	78	86	60	54
Gemfibrozil	<1	14	<1	82	79	<1
Hydrocodone	19	81	*	*	77	2
Ibuprofen (Advil)	28	77	<1	80	68	8
Iopromide	36	90	<1	86	86	38
Lindane (γ-BHC)	24	71	18	86	75	18
Meprobamate	<1	32	3	70	71	5
Metolachlor	19	78	41	74	80	<1
Musk Ketone	63	76	63	>93	78	45
Naproxen	70	87	<1	17	25	6
Oxybenzone	84	96	78	88	95	83
Pentoxifylline	<1	66	7	66	70	2
Progesterone	76	92	48	52	86	60
Sulfamethoxazole	20	71	19	56	20	29
TCEP	32	82	22	84	69	11
Testosterone	<1	50	<1	76	93	45
Triclosan	93	96	88	93	92	92
Trimethoprim	<1	41	20	76	77	8

* Target Compound Not Analyzed

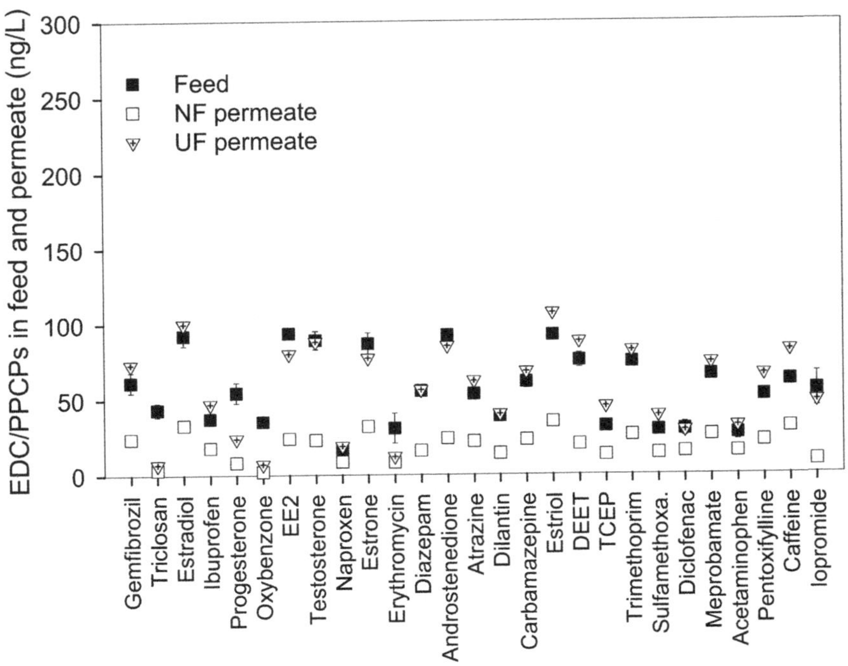

Figure 9.1 Concentration comparison of EDCs/PPCPs in feed and NF/UF permeates (CRW).

The removal of EDCs/PPCPs by UF was somewhat surprising, because the molecular weights of the selected compounds were all less than the nominal pore size of the UF membrane. In all three waters, UF membranes tended to remove the more hydrophobic compounds (e.g., fluorine, DDT and benzo(a)pyrene) while other more polar compounds (e.g., iopromide) exhibited lower removals. This suggested that adsorption of EDCs/PPCPs onto virgin membrane surfaces may be a factor in removal. A weak removal trend due to hydrophobic adsorption as a function of octanol–water partition coefficient, log K_{ow}, was observed between the hydrophobic compounds and the virgin membranes for CRW and ORW. PVW failed to exhibit such a trend indicating the presence of a second mechanism, which might be due to the interaction of the target compounds with natural organic matter, a mechanism previously proposed to explain the removal of diclofenac and ibuprofen by UF process (Burba, Geltenpoth, and Nolte 2005). The nature of the water matrix will also affect the compound removal during membrane processes (Snyder et al. 2006; Yoon et al. 2006).

Figure 9.2 compares the average percentage rejection/retention of EDCs/PPCPs by UF versus the percentage by NF. The results show that 10 of the 27 target compounds (those with labels on the graph and the two in that group) exhibited greater than 20% rejection by both UF and NF. Although initial concentrations and source water chemistry conditions are different, Figure 9.2 depicts compound retention divided into two general categories. In the first group are compounds that are relatively well retained by both types of membranes; NF (40-95%) and UF (25-90%). Such compounds shown close to the 1:1 in this figure, suggest similar removal mechanisms exhibited by both UF and NF membranes. Common structural properties for these compounds generally include aliphatic substituted aromatic ring structures, high pK_a values, and/or high log K_{ow} values. The undissociated hydrophobic compounds at pH conditions lower than their pK_a values are likely readily adsorbed onto the hydrophobic membranes. In summary, hydrophobic interaction seems to be the dominating driving force for the removal of these types of compounds.

The second group is not labeled in the figure and includes compounds poorly retained: UF (0 to < 25%) and NF (10 to <60%) (Figure 9.2). In general, most of these compounds have low log K_{ow} values due to aliphatic/aromatic nitrogen, carbonyl, phosphate, amine, or hydroxyl functional groups. The NF membrane had greater retention than the UF membrane. Size exclusion likely contributed to the retention for the NF membrane (MWCO = 600 Daltons) as compared to the UF membrane (MWCO = 8,000 Daltons). The retention of iopromide (MW = 791 Daltons) by the NF membrane was the highest among the second group compounds (58%), although its log K_{ow} of -2.1 is the lowest.

The results suggest that the NF membrane retained many EDCs/PPCPs due to both hydrophobic adsorption and size exclusion, while the UF membrane retained typically hydrophobic EDCs/PPCPs due mainly to adsorption. Adsorption to membrane surfaces will be highly dependent upon membrane fouling and water quality. In the experiments shown here, virgin membranes were used. Pilot and full-scale experiments using virgin and fouled membranes were shown in the following sections of this chapter.

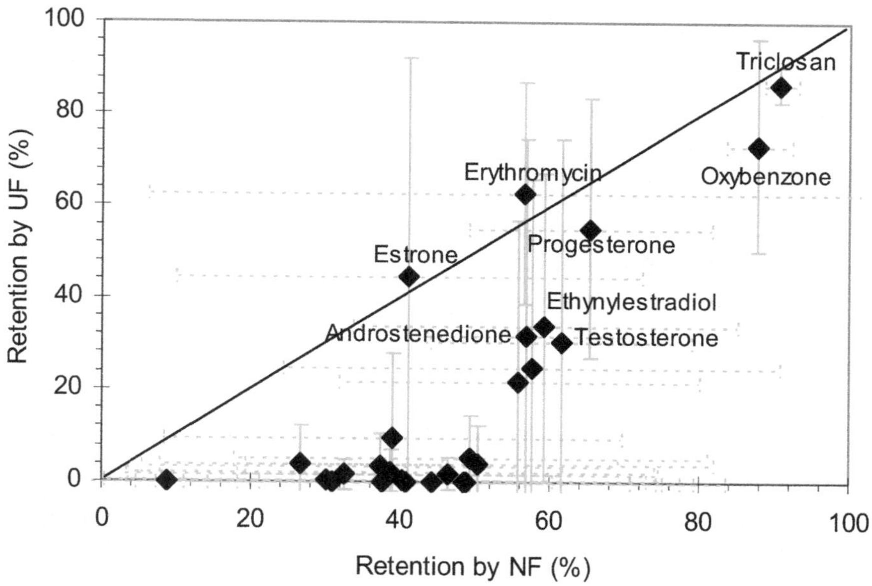

Figure 9.2 Summary of average percentage retention across four waters spiked with EDCs/PPCPs (SRW, CRW, ORW, and PVW) by the NF and UF membranes.

PILOT-SCALE EVALUATIONS

All pilot membrane evaluations were performed in dynamic flow-through skids. The experimental methodologies and procedures are described in detail in Chapter 3, while a general explanation of the experimental conditions is provided here. Membrane types and manufacturers, as well as feed water types, are provided in Table 9.2. The most significant differences between pilot and batch tests are that pilot test membranes had continuous and lengthy exposure to organics prior to studying EDC/PPCP removal, and the RO systems have cross-flow sheer stresses that help remove particles. The batch tests, run under dead-end conditions, used new membrane surfaces that exhibit significant adsorption of hydrophobic EDCs/PPCPs (Snyder et al. 2006; Yoon et al. 2006).

Table 9.2
Pilot membrane experiments

Membrane Type	Manufacturer	Feed Water	Spiked
UF	Ionics	Secondary Effluent	Yes
MBR	US Filter	Primary Effluent	No
RO	Koch	Saline Groundwater	Yes
UF/RO	Zenon/Saehan	Tertiary Effluent	No
MBR/RO	Zenon/Saehan	Primary Effluent	No
MBR/RO	Osmonics/Saehan	Raw Wastewater	No
MF/RO/EDR	US Filter/Dow/Aquamite	Tertiary Effluent	No
MF/RO	US Filter/Hydranautics	Tertiary Effluent	No
MF/RO	US Filter/Hydranautics	Secondary Effluent	No

Source: Snyder et al. 2006a. Reprinted from *Desalination*, 202, Snyder, S. A., S. Adham, A. M. Redding, F. S. Cannon, J. DeCarolis, J. Oppenheimer, E. C. Wert, and Y. Yoon. Role of Membranes and Activated Carbon in the Removal of Endocrine Disruptors and Pharmaceuticals. 156-181. Copyright 2006, with permission from Elsevier.

Ultrafiltration Using Secondary Wastewater Effluent

An Ionics UF pressure-vessel skid using secondary effluent from a WWTP was evaluated for EDC/PPCP removal. Target compounds were spiked into a head tank that was rapidly filled and mixed using secondary wastewater effluent. Results indicate poor removal of the majority of detectable target compounds by UF during this pilot evaluation (Table 9.3). Overall, a relatively a poor relationship was observed between percent removal and compound log K_{ow} (Figure 9.3). There appears to be a relationship between UF removal and log K_{ow} for neutral and positively charged compounds, while log K_{ow} was not predictive for removal of negative compounds. However, as discussed previously, log K_{ow} values for species with pKa values within the range of pH evaluated here may misrepresent the hydophobicity of these compounds as some reported log K_{ow} values are based on the unionized form of these compounds.

It is interesting to note that steroids were well-removed by UF. This may be due to their unique interactions with natural organic matter, which apparently is water specific since different removal patterns were observed in bench-scale experiments presented in Table 9.1. It also is possible that anti-scalant used in this pilot skid may have impacted the removal of certain compounds. The impact of anti-scalant alone was not evaluated.

Table 9.3
Results of ultrafiltration pilot testing in secondary effluent

Target Compound	Secondary Effluent (ng/L) (n=3)	Ultrafiltration Permeate (ng/L) (n=3)	Percent Removal (%)
Acetaminophen	18	17	6
Androstenedione	77	22	72
Atrazine	61	52	15
Benzo[a]pyrene	87	<10	>89
Caffeine	85	79	7
Carbamazepine	191	161	16
DDT	67	<10	>85
DEET	249	228	9
Diazepam	58	9	84
Diclofenac	38	37	2
Dilantin	130	98	25
Erythromycin-H_2O	289	245	15
Estradiol	87	<1	>99
Estriol	108	64	41
Estrone	98	9	91
Ethinylestradiol	78	<1	>99
Fluorene	38	<10	>74
Fluoxetine	45	14	69
Galaxolide	1601	1718	<1
Gemfibrozil	82	89	<1
Hydrocodone	105	90	14
Ibuprofen	39	36	8
Iopromide	75	79	<1
Lindane	68	<10	>85
Meprobamate	561	529	6
Metolachlor	108	47	56
Musk Ketone	147	93	37
Naproxen	24	21	10
Oxybenzone	32	5	85
Pentoxifylline	49	44	11
Progesterone	64	<1	>98
Sulfamethoxazole	66	63	5
TCEP	169	156	7
Testosterone	81	23	71
Triclosan	32	4	88
Trimethoprim	138	113	18

Source: Snyder et al. 2006a. Reprinted from *Desalination*, 202, Snyder, S. A., S. Adham, A. M. Redding, F. S. Cannon, J. DeCarolis, J. Oppenheimer, E. C. Wert, and Y. Yoon. Role of Membranes and Activated Carbon in the Removal of Endocrine Disruptors and Pharmaceuticals. 156-181. Copyright 2006, with permission from Elsevier.

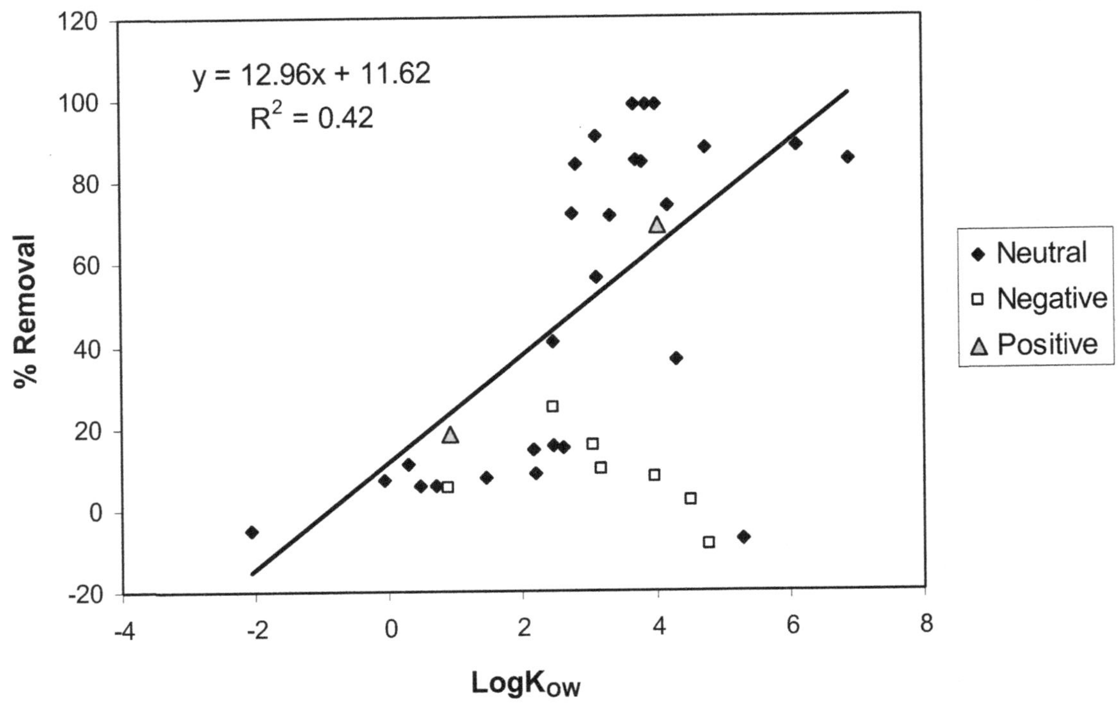

Figure 9.3 Dependence of percent removal on log K_{ow} during UF pilot testing

Membrane Bioreactor Using Primary WWTP Effluent

A membrane bioreactor system (US Filter Jet Tech) was tested at pilot-scale using primary effluent as the feed water. No spiking was performed for this experiment; therefore, compounds shown are limited to those detectable in the primary effluent. Smaller sample volumes were processed for this experiment, since raw wastewater contains far greater concentrations of target analytes and creates extracts that are much more challenging for instrumental analysis. Therefore, reporting limits are proportionally higher than in evaluations using less challenging waters.

The results of this MBR experiment are shown in Table 9.4. Percent removal shown in Table 9.4 is for the MBR process only and not through the entire WWTP. The difference between the raw wastewater influent concentration and primary effluent indicates that hydrophobic adsorption can result in significant target concentration reduction. These data show that some compounds are well-removed by the MBR while others are not. In some cases, particularly meprobamate, concentrations appear to increase through the MBR. This phenomenon has been mentioned previously and has been documented on several occasions; however the specific cause for this apparent increase has not yet been determined.

Table 9.4
Removal of target compounds by a pilot MBR

Target Compound	WWTP Influent (ng/L)	Primary Effluent (ng/L)	MBR Permeate (ng/L)	Percent Removal (%)
Caffeine	72200	68	<10	>85
Carbamazepine	189	281	<10	>96
DEET	150	213	171	20
Diclofenac	<100	16	<10	>38
Dilantin	210	192	184	4
Erythromycin - H_2O	1050	800	34	96
Fluoxetine	<100	44	<10	>77
Gemfibrozil	2210	74	<10	>86
Hydrocodone	118	168	<10	>94
Ibuprofen	12000	27	43	<1
Meprobamate	966	652	1340	<1
Naproxen	12500	70	<10	>86
Sulfamethoxazole	1110	23	<10	>57
Triclosan	1280	17	<10	>41
Trimethoprim	693	42	<10	>76

Source: Snyder et al. 2006a. Reprinted from *Desalination*, 202, Snyder, S. A., S. Adham, A. M. Redding, F. S. Cannon, J. DeCarolis, J. Oppenheimer, E. C. Wert, and Y. Yoon. Role of Membranes and Activated Carbon in the Removal of Endocrine Disruptors and Pharmaceuticals. 156-181. Copyright 2006, with permission from Elsevier.

Reverse Osmosis with Virgin and Fouled Membranes Using Saline Groundwater

A reverse osmosis system was pilot tested using saline groundwater with total dissolved solids of approximately 5,000 mg/L. A feed tank containing the groundwater was spiked and homogenized with 100-300 ng/L of a limited group of target compounds prior to testing. These RO evaluations were conducted very early in the project, before a final target list and analytical method had been developed. Therefore, a limited subset of compounds was used and the method reporting limits were 25 ng/L (as compared to 1 ng/L in most experiments). Results from the spiking study using virgin and fouled membranes are shown in Tables 9.5 and 9.6. Similarly to the UF pilot described previously, the antiscalant appeared to remove a significant portion of the phenolic steroids (i.e. estradiol, estrone, ethynyl estradiol). The concentrations of steroids decreased by up to 68% after the addition of anti-scalants. All target analytes were well rejected by the RO membranes and membrane fouling had a minimal effect on removal. The detections of pentoxifylline and iopromide in the permeate from the virgin and fouled membranes, respectively, were not expected and there is no clear reason why these compounds were observed in the permeate.

Table 9.5
RO treatment using virgin membranes

Target Compound	Feed Tank (ng/L)	Post Antiscalant (ng/L)	RO Permeate (ng/L)	Brine Recycle (ng/L)
Androstenedione	284	306	<25	315
Caffeine	311	324	52	344
Diclofenac	26	32	<25	31
Dilantin	259	275	<25	287
Estradiol	125	66	<25	57
Estriol	128	78	<25	58
Estrone	167	57	<25	78
Ethynylestradiol	125	65	<25	58
Fluoxetine	263	284	<25	499
Gemfibrozil	230	211	<25	218
Ibuprofen	259	244	<25	251
Iopromide	165	170	<25	158
Naproxen	118	129	<25	119
Oxybenzone	218	176	<25	192
Pentoxifylline	458	483	45	471
Progesterone	285	324	<25	312
Triclosan	246	185	<25	180
Trimethoprim	265	294	<25	268

Source: Snyder et al. 2006a. Reprinted from *Desalination*, 202, Snyder, S. A., S. Adham, A. M. Redding, F. S. Cannon, J. DeCarolis, J. Oppenheimer, E. C. Wert, and Y. Yoon. Role of Membranes and Activated Carbon in the Removal of Endocrine Disruptors and Pharmaceuticals. 156-181. Copyright 2006, with permission from Elsevier.

Table 9.6
RO treatment using fouled membranes

Target Compound	Feed Tank (ng/L)	Post Antiscalant (ng/L)	RO Permeate (ng/L)	Brine Recycle (ng/L)
Androstenedione	247	250	<25	243
Caffeine	196	193	<25	219
Dilantin	239	242	<25	225
Estradiol	27	<25	<25	<25
Estrone	83	<25	<25	<25
Ethynylestradiol	51	<25	<25	<25
Fluoxetine	564	441	<25	451
Gemfibrozil	234	234	<25	221
Ibuprofen	302	275	<25	284
Iopromide	125	115	72	133
Naproxen	91	73	<25	77
Oxybenzone	221	34	<25	<25
Pentoxifylline	169	154	<25	160
Progesterone	250	251	<25	250
Triclosan	166	105	<25	90
Trimethoprim	278	309	<25	371

Source: Snyder et al. 2006a. Reprinted from *Desalination*, 202, Snyder, S. A., S. Adham, A. M. Redding, F. S. Cannon, J. DeCarolis, J. Oppenheimer, E. C. Wert, and Y. Yoon. Role of Membranes and Activated Carbon in the Removal of Endocrine Disruptors and Pharmaceuticals. 156-181. Copyright 2006, with permission from Elsevier.

Ultrafiltration followed by Reverse Osmosis Using Secondary Wastewater

A Zenon ZeeWeed 1000 submerged UF system along with a Saehan RO pilot system were evaluated using secondary wastewater effluent. Compound spiking was not performed during this experiment; therefore, only the target compounds detected in the secondary effluent are shown. Results of contaminant removal using this pilot are shown in Table 9.7. The steroid estrone was one of the few compounds than was partially removed by the UF membrane, which matched previous UF results. In general, UF had little effect on target compounds while RO removed nearly all target analytes to below or very near the MRLs. Interestingly, iopromide was again detected in the RO permeate, which was unexpected due to the relatively large molecular weight (791 Daltons) of this contaminant. The RO concentrate stream showed an approximately a 3-fold contaminant concentration increase as compared to feed water concentrations, which is as expected considering the RO system was operating at approximately 75% recovery. Some target compounds, such as acetaminophen and fluoxetine were not detected in the secondary effluent, but were detected in the RO concentrate. Even though a target compound may not be detected in the feed water, it may appear in the brine due to concentration through the system.

Table 9.7
EDC/PPCP removal by UF/RO pilot

Target Compound	Secondary Effluent (ng/L)	UF Permeate (ng/L)	RO Permeate (ng/L)	RO Retentate (ng/L)
Acetaminophen	<20	<10	<1.0	16
Caffeine	<20	14	1.8	298
Carbamazepine	110	147	<1.0	278
DEET	104	103	<1.0	293
Diclofenac	<20	37	<1.0	59
Dilantin	126	191	<1.0	416
Erythromycin-H_2O	336	357	<1.0	940
Estrone	35	<10	<1.0	78
Fluoxetine	<20	<10	<1.0	17
Galaxolide	968	816	<10.0	2180
Gemfibrozil	100	142	<1.0	329
Hydrocodone	87	89	<1.0	215
Iopromide	<20	58	1.1	89
Meprobamate	693	715	<1.0	1610
Musk Ketone	97	106	<10.0	329
Naproxen	<20	17	<1.0	33
Oxybenzone	48	26	<1.0	20
Sulfamethoxazole	90	56	1.2	121
TCEP	189	219	1.4	426
Triclosan	29	<10	<1.0	14
Trimethoprim	186	158	<1.0	403

Source: Snyder et al. 2006a. Reprinted from *Desalination*, 202, Snyder, S. A., S. Adham, A. M. Redding, F. S. Cannon, J. DeCarolis, J. Oppenheimer, E. C. Wert, and Y. Yoon. Role of Membranes and Activated Carbon in the Removal of Endocrine Disruptors and Pharmaceuticals. 156-181. Copyright 2006, with permission from Elsevier.

Membrane Bioreactor Followed by Reverse Osmosis Pilot

Tables 9.8 and 9.9 present the results of the MBR/RO investigations. The data show that while the MBR is effective for reducing the concentration of many EDCs/PPCPs, several compounds are unaffected and very few compounds are reduced to below the corresponding MRLs. The observed removal is likely related to biodegradability of each individual compound. In each experiment, naproxen, acetaminophen, ibuprofen, and caffeine experienced significant removal through the MBR. These compounds were previously shown to be rapidly biodegraded (Snyder et al. 2004). While individual differences were observed for removal efficiency between the two MBR systems evaluated, the two sampling events at Facility #2 demonstrate that influent and effluent concentrations can be quite variable.

Table 9.8
Results of MBR/RO – Facility #1

	WWTP Influent (ng/L)	Primary Effluent (ng/L)	MBR Permeate (ng/L)	RO Permeate (ng/L)
Acetaminophen	21950	4095	<1.0	<1.0
Androstenedione	<100	60	<1.0	<1.0
Atrazine	<100	<1.0	<1.0	<1.0
Caffeine	58550	6775	<10	<10
Carbamazepine	299	138	171	<1.0
DEET	690	168	46	2.3
Diazepam	<100	1.1	2.4	<1.0
Diclofenac	<100	5.6	17	<1.0
Dilantin	143	21	67	<1.0
Erythromycin-H_2O	479	9.4	54	<1.0
Estradiol	<100	<1.0	<1.0	<1.0
Estriol	226	67	<1.0	<1.0
Estrone	<100	36	<1.0	<1.0
Ethynylestradiol	<100	<1.0	<1.0	<1.0
Fluoxetine	<100	4.4	9.8	<1.0
Gemfibrozil	<100	331	270	<1.0
Hydrocodone	<100	32	39	<1.0
Ibuprofen	70350	641	5.3	<1.0
Iopromide	<100	<1.0	3.5	<1.0
Meprobamate	520	92	216	1.3
Naproxen	21000	599	1.3	<1.0
Oxybenzone	896	181	9.4	<1.0
Pentoxifylline	<100	6.9	<1.0	<1.0
Progesterone	<100	22	<1.0	<1.0
Sulfamethoxazole	234	103	33	<1.0
TCEP	464	151	171	6.5
Testosterone	<100	23	<1.0	<1.0
Triclosan	4030	176	11	<1.0
Trimethoprim	699	144	7.4	<1.0

Source: Snyder et al. 2006a. Reprinted from *Desalination*, 202, Snyder, S. A., S. Adham, A. M. Redding, F. S. Cannon, J. DeCarolis, J. Oppenheimer, E. C. Wert, and Y. Yoon. Role of Membranes and Activated Carbon in the Removal of Endocrine Disruptors and Pharmaceuticals. 156-181. Copyright 2006, with permission from Elsevier.

Table 9.9
Results of two MBR/RO sampling events – Facility #2

Target Compound	Event #1			Event #2		
	MBR Feed	MBR Permeate	RO Permeate	MBR Feed	MBR Permeate	RO Permeate
	ng/L	ng/L	ng/L	ng/L	ng/L	ng/L
Acetaminophen	199	<5.0	<1.0	392	<10	<1.0
Androstenedione	30	<5.0	<1.0	25	2.6	<1.0
Caffeine	2270	223	<10	7140	1650	<10
Carbamazepine	243	199	<1.0	367	410	<1.0
DEET	74	16	<1.0	365	20	<1.0
Diclofenac	28	32	<1.0	32	38	<1.0
Dilantin	236	206	<1.0	259	310	<1.0
Erythromycin-H_2O	34	50	<1.0	50	90	<1.0
Estradiol	8.1	<5.0	<1.0	17	<1.0	<1.0
Estrone	72	13	<1.0	70	21	<1.0
Ethynylestradiol	<5.0	<1.0	<1.0	32	<1.0	<1.0
Fluoxetine	15	9.0	<1.0	17	23	<1.0
Gemfibrozil	660	357	<1.0	701	698	<1.0
Hydrocodone	76	54	<1.0	81	85	<1.0
Ibuprofen	1980	33	<1.0	1990	169	<1.0
Iopromide	233	186	<1.0	564	1230	<1.0
Meprobamate	251	205	<1.0	309	339	<1.0
Naproxen	1710	66	<1.0	2660	467	<1.0
Oxybenzone	37	5.7	<1.0	50	16	<1.0
Progesterone	<5.0	<1.0	<1.0	6.4	<1.0	<1.0
Sulfamethoxazole	883	667	<1.0	1490	1800	<1.0
TCEP	218	180	<10	291	334	30
Triclosan	44	17	<1.0	50	33	<1.0
Trimethoprim	92	101	<1.0	199	250	<1.0

Source: Snyder et al. 2006a. Reprinted from *Desalination*, 202, Snyder, S. A., S. Adham, A. M. Redding, F. S. Cannon, J. DeCarolis, J. Oppenheimer, E. C. Wert, and Y. Yoon. Role of Membranes and Activated Carbon in the Removal of Endocrine Disruptors and Pharmaceuticals. 156-181. Copyright 2006, with permission from Elsevier.

The removal of target analytes by MBR alone in the pilot MBR and MBR/RO are compared with predicted biodegradation potential from EPI BIOWIN in Table 9.10. Some consistencies in removal were observed between the pilot MBR and MBR/RO systems. Caffeine and naproxen were well removed by each MBR system. The BIOWIN model also predicts rapid biodegradation for these two compounds. Approximately half of the target analytes have shown similar removal by the three MBR pilot systems investigated. The variability for the remaining compounds among different waters could be related to water characteristics and the efficiency of the biological systems, which poses a great challenge in empirical predictions. Also, the BIOWIN model predicts only metabolism rate without consideration of adsorption. Considering the large biomass generated during these processes, adsorption is likely to play a role in apparent compound removal. BIOWIN was able to accurately predict the biodegradation behavior for approximately half of the observed compound removal.

Table 9.10
Predicted biodegradation potential and actual removal of target analytes by MBR systems

	MBR	MBR/RO		
	Only	Facility #1	Facility #2	BIOWIN
Acetaminophen	NA	100	> 97	Biodegrades fast
Androstenedione	NA	> 98	> 86	Does not biodegrade fast
Atrazine		NA		Does not biodegrade fast
Caffeine	> 85	100	84	Biodegrades fast
Carbamazepine	> 96	0	9	Biodegrades fast
DEET	20	73	86	Biodegrades fast
Diazepam		0		Biodegrades fast
Diclofenac	> 38	0	0	Does not biodegrade fast
Dilantin	4	0	6	Biodegrades fast
Erythromycin-H2O	96	0	0	Does not biodegrade fast
Estradiol		NA	> 66	Biodegrades fast
Estriol		> 99		Biodegrades fast
Estrone		> 97	76	Biodegrades fast
Ethynyl estradiol		NA	> 97	Does not biodegrade fast
Fluoxetine	> 77	0	20	Does not biodegrade fast
Gemfibrozil	> 86	18		Biodegrades fast
Hydrocodone	> 94	0	14	Biodegrades fast
Ibuprofen	0	99	95	Biodegrades fast
Iopromide		0	10	Does not biodegrade fast
Meprobamate	0	0	9	Biodegrades fast
Naproxen	>86	100	89	Biodegrades fast
Oxybenzone	NA	95	76	Biodegrades fast
Pentoxifylline		> 86		Biodegrades fast
Progesterone		> 95	> 84	Does not biodegrade fast
Sulfamethoxazole	> 57	68	12	Does not biodegrade fast
TCEP		0	9	Biodegrades fast
Testosterone		> 96		Does not biodegrade fast
Triclosan	> 41	94	48	Does not biodegrade fast
Trimethoprim	> 76	95	0	Biodegrades fast

Data from Snyder et al. 2004

Microfiltration Followed by Reverse Osmosis and Electrodialysis Reversal (EDR)

Results of EDC/PPCP removal using MF/RO and MF/EDR systems are provided in Table 9.11. The concentrations of several compounds in the tertiary effluent were below the method reporting limits. The tertiary effluent was the influent for MF. Of the detectable compounds, none were well-removed by EDR, indicating that EDR is not effective in removing the selected target analytes. Conversely RO demonstrated excellent removal efficiency for all target compounds.

Table 9.11
Results from MF/RO and MF/EDR testing

Target Compound	WWTP Influent ng/L	Tertiary Effluent ng/L	MF Permeate ng/L	EDR Effluent ng/L	RO Permeate ng/L
Acetaminophen	14200	2.5	2.4	3.4	<1.0
Androstenedione	52	5.8	5.2	5.2	<1.0
Caffeine	32500	<10	<10	<10	<10
Carbamazepine	78	19	17	18	<1.0
DEET	154	122	100	112	4.2
Dilantin	51	52	31	47	<1.0
Erythromycin-H_2O	79	<1.0	<1.0	<1.0	<1.0
Estradiol	33	<1.0	<1.0	<1.0	<1.0
Estriol	137	<5.0	<5.0	<5.0	<5.0
Estrone	49	<1.0	<1.0	<1.0	<1.0
Fluoxetine	10	8.5	4.7	5.8	<1.0
Galaxolide	931	544	617	587	<10
Gemfibrozil	1220	<1.0	<1.0	<1.0	<1.0
Hydrocodone	35	<1.0	<1.0	<1.0	<1.0
Ibuprofen	2270	6.0	2.7	5.4	<1.0
Iopromide	17	42	34	51	<1.0
Meprobamate	124	75	67	71	<1.0
Musk Ketone	119	65	45	45	<10
Naproxen	4480	<1.0	<1.0	<1.0	<1.0
Oxybenzone	657	5.8	4.9	3.8	<1.0
Sulfamethoxazole	360	<1.0	<1.0	<1.0	<1.0
TCEP	244	133	127	127	2.3
Testosterone	47	<1.0	<1.0	<1.0	<1.0
Triclosan	564	1.2	1.2	<1.0	<1.0
Trimethoprim	213	<1.0	<1.0	<1.0	<1.0

Snyder et al. 2006a. Reprinted from Desalination, 202, Snyder, S. A., S. Adham, A. M. Redding, F. S. Cannon, J. DeCarolis, J. Oppenheimer, E. C. Wert, and Y. Yoon. Role of Membranes and Activated Carbon in the Removal of Endocrine Disruptors and Pharmaceuticals. 156-181. Copyright 2006, with permission from Elsevier.

FULL-SCALE EVALUATIONS

Full-Scale Microfiltration Followed by Reverse Osmosis and UV Advanced Oxidation

A 10 million gallon per day (MGD) MF/RO water reuse facility equipped with a UV advanced oxidation process (AOP) was evaluated for EDC/PPCP removal. The facility is provided non-nitrified tertiary treated wastewater. Chlorine is added to reduce biofouling of the membranes; therefore, a chloramine residual is generally present. Ascorbic acid was added to sample bottles prior to sample collection to quench residual oxidants.

This MF/RO facility used US Filter submerged hollow-fiber MF membranes as a pretreatment for the RO process. Hydranautics ESPA 2 membranes were used for the RO process. The RO permeate is exposed to UV light in the presence of hydrogen peroxide for advanced oxidation. Feed water was not spiked and samples were collected between each unit process.

Results from the MF/RO/AOP facility are provided in Table 9.12. MF did not significantly remove detectable target compounds. However, RO was very efficient in removing the target compounds, although a few compounds were detectable at trace levels in the permeate. The UV/AOP provided additional removal for those compounds not completely removed by RO. The degree of removal demonstrated in the AOP process is as expected from previous UV AOP experiments conducted previously (Chapter 8). The trace amount of oxybenzone (2.1 ng/L) is likely a result of contamination during sampling, as oxybenzone is used in a large variety of personal care products.

Full-Scale Microfiltration Followed by Double-Pass Reverse Osmosis

A 37 MGD water reuse facility equipped with MF and double-pass RO was evaluated for EDC/PPCP removal. The facility is provided with non-disinfected secondary-treated wastewater and uses a US Filter Memcor pressure-vessel MF membrane system as a pretreatment for Hydranautics ESPA 2 RO membranes. Feed water was not spiked and samples were collected between each unit process. All samples were collected in duplicate or triplicate, so results shown are averages (Table 9.13). Variability in replicate measurements generally was less than 10%. The results were consistent with other observations indicating that MF does not provide appreciable contaminant removal of commonly detected target analytes. Double pass RO was shown to be very efficient in the rejection of target analytes, with removal to below detection after the second stage.

Table 9.12
Results from full-scale MF/RO/UV-AOP plant

Target Compound	Tertiary Effluent ng/L	MF Permeate ng/L	RO Permeate ng/L	UV-AOP Effluent ng/L
Caffeine	1037	994	<10	<10
Carbamazepine	258	237	<1.0	<1.0
DEET	3167	2430	4.4	2.2
Diclofenac	49	59	<1.0	<1.0
Dilantin	175	180	<1.0	<1.0
Erythromycin-H_2O	191	181	<1.0	<1.0
Estradiol	11	14	<1.0	<1.0
Estrone	137	158	<1.0	<1.0
Galaxolide	1130	1084	11	<10
Gemfibrozil	2740	3750	2.4	1.1
Hydrocodone	54	51	<1.0	<1.0
Ibuprofen	379	500	<1.0	<1.0
Iopromide	127	131	<1.0	<1.0
Meprobamate	308	260	<1.0	<1.0
Musk Ketone	61	68	<10	<10
Naproxen	516	621	1.2	<1.0
Oxybenzone	121	77	6.1	2.1
Pentoxifylline	39	33	<1.0	<1.0
Sulfamethoxazole	1050	1030	1.9	<1.0
TCEP	407	370	7.4	2.6
Triclosan	136	64	<1.0	<1.0
Trimethoprim	248	227	<1.0	<1.0

Snyder et al. 2006a. Reprinted from *Desalination*, 202, Snyder, S. A., S. Adham, A. M. Redding, F. S. Cannon, J. DeCarolis, J. Oppenheimer, E. C. Wert, and Y. Yoon. Role of Membranes and Activated Carbon in the Removal of Endocrine Disruptors and Pharmaceuticals. 156-181. Copyright 2006, with permission from Elsevier.

Table 9.13
Average results from full-scale MF double-pass RO

Target Compound	Secondary Effluent ng/L	MF Permeate ng/L	1st Stage RO Permeate ng/L	2nd Stage RO Permeate ng/L
Acetaminophen	16	10	<1.0	<1.0
Caffeine	3460	6125	16	1.2
Carbamazepine	174	271	<1.0	<1.0
DEET	2020	3365	3.4	<1.0
Diclofenac	56	49	<1.0	<1.0
Dilantin	207	336	<1.0	<1.0
Erythromycin-H_2O	312	507	<1.0	<1.0
Estrone	85	63	<1.0	<1.0
Fluoxetine	12	23	<1.0	<1.0
Gemfibrozil	2885	3040	2.7	<1.0
Hydrocodone	62	104	<1.0	<1.0
Ibuprofen	354	422	<1.0	<1.0
Iopromide	1670	1810	2.3	<1.0
Meprobamate	230	341	<1.0	<1.0
Naproxen	1068	1205	2.0	<1.0
Oxybenzone	31	60	1.9	<1.0
Pentoxifylline	67	109	<1.0	<1.0
Sulfamethoxazole	341	805	2	<1.0
TCEP	300	467	1.9	1.3
Triclosan	620	424	<1.0	<1.0
Trimethoprim	248	409	<1.0	<1.0

Snyder et al. 2006a. Reprinted from *Desalination*, 202, Snyder, S. A., S. Adham, A. M. Redding, F. S. Cannon, J. DeCarolis, J. Oppenheimer, E. C. Wert, and Y. Yoon. Role of Membranes and Activated Carbon in the Removal of Endocrine Disruptors and Pharmaceuticals. 156-181. Copyright 2006, with permission from Elsevier.

RESULT SUMMARY AND DISCUSSION

Membrane systems can be an effective technology for reducing the concentration of a diverse set of EDC and PPCP contaminants during both drinking water and wastewater treatment. The degree of removal is directly related to membrane characteristics, and to some extent the molecular properties of the particular contaminant. The experiments discussed here demonstrate that MF and UF membranes (without substantial pretreatment steps) offer little removal effectiveness in full-scale operations. This limited data set indicates a potential for the removal of steroids using MF/UF membranes, especially when biologically active (MBR). It is also important to note that bench-scale testing using virgin membranes did show higher removals; however, this appears to be a limitation of bench-scale membrane testing using virgin membranes and does not predict full-scale removal accurately. While additional studies are needed to validate this observation at additional full-scale utilities, the removal of steroids is valuable since these compounds have been shown to induce biological effects at even minute concentrations (Jobling et al. 1998; Snyder et al. 2001; Palace et al. 2002; van Aerle et al. 2002).

In contrast, RO membranes are capable of removing nearly all compounds investigated to levels less than method reporting limits. Trace levels of some contaminants were still detectable in RO permeates. The compounds that passed through the RO membranes were not consistent, and no clear relationship between molecular structure and membrane passage could be established. Initially it appeared that the flame-retardant TCEP and the stimulant caffeine had permeated the RO membranes evaluated; however, closer examination of the data demonstrated that both of these compounds were ubiquitous in field and laboratory blanks. The MRL for caffeine and TCEP were adjusted to approximately twice the average concentration detected in blanks. Once the MRL was raised to compensate for contamination, no significant break-through could be observed. Permeation of the RO membrane could be the result of diffusion into and through the membrane, short-circuiting of the membrane, or support media failure (i.e. adhesive failure). The full-scale double-pass RO system showed that the second pass was able to remove compounds that were not entirely removed during the first pass. Without question, a multi-barrier approach (e.g., RO/AOP or double pass RO) is most successful in the removal of trace contaminants.

Table 9.14 places each target compound in a percent removal bin based upon results obtained through bench, pilot, and full-scale studies. The number of waters evaluated by each membrane type is also listed. It must be noted that while RO consistently falls into the >80% category in Table 9.14, the observed removal using RO was generally >99%. MF and UF systems may be applicable for removing those compounds with higher log K_{ow} due to particle binding and subsequent physical removal. MBR systems using UF membranes provided additional removal of biodegradable compounds (i.e. acetaminophen, naproxen, ibuprofen, and caffeine). NF and RO systems were shown to reject most target compounds. The tighter membranes provide an excellent treatment process for the removal of EDCs/PPCPs. The detection of trace levels of certain contaminants near the method reporting limits after NF/RO processes is primarily a function of the sensitivity of the analytical methods. However, additional oxidation or adsorption processes will provide even greater removal potential. Even with multiple barriers, each of which are highly efficient for contaminant removal, it is likely that sensitive analytical techniques will still be able to detect several trace contaminants in the finished water.

Table 9.14
Summary of percent removal by membranes

Membrane Size # of Systems Tested	Percent Removal MF n=3	UF n=5	UF/MBR n=4	NF n=3	RO n=9
Acetaminophen (Tylenol)	<20	<20	>80	20-50	>80
Androstenedione	<20	20-50	>80	50-80	>80
Atrazine	*	<20	*	50-80	*
Benzo(a)pyrene	*	>80	*	>80	*
Caffeine	<20	<20	>80	50-80	>80
Carbamazepine	<20	<20	20-50	50-80	>80
DDT	*	>80	50-80	>80	*
DEET	<20	<20	50-80	50-80	>80
Diazepam (Valium)	*	20-50	<20	50-80	>80
Diclofenac	<20	<20	<20	50-80	>80
Dilantin	<20	<20	<20	50-80	>80
Erythromycin	<20	20-50	20-50	>80	>80
Estradiol	<20	20-50	50-80	50-80	>80
Estriol	*	<20	>80	50-80	>80
Estrone	<20	20-50	>80	50-80	>80
Ethinyl Estradiol	*	20-50	>80	50-80	>80
Fluorene	*	>80	*	>80	*
Fluoxetine (Prozac)	20-50	>80	20-50	>80	>80
Galaxolide	<20	20-50	*	50-80	>80
Gemfibrozil	<20	<20	20-50	50-80	>80
Hydrocodone	<20	<20	20-50	50-80	>80
Ibuprofen (Advil)	<20	<20	50-80	50-80	>80
Iopromide	<20	<20	<20	>80	>80
Lindane (γ-BHC)	*	20-50	*	50-80	*
Meprobamate	<20	<20	<20	50-80	>80
Metolachlor	*	20-50	*	50-80	*
Musk Ketone	<20	20-50	*	>80	>80
Naproxen	<20	<20	>80	20-50	>80
Oxybenzone	<20	50-80	>80	>80	>80
Pentoxifylline	<20	<20	>80	50-80	>80
Progesterone	*	50-80	>80	50-80	>80
Sulfamethoxazole	<20	20-50	20-50	50-80	>80
TCEP	<20	<20	<20	50-80	>80
Testosterone	*	20-50	>80	50-80	*
Triclosan	20-50	>80	50-80	>80	>80
Trimethoprim	<20	<20	20-50	50-80	>80

* Not detected

REFERENCES

Bellona, C., J. E. J. E. Drewes, P. Xu, and G. Amy. 2004. Factors affecting the rejection of organic solutes during NF/RO treatment--a literature review. *Water Res.*, 38(12): 2795-2809.

Burba, P., H. Geltenpoth, and J. Nolte. 2005. Ultrafiltration behavior of selected pharmaceuticals on natural and synthetic membranes in the presence of humic-rich hydrocolloids. *Analytical and Bioanalytical Chemistry*, 382: 1934-1941.

Huang, C.-H., and D. L. Sedlak. 2001. Analysis of estrogenic hormones in municipal wastewater effluent and surface water using enzyme-linked immunosorbent assay and gas chromatography/tandem mass spectrometry. *Environ. Toxicol. Chem.*, 20(1): 133-139.

Jobling, S., M. Noylan, C. R. Tyler, G. Brighty, and J. P. Sumpter. 1998. Widespread Sexual Disruption in Wild Fish. *Environ. Sci. Technol.*, 32: 2498-2506.

Nghiem, L. D., J. McCutheon, A. I. Schafer, and M. Elimelech. 2004. The role of endocrine disrupters in water recycling: risk or mania? *Water Sci. Technol.*, 50(2): 215-220.

Palace, V. P., R. E. Evans, K. Wautier, C. Baron, L. Vandenbyllardt, W. Vandersteen, and K. Kidd. 2002. Induction of vitellogenin and histological effects in wild fathead minnows from a lake experimentally treated with the synthetic estrogen, ethynylestradiol. *Water Quality Research Journal of Canada* 37(3): 637-650.

Snyder, S. A., S. Adham, A. M. Redding, F. S. Cannon, J. DeCarolis, J. Oppenheimer, E. C. Wert, and Y. Yoon. 2006. Role of Membranes and Activated Carbon in the Removal of Endocrine Disruptors and Pharmaceuticals. *Desalination*, 202: 156-181.

Snyder, S. A., J. Leising, P. Westerhoff, Y. Yoon, H. Mash, and B. J. Vanderford. 2004. Biological attenuation of EDCs and PPCPs: Implications for water reuse. *Ground Water Monit. R.*, 24(2): 108-118.

Snyder, S. A., D. L. Villeneuve, E. M. Snyder, and J. P. Giesy. 2001. Identification and quantification of estrogen receptor agonists in wastewater effluents. *Environ. Sci. Technol.*, 35(18): 3620-3625.

van Aerle, R., N. Pounds, T. H. Hutchinson, S. Maddix, and C. R. Tyler. 2002. Window of Sensitivity for the Estrogenic Effects of Ethinylestradiol in Early Life-Stages of Fathead Minnow, Pimephales promelas. *Ecotoxicology* 11(6): 423-434.

Yoon, Y., P. Westerhoff, S. A. Snyder, and E. C. Wert. 2006. Nanofiltration and ultrafiltration of endocrine disrupting compounds, pharmaceuticals and personal care products. *J. Membr. Sci.*, 270: 88-100.

CHAPTER 10
MAGNETIC ION-EXCHANGE

BASIC THEORY

Magnetic ion-exchange (MIEX®) is a process that utilizes magnetic polymer microspheres as an ion exchange resin. The resin is composed of a polymer and inorganic magnetic material such as Fe_3O_4, Fe_2O_3, nickel, and cobalt. The resin exhibits both the characteristics of ionic exchange and magnetism. MIEX® is a high capacity ion exchange resin that is used as a continuous feed suspension in a stirred contactor. The magnetic component of this resin aids in settling by aggregating particles for more rapid settling. In environmental applications, the resin is made via a quaternization reaction of the amine groups in the polymer which has affinity for anions. The base anion exchange resin in MIEX® can remove inorganic such as sulfate, bicarbonate, bromide as well as bulk organic acids (e.g., NOM with acidic functional groups). The resin is introduced to water as a slurry such that doses are expressed as mL resin/L water.

The process has previously been shown to be effective for reducing the concentration of dissolved organic carbon (DOC) from water (Singer and Bilyk 2002; Fearing et al. 2004; Son et al. 2005; Wert et al. 2005). In this investigation, MIEX® was evaluated for removal of several endocrine disrupting chemicals (EDCs) and pharmaceuticals. Both static bench-scale and dynamic pilot-scale tests were completed using Colorado River water spiked with target compounds.

BENCH-SCALE EVALUATION

MIEX® resin from an on-site pilot plant was used for jar testing. Since the MIEX® resin used for bench-scale testing was cycled several times through the pilot system, the binding capacity was likely decreased and more representative of a resin that would be in service as compared to virgin material. MIEX® concentrations of 5, 15, and 20 mL/L were evaluated with a 10-minute contact time. MIEX® concentrations of 5, 10, 15, and 20 mL/L were evaluated with a 20-minute contact time. These experiments were designed to determine optimal conditions for pilot plant testing.

Bench-Scale Results

During MIEX® bench testing, 18 of 28 target compounds spiked showed less than 20% removal when using a contact time of 10 minutes. Those compounds showing appreciable removal followed a response curve related to the MIEX® concentration (Figure 10.1). Only triclosan and diclofenac were removed by greater than 90%. Both triclosan and diclofenac were negatively charged at the ambient pH and thus were expected to exhibit appreciable removal by anionic exchange. When the contact time was increased to 20 minutes, removal of the same contaminants showed a meager increase in removal (Figure 10.2). Increasing the MIEX® dose and contact time did not result in linear increases in removal.

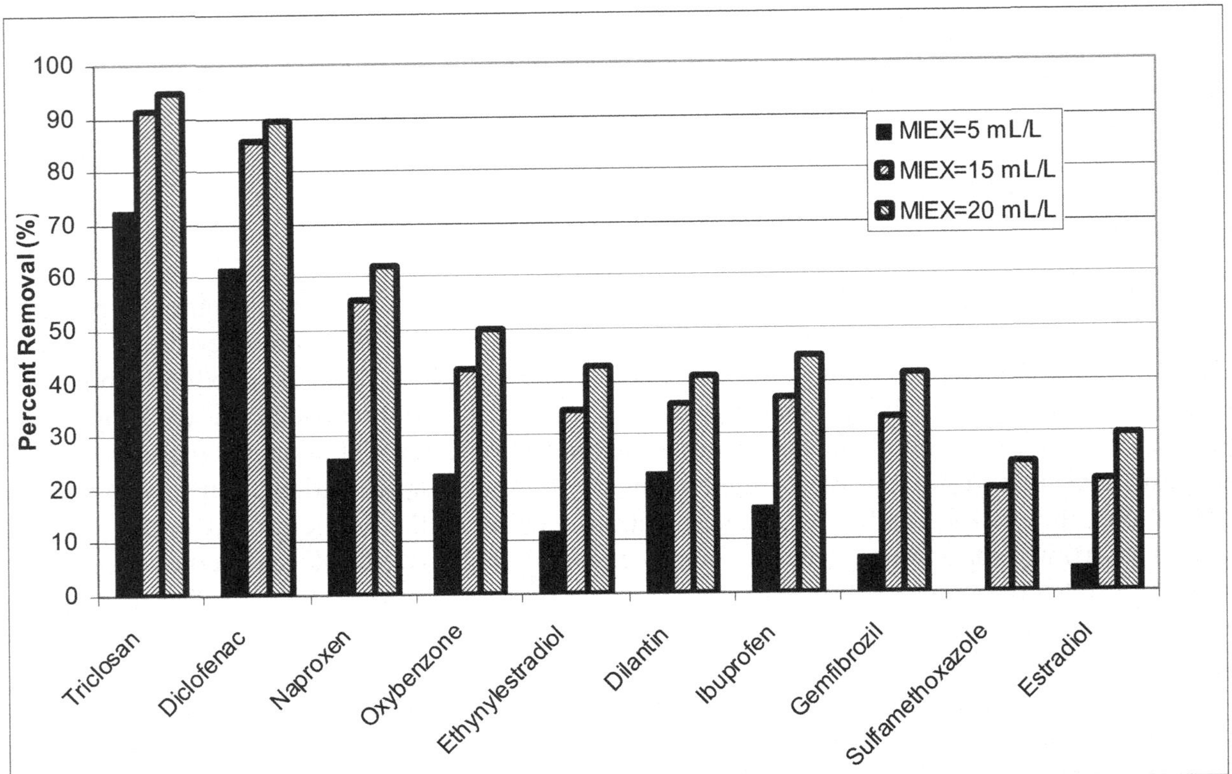

Figure 10.1 Bench-scale percent removal of target compounds by MIEX® treatment with a contact time of 10 minutes

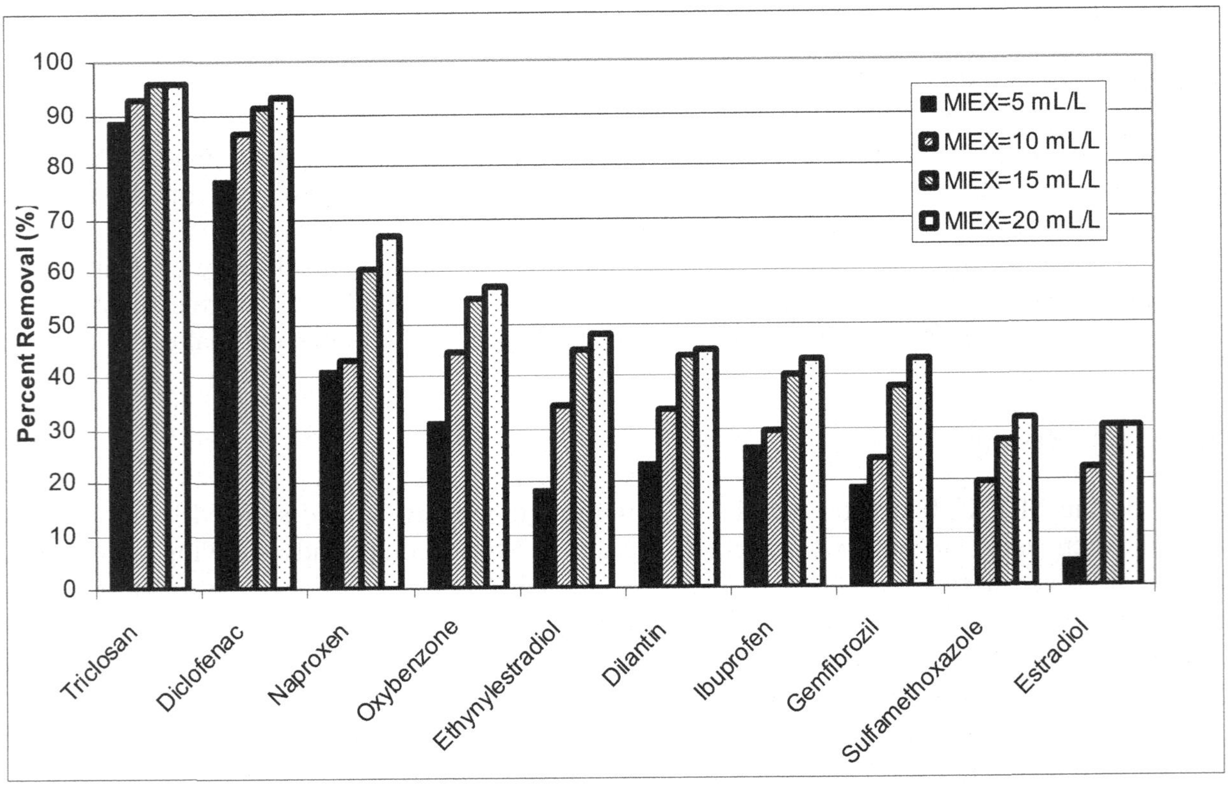

Figure 10.2 Bench-scale percent removal of target compounds by MIEX® treatment with a contact time of 20 minutes

PILOT-SCALE EVALUATION

Pilot plant testing was conducted at SNWA using Colorado River water spiked with target compounds. A MIEX® concentration of 20 mL/L was selected to represent a best case scenario based on bench-scale results. Samples were collected after each contactor chamber having total elapsed contact times of 10 and 20 minutes.

Pilot Plant Results

During pilot plant testing, 16 of 28 target compounds poorly removed during bench-scale testing showed less than 30% removal at pilot-scale. Select compound removal is shown in Figure 10.3. Progesterone and estrone showed slightly higher removal during pilot-scale experiments. Variation from the bench-scale test may have occurred due to less consistent concentrations of target compounds spiked into raw water as a result of dynamic spike feed.

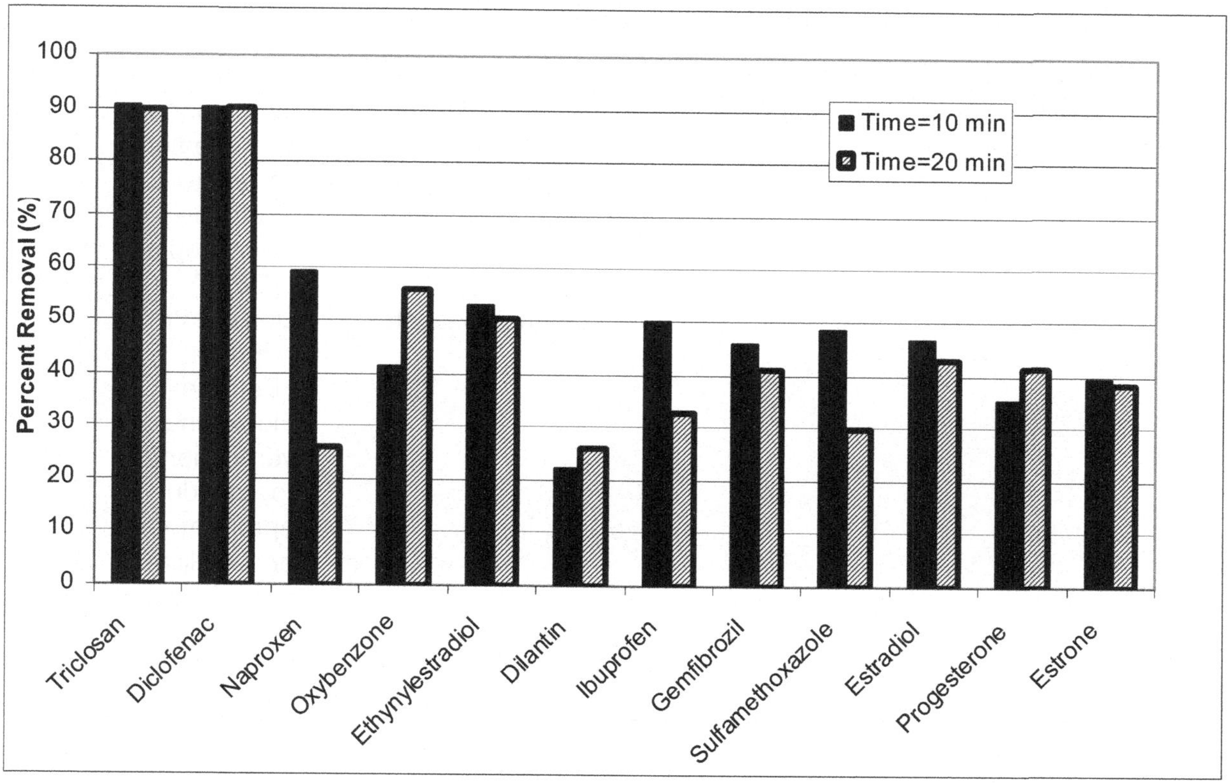

Figure 10.3 Pilot-scale percent removal of target compounds by MIEX® (20 mL/L)

RESULT SUMMARY AND DISCUSSION

Overall, poor removal was observed for MIEX® treatment compared to adsorption by activated carbon and oxidation. A summary of the percent removal of all target compounds evaluated is shown in Table 10.1. Removal was better among compounds carrying negative charge at the ambient water pH of 8.2. The two compounds, diclofenac and triclosan, exhibiting over 90% removal were both negatively charged at pH 8.2. The other five compounds that were either fully or partially depprononated had over 40% removal. The low removal (28%) of sulfamethoxazole, however, was not expected since at pH 8.2, it is present as a mixture of both neutral and negatively charged species. Unlike the other negatively charged compounds, the negative charge of sulfamethoxazole comes from the deprotonated secondary amine. This may partially hinder the electrostatic interaction. As expected, fluoxetine and trimethoprim were poorly removed due to their positive charge.

Table 10.1
Summary of percent removal by MIEX

> 80% Removal	50-80% Removal	20-50% Removal	< 20% Removal
Diclofenac	Naproxen	Dilantin	Acetaminophen
Triclosan		Estradiol	Androstenedione
		Ethynylestradiol	Atrazine
		Gemfibrozil	Caffeine
		Ibuprofen	Carbamazepine
		Oxybenzone	DEET
		Sulfamethoxazole	Diazepam
			Estriol
			Estrone
			Fluoxetine
			Hydrocodone
			Iopromide
			Meprobamate
			Pentoxifylline
			Progesterone
			TCEP
			Testosterone
			Trimethoprim

Most neutral compounds were poorly removed, yet a significant percentage of removal can be attained for compounds with large log K_{ow} values. This observation is confirmed by a recent study that also observed removal of hydrophobic compounds in wastewater secondary effluent by MIEX® (Zhang et al. 2006). Further examination on the removal of neutral compounds disclosed a weak dependence of percent removal on log K_{ow}, as illustrated in Figure 10.4, indicating that MIEX® can provide some removal of non-ionic species. The log K_{ow} appeared to also influence the removal of negatively charged contaminants; however, the dependence was shifted due to the electrostatic interactions (Figure 10.4). Van der Waals interactions from dipole or induced dipole moments between the analytes and polymer resins also may contribute to this type of dependence.

Dose response curves, obtained for compounds with removal data, showed that removal is related to MIEX® dose. Contact time seemed to be less important, as the 20 min contact time generally improved removal by only 20% or less. MIEX® proved to be of limited use for the removal of EDCs and PPCPs, as most compounds were not well removed. However, for specific compounds and multi-barrier processes, MIEX® may be applicable for microcontaminant removal. Additionally, this process has been shown very efficient in removing several important ionic DBP precursors, such as NOM and bromide ion, and is superior to traditional and enhanced coagulation for this purpose (Hammann, Bourke, and Topham 2004; Johnson and Singer 2004; Boyer and Singer 2005; Wert et al. 2005).

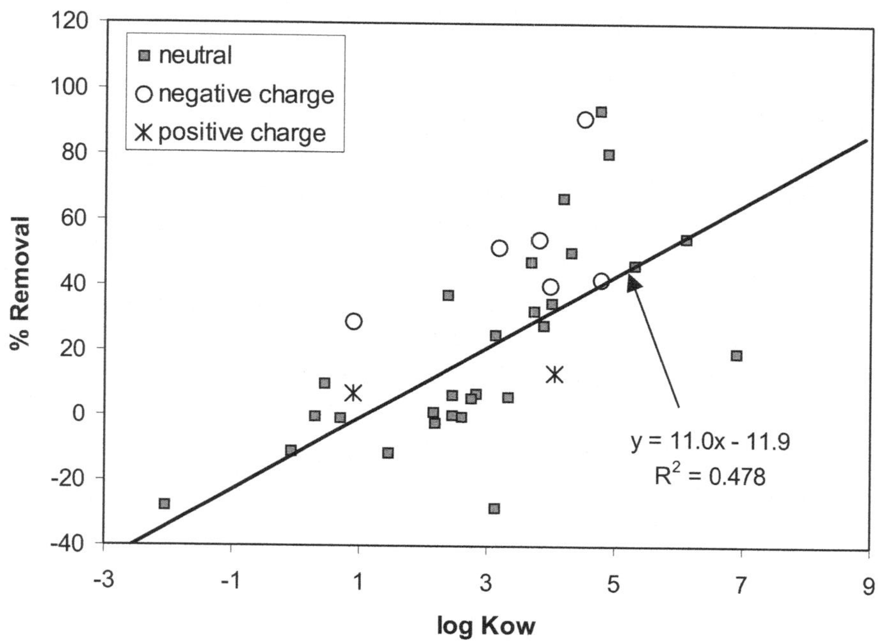

Figure 10.4 Percentage removal of target compounds in relation to log K_{ow} for neutral, positive and negatively charged species.

The linear line represents the relationship for neutral species only.

REFERENCES

Boyer, T. H., and P. C. Singer. 2005. Bench-scale testing of a magnetic ion exchange resin. for removal of disinfection by-product precursors. *Water Res.*, 39(7): 1265-1276.

Fearing, D. A., J. Banks, S. Guyetand, C. Monfort Eroles, B. Jefferson, D. Wilson, P. Hillis, A. T. Campbell, and S. A. Parsons. 2004. Combination of ferric and MIEX for the treatment of a humic rich water. *Water Res.*, 38(10): 2551-2558.

Hammann, D., M. Bourke, and C. Topham. 2004. Evaluation of a magnetic ion exchange resin to meet DBP regs at the Village of Palm Springs. *Journal American Water Works Association*, 96(2): 46-50.

Johnson, C. J., and P. C. Singer. 2004. Impact of a magnetic ion exchange resin on ozone demand and bromate formation during drinking water treatment. *Water Res.*, 38(17): 3738-3750.

Singer, P. C., and K. Bilyk. 2002. Enhanced coagulation using a magnetic ion exchange resin. *Water Res.*, 36(16): 4009-4022.

Son, H. J., Y. D. Hwang, J. S. Roh, K. W. Ji, P. S. Sin, C. W. Jung, and L. S. Kang. 2005. Application of MIEX pre-treatment for ultrafiltration membrane process for NOM removal and fouling reduction *Water Sci. Technol.*, 5(5): 15-24.

Wert, E. C., J. C. Edwards-Brandt, P. C. Singer, and G. C. Budd. 2005. Evaluating magnetic ion exchange resin (MIEX)® pretreatment to increase ozone disinfection and reduce bromate formation. *Ozone-Science & Engineering*, 27(5): 371-379.

Zhang, R., S. Vigneswaran, H. H. Ngo, and H. Nguyen. 2006. Magnetic ion exchange (MIEX®) resin as a pre-treatment to a submerged membrane system in the treatment of biologically treated wastewater. *Desalination*, 192(1-3): 296-302.

CHAPTER 11
BIOLOGICAL PROCESSES

BASIC THEORY

Biological processes rely on the microbial activity of non-pathogenic bacteria to degrade organic chemicals. Because trace organics usually are insufficient in concentration to provide a primary substrate for the growth of microorganisms, trace organics are co-metabolized. In most cases the DOC present in the water provides the primary substrate for microbial activity, and trace organics are co-metabolized as a secondary substrate. Secondary substrates typically degrade at a rate related to the amount of biomass present in a system (biomass supported by the primary substrate). In addition to degradation of trace organics, which require enzymes to react with the organics, trace organics can adsorb or partition into cellular materials (termed biosorption). Biosorption can occur alone or as a first step prior to biodegradation.

Some studies on the biological degradation of EDCs and PPCPs have been reported, but the majority of these reports focused on more traditional contaminants now known to have endocrine effects (i.e., pesticides, herbicides, and polycyclic aromatic hydrocarbons). Akkanen *et* al. {Akkanen, 2001 #1188} investigated the effects of dissolved organic matter (DOM) and water quality on the bioavailability of atrazine, pyrene, and benzo(a)pyrene through the formation DOM-contaminant complex formation. In that study, the bioavailability of atrazine was not affected by DOM or water quality, presumably due to the hydophilicity of atrazine. Although equilibrium dialysis showed binding of pyrene to DOM, the bioavailability of pyrene was not significantly affected by DOM. In contrast, the bioavailability of benzo(a)pyrene was significantly reduced by the presence of DOM. Soil-aquifer treatment (SAT) was studied to investigate the fate of pharmaceuticals present in tertiary-treated wastewater (gemfibrozil, clofibric acid, fenofibrate, carbamazepine, primidone, diclofenac, ibuprofen, ketoprofen, naproxen, fenoprofen, propyphenazone, meclofenamic acid, and tolfenamic acid) {Drewes, 2001 #1195}. SAT was shown to efficiently remove certain anti-inflammatory and lipid-regulating drugs. However, the antiepileptic drugs carbamazepine and primidone persisted in groundwater and no clear indication of degradation was found during travel times of more than 6 years (Snyder et al. 2004). One study investigated the occurrence of pharmaceuticals, emphasizing tri-iodinated benzene derivatives used as X-ray contrast media, in domestic effluents and their fate during subsequent groundwater recharge {Drewes, 2001 #1196}. Organic iodine compounds in secondary treated effluents were not removed by advanced wastewater treatment using ozone. A concentration range of 8-15 μg iodine/L was observed in groundwater recharge systems after travel times of 8 to 10 years.

This chapter includes experiments designed to address EDC/PPCP removal under drinking water conditions along with field studies that employ reclaimed wastewaters in order to cover a wide breadth of potential biodegradation scenarios. The largest difference between drinking water conditions (aerobic, low organics content) and reclaimed wastewater (possibly anaerobic, higher organics content) is related to the type of microorganics (aerobic, anaerobic) and level of biodegradable primary organic substrate content.

BENCH-SCALE EVALUATIONS

Biologically-Active Sand Experiments

In these tests, the biodegradation of EDCs and PPCPs by BAS was investigated by spiking Colorado River water (CRW) with EDCs and PPCPs. Approximately 10% of the DOC in CRW was removed by biologically-active sand (BAS). The initial DOC was 2.52 mg/L. After two and five days of incubation, the DOC levels remaining in the Biodegradable dissolved organic carbon (BDOC) reactors were 2.27 mg/L (n=6) and 2.20 mg/L (n=6), respectively. The BDOC material (0.2 to 0.4 mg/L) was assumed to serve as the primary substrate for support of microbial activity on the BAS (Snyder et al. 2004).

Removals of target compounds in BDOC batch experiments are summarized in Table 11.1. Significant removal for some compounds was observed in samples with pre-acidified BAS (five day exposure in CRW). In this study, the compound removal with pre-acidified BAS was likely due to adsorption onto the sand or the biomass for compounds with no pH dependent speciation change. If the analyte charge changed as a result of pH change, it cannot be ruled out the observed removal was the result of the change in ionization state. For example, the removal of trimethoprim (pKa = 7.1) was 55% in the pre-acidified BAS and 19% and 24% in the two and five day BAS samples. This corresponds with higher adsorption at pH 2.0 (pre-acidified BAS) than at pH 8.1 (2 and 5 day experiments), and is probably due to the electrostatic interaction of the protonated trimethoprim (low pH) with the sand and/or biomass at pH 2.0. At pH 8.1 only a small percentage of the trimethoprim is protonated, with the majority being non-ionized. Biosorption (sand and biomass) appears to be significant for some target compounds. The sustainability of biosorption alone (without biodegradation) is unknown. If biodegradation is assumed to be the dominant mechanism, compounds lost from filtered CRW (with BAS) would be considered biologically labile (i.e., easily biodegradable) (Snyder et al. 2004).

PILOT-SCALE EVALUATIONS

Biological Filtration - SNWA

During pilot plant testing at SNWA, one pilot filter contained biologically active carbon (BAC) and another contained biologically active anthracite (BAF). The SNWA pilot used an empty bed contact time (EBTC) of 2 minutes, representative of full-scale treatment plants operated by SNWA. The SNWA filters are not used for adsorption, but simply for conventional filtration. Results showed that 11 target compounds were greater than 20% removed by the BAF (Table 11.2). However, 18 of the target compounds were greater than 90% removed by the BAC filter. The additional removal observed with the BAC filter was likely due to either the adsorptive properties of the carbon or additional biological removal. BAC has more surface area for biofilm development than anthracite media. Since non-biological GAC was not pilot tested simultaneously, it is difficult to conclusively determine the mechanism for observed removal. However, biologically active GAC was evaluated at full-scale (Chapter 5). Based on full-scale observations, it is likely that removal shown in the pilot-scale BAC for the less biodegradable contaminants was largely due to adsorption rather than biodegradation.

Table 11.1
Summary of removal in BDOC reactors relative to acidified controls*

Process Evalauted	Adsorption + Biodegradation	Adsorption + Biodegradation	Adsorption
pH Condition	Ambient	Ambient	pH=2.0
Exposure Time	2 days	5 days	5 days
Target Compound	Percent Removal		
Acetaminophen	94	99	79
Androstenedione	99	99	96
Atrazine	58	55	54
Caffeine	76	92	77
Carbamazepine	57	54	56
DEET	40	57	37
Diazepam	83	82	83
Diclofenac	71	74	67
Dilantin	79	77	78
Erythromycin-H_2O	84	79	83
Estradiol	98	99	85
Estriol	81	99	81
Estrone	95	99	62
Ethynylestradiol	76	79	73
Fluoxetine	98	99	98
Gemfibrozil	82	99	54
Hydrocodone	48	47	49
Ibuprofen	94	99	66
Iopromide	31	33	28
Meprobamate	37	48	36
Naproxen	85	98	80
Oxybenzone	91	99	83
Pentoxifylline	92	99	91
Sulfamethoxazole	80	77	81
TCEP	54	53	54
Testosterone	99	99	92
Triclosan	97	99	97
Trimethoprim	19	24	55

* Concentrations from duplicate experiments were averaged to obtain the percentage removals; differences were <10% on average

Source: Snyder et al. 2004. Copyright 2004 National Ground Water Association. Reprinted with permission.

Table 11.2
Summary of removal during BAC and BAF pilot-scale evaluations

Source Water	Colorado River Water		
Filter Type		BAF	BAC
EBCT [min]		2	2
Target Compound	Influent Concentration (ng/L)	Percent Removal	
Acetaminophen	39	6	95
Androstenedione	157	41	97
Atrazine	142	15	83
Benzo[a]pyrene	94	89	89
Caffeine	102	8	93
Carbamazepine	112	5	90
DDT	77	>87	85
DEET	147	8	80
Diazepam	110	15	84
Diclofenac	108	13	75
Dilantin	93	19	80
Erythromycin-H_2O	111	31	78
Estradiol	147	1	94
Estriol	143	<1	92
Estrone	150	<1	95
Ethynylestradiol	145	1	91
Fluorene	167	28	>94
Fluoxetine	79	97	>99
Galaxolide	48	19	74
Gemfibrozil	135	13	74
Hydrocodone	74	14	92
Ibuprofen	118	30	83
Iopromide	120	11	42
Lindane (γ-BHC)	150	7	91
Meprobamate	133	4	71
Metolachlor	130	4	79
Musk Ketone	170	10	83
Naproxen	50	3	82
Oxybenzone	66	14	98
Pentoxifylline	98	13	90
Progesterone	142	52	99
Sulfamethoxazole	77	5	63
TCEP	94	10	80
Testosterone	144	35	96
Triclosan	85	37	97
Trimethoprim	102	20	94

Riverbank Filtration – Louisville Kentucky

A riverbank filtration (RBF) pilot-plant at Louisville Water Company, Kentucky, was utilized to evaluate the removal of EDCs/PPCPs. This pilot-plant uses three sequential biologically-active sand columns described in detail in Chapter 3. The experiment was conducted over a period of 36 days, wherein 90% of a fluoride tracer had been recovered from the final column (Figure 11.1). Samples from all three filters were collected prior to spiking in order to determine background concentrations of target compounds. Spiking began on January 15th, 2004 and ended on February 20th, 2004. Three sets of samples from each filter were collected; the initial background samples, samples on February 12th (90% of the fluoride tracer had breached filters 1 & 2), and again on February 20th (90% of the fluoride tracer had passed through filter 3). Table 11.3 provides available water quality data for the experiment. Only LC-MS/MS target compounds were used for this experiment.

Although samples from all three filters were collected on January 15th, February 12th, and February 20th, the filter 1 and filter 2 samples from February 12th were accidentally destroyed during sample processing. Table 11.4 presents target compound concentrations for the RBF pilot plant experiment, with raw and spiked raw data presented as averages (n = 2 and 6, respectively). In general, the majority of removal occurred within the first column with 14 of the 29 target contaminants removed to less than detection. The remaining compounds were removed to various extents through the RBF columns.

The trends for biodegradation seen in the RBF pilot-plant are in good agreement with those obtained in biological degradation batch tests conducted and published previously by this team {Snyder, 2004 #1361}. Table 11.5 provides the percent removal for the RBF pilot columns (Feb. 20th data) and compares these with biologically-active sand batch tests. The results in Table 11.5 are provided as percent removal from the raw water based on the effluent concentration from each column as shown in Table 11.4. In other words, the percent removals are based on discrete comparisons to the raw water and are not cumulative. In some cases the percent removal appears to decrease from column 2 to column 3 (i.e., TCEP); however, this is most likely the result of analytical variation for compounds where large concentration changes were not observed. In general, results from the RBF pilot were in good agreement with batch testing; however, some discrepancies were observed.

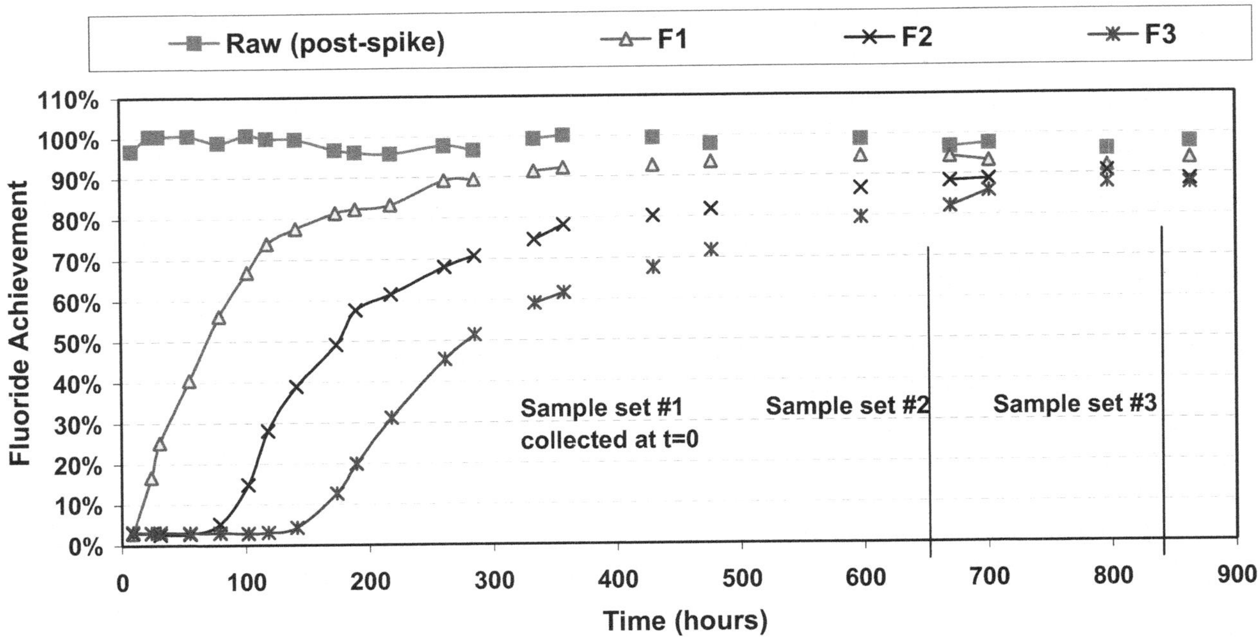

Figure 11.1 Fluoride tracer data from RBF pilot-plant (F = filter#)

Table 11.3
Water quality data for RBF experiments

Date	Time	Location	pH	Turbidity	Hardness	Alkalinity	UV_{254}
1/15/2004	9:00	Raw (pre-spike)	7.87	69.1	128	78	0.184
	9:00	Filter 1 Effluent	7.74	3.5	124	78	0.073
	9:00	Filter 2 Effluent	8.07	0.84	130	84	0.054
	9:00	Filter 3 Effluent	8.05	0.14	128	84	0.048
2/12/2004	18:00	Raw (post-spike)	7.76	98	110	58	0.067
	18:00	Filter 1 Effluent	7.54	1.89	119	62	0.053
	18:00	Filter 2 Effluent	8.05	0.19	126	70	0.055
	18:00	Filter 3 Effluent	8.06	0.09	132	74	0.049
2/20/2004	9:00	Raw (post-spike)	7.83	21	135	79	0.045
	9:00	Filter 1 Effluent	7.52	0.45	140	79	0.035
	9:00	Filter 2 Effluent	8.04	0.21	155	94	0.035
	9:00	Filter 3 Effluent	8.07	0.09	146	96	0.033

Table 11.4
River bank filtration pilot (units = ng/L)

	AVE Raw	AVE Spiked	15th Jan 2004 C-1	15th Jan 2004 C-2	15th Jan 2004 C-3	12th Feb 2004 C-3	20th Feb 2004 C-1	20th Feb 2004 C-2	20th Feb 2004 C-3
Acetaminophen	14	52	<1.0	<1.0	<1.0	<1.0	<1.0	<1.0	<1.0
Androstenedione	<1.0	121	<1.0	<1.0	<1.0	<1.0	<1.0	<1.0	<1.0
Atrazine	22	113	16	18	18	104	124	102	108
Caffeine	48	138	1.8	3.1	3.0	2.2	6.3	4.2	2.7
Carbamazepine	1.6	119	1.5	2.0	1.9	93	123	93	103
DEET	17	144	22	44	36	26	17	8.8	12
Diazepam	<1.0	99	<1.0	<1.0	<1.0	19	107	58	35
Diclofenac	<1.0	64	<1.0	<1.0	<1.0	<1.0	24	<1.0	<1.0
Dilantin	8.7	105	1.7	1.4	1.9	78	126	81	88
Erythromycin-H_2O	2.1	83	<1.0	<1.0	<1.0	2.1	40	1.5	2.1
Estradiol	<1.0	146	<1.0	2.1	<1.0	<1.0	<1.0	<1.0	<1.0
Estriol	1.6	149	<1.0	<1.0	<1.0	<1.0	1.1	<1.0	1.1
Estrone	<1.0	169	<1.0	<1.0	<1.0	<1.0	<1.0	<1.0	<1.0
Ethynylestradiol	<1.0	164	<1.0	<1.0	<1.0	<1.0	8.1	<1.0	<1.0
Fluoxetine	<1.0	45	<1.0	<1.0	<1.0	2.1	<1.0	<1.0	<1.0
Gemfibrozil	5.5	125	<1.0	<1.0	<1.0	<1.0	<1.0	<1.0	<1.0
Hydrocodone	<1.0	81	<1.0	<1.0	<1.0	<1.0	75	2.6	<1.0
Ibuprofen	7.3	78	<1.0	<1.0	<1.0	1.2	<1.0	<1.0	<1.0
Iopromide	1.7	106	3.2	1.5	<1.0	5.4	41	8.5	6.2
Meprobamate	1.4	115	4.1	2.1	<1.0	54	54	33	30
Naproxen	5.0	50	<1.0	<1.0	<1.0	<1.0	<1.0	<1.0	<1.0
Oxybenzone	<1.0	35	<1.0	<1.0	<1.0	<1.0	<1.0	<1.0	<1.0
Pentoxifylline	<1.0	97	<1.0	<1.0	<1.0	<1.0	<1.0	<1.0	<1.0
Progesterone	<1.0	70	<1.0	<1.0	<1.0	<1.0	<1.0	<1.0	<1.0
Sulfamethoxazole	2.0	20	10	7.2	3.7	86	82	68	71
TCEP	59	144	53	103	85	49	64	44	53
Testosterone	<1.0	124	<1.0	<1.0	<1.0	1.1	<1.0	<1.0	<1.0
Triclosan	2.2	43	<1.0	<1.0	<1.0	<1.0	3.4	<1.0	<1.0
Trimethoprim	1.2	131	<1.0	<1.0	<1.0	<1.0	<1.0	<1.0	<1.0

Table 11.5
Comparison of riverbank pilot to batch tests (% removal)

	Column 1	Column 2	Column 3	Batch
Acetaminophen	>98	>98	>98	>99
Androstenedione	>99	>99	>99	>99
Atrazine	-10	10	4	58
Caffeine	95	97	98	92
Carbamazepine	-3	22	13	57
DEET	87	93	91	57
Diazepam	-8	42	65	83
Diclofenac	67	>99	>99	74
Dilantin	-11	28	22	79
Erythromycin	56	98	98	84
Estradiol	>99	>99	>99	>99
Estriol	99	>99	99	>99
Estrone	>99	>99	>99	>99
Ethynylestradiol	95	>99	>99	79
Fluoxetine	>99	>99	>99	>99
Gemfibrozil	>99	>99	>99	>99
Hydrocodone	8	97	>99	49
Ibuprofen	>99	>99	>99	>99
Iopromide	64	93	95	33
Meprobamate	53	71	74	48
Naproxen	>98	>98	>98	>98
Oxybenzone	>97	>97	>97	>99
Pentoxifylline	>99	>99	>99	>99
Progesterone	>99	>99	>99	NM
TCEP	18	44	32	54
Testosterone	>99	>99	>99	>99
Triclosan	92	>98	>98	>99
Trimethoprim	>99	>99	>99	55

Simulated Aquifer Storage and Recovery (ASR) (The following pages were previously published in Snyder et al. 2004).

Results for the simulated ASR experiment are shown in Figure 11.2 for specific target compounds and fluoride. Fluoride acted as a conservative tracer and indicated the travel time within the column system to be approximately 24 days. This contact time was sufficiently long enough to attenuate and remove oxybenzone, ethynyl estradiol, and testosterone. TCEP, carbamazepine, and DEET showed minimal attenuation with similar breakthrough profiles as fluoride. The profile for meprobamate showed an initial 50% reduction followed by a slower attenuation.

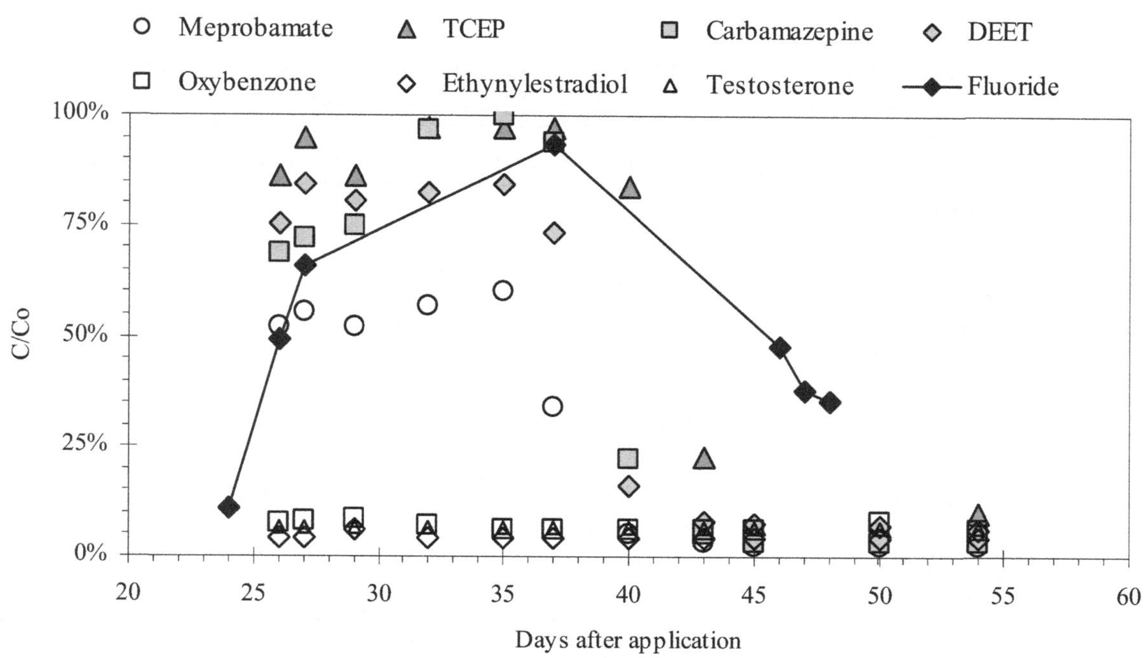

Figure 11.2 Simulated aquifer storage and recovery. Source: Snyder et al. 2004. Copyright 2004 National Ground Water Association. Reprinted with permission.

FIELD STUDIES

Two golf courses irrigated with reuse water were studied to determine if turf applications led to contamination of nearby groundwater. A wastewater treatment facility that uses rapid infiltration basins seasonally was also tested to determine the degree of attenuation of detectable target compounds along a subsurface flow path. Field observations were limited to detectable compounds and various non-biological removal or concentration effects that may impact data interpretations are discussed. Additional details on these field sites can be found in Snyder et al., 2004.

The field sites are explained in detail in Chapter 3 and analytical results are summarized in Tables 11.6 – 11.8. In all cases, only those compounds detectable in at least one sample from each individual site are shown. The Desert Rose results (Table 11.6) indicate rapid attenuation of some target compounds (meprobamate, dilantin, carbamazepine, and DEET) with questionable removal of other compounds that were near the method reporting limits (acetaminophen, caffeine, and fluoxetine). Oxybenzone appears to increase in Desert Rose wells; however, this may be an artifact of sunscreen use by sample collectors. Wildhorse golf course results are presented in Table 11.7. Based on monitoring data, the feed to this golf course was assumed similar to that of the Henderson WWTF ponds. For the purposes of approximation, this assumption was considered justifiable given the broad similarity of waters at the Desert Rose feed and the Henderson WWTF ponds shown in Tables 11.6 and 11.8, as well as previous monitoring of regional WWTFs (data not shown). While some compounds have been shown to photodegrade, it is unlikely that photodegradation plays a significant role here, since influent flows are relatively rapid and the water has high turbidity.

Table 11.6
Analytical results from Desert Rose Golf Course monitoring (ng/L)*
(Data from Snyder et al. 2004)

	DRose Feed	DRose	DRose 10G	DRose 11G
Acetaminophen	<10	24	<10	<10
Caffeine	18	<10	<10	<10
Carbamazepine	128	31	<10	15
DEET	193	55	<10	12
Dilantin	172	<10	<10	<10
Fluoxetine	20	<10	<10	<10
Meprobamate	641	60	<10	<10
Oxybenzone	<10	391	2580	380
Chloride (mg/l)	222	484	770	578

* DRose, DRose 10G, and DRose 11G are monitoring wells. DRose is directly down gradient from the reservoir pond that receives irrigation feed from a nearby City of Las Vegas wastewater treatment facility.

Table 11.7
Analytical results from Wildhorse Golf Course monitoring (ng/L)*
(Data from Snyder et al. 2004)

	Soil 15cm	Soil 45cm	Soil 75cm	Soil 105 cm	FR-4GV	FR-1GV	MW8-GV
Acetaminophen	<10	<10	<10	15	<10	<10	<10
Caffeine	<10	36	18	112	<10	<10	<10
Carbamazepine	53	13	<10	<10	<10	<10	<10
DEET	<10	22	25	15	<10	<10	<10
Estriol	<25	<25	<25	175	<10	<10	<10
Oxybenzone	476	852	515	1590	<10	<10	<10
Progesterone	<10	<10	14	27	<10	<10	<10
Triclosan	<25	<25	53	298	<25	<25	<25
Chloride (mg/L)	863	668	NM	4692	1235	733	351

Note: drinking water system chloride concentration is ~78 mg/l
* FR-4GV, FR-1GV, and MW8-GV are monitoring wells down gradient from the Wildhorse golf course. Soil samples are from 4 tensiometers installed at a sampling and measuring site.

Table 11.8
Analytical results from Henderson WWTF (ng/L)* (Data from Snyder et al. 2004)

	HPond 4	HMW 14	HPond 9	PC105
Carbamazepine	182	206	215	237
DEET	21	94	27	90
Dilantin	201	190	187	219
Erythromycin	87	<10	86	<10
Gemfibrozil	1670	703	1780	951
Ibuprofen	34	33	28	46
Iopromide	11	<10	12	<10
Meprobamate	1270	773	1280	80
Oxybenzone	<10	<10	<10	20
Progesterone	74	63	83	<10
Sulfamethoxazole	<10	24	<10	72
Triclosan	<10	<10	<10	24
Trimethoprim	12	<10	52	<10
Chloride (mg/L)	369	450	331	425

* HPond 4 and HPond 9 are evaporation/percolation ponds for the Henderson WWTF, while HMW 14, and PC105 are monitoring wells down gradient from these ponds

By assuming similar feed, it appears that detectable target compounds are rapidly removed within the first 15 cm of soil with the possible exceptions of carbamazepine and oxybenzone. It must be noted that contact with disinfectant chloramines in the distribution system may have altered the composition of target compounds in the feed water as compared to the WWTF ponds. Data presented in Table 11.8 indicate that meprobamate and gemfibrozil were both degraded by more than 50% during subsurface flow. Trimethoprim and erythromycin appear to have been degraded, though the extent is difficult to quantify because of lower concentration in source water. Sulfamethazole, dilantin, carbamazepine, DEET, progesterone, and ibuprofen were not significantly degraded along the flow path.

Golf course analyses must be interpreted more carefully in light of other removal/dilution processes. To estimate target compound attenuation within the saturated zone it is necessary not only to establish transit times, initial concentrations, and final concentrations, but also to identify any other processes that potentially might have impacted contaminant concentrations. At the study sites, processes that could impact contaminant concentrations included evapotranspiration (ET) that can concentrates solutes, and dilution by other waters that lack EDCs and PPCPs. Dilution by native groundwater was found to be negligible, but some dilution by non-reuse (potable) water probably took place around the Wildhorse monitoring wells, which are located among single-family residences, which use potable for landscape irrigation. Table 11.9 summarizes some of these conditions for the field sites investigated.

Chloride ion (Cl^-) data were used to quantify the effects of dilution and ET, under the assumption that Cl^- mass is conserved. This is realistic given the high solubility of chloride salts. Changes in Cl^- concentration between samples along the same flow path therefore arise either from loss of water due to ET, or through mixing with waters that contain different Cl^- concentrations. If conservative, EDCs and PPCPs would behave similarly. However, any attenuation of these compounds would manifest itself as a change in the ratio of that compound's concentration to Cl^- concentration.

Beneath the golf courses, where groundwater was presumed to derive solely from irrigation application, Cl^- increases result from consumptive water use by vegetation and target compound attenuation can be measured directly. At the WWTF; however, Cl^- bearing groundwater is present that derives in part from industrial application and in part from native groundwater {Carlsen, 1991 #1194}. These studies reported Cl^- concentrations up gradient of the Henderson ponds from a low of ~1100 to a maximum of ~6100 mg/L, with an average of ~3500 mg/L. Groundwater away from the impacted industrial area ranged from ~400 to ~1400 mg/L and averaged ~750 mg/L.

Prior to assessing EDC and PPCP attenuation, the effect of ET and dilution was determined. Table 11.9 indicates that Cl^- increases between the ponds and monitor wells are ~22% and ~29%. These increases may result from ET or from mixing with groundwater. They are consistent with calculated evaporative losses based upon flow to the ponds and an evaporation rate of 6 ft/y over a 10-acre (~4 ha) pond area.

Table 11.9
Best estimates of hydraulic properties and transit times*

Hydraulic Property	Desert Rose	Wildhorse	WWTF
Gradient (ft/ft)	0.022	0.02	0.4
Effective Porosity (%)	20	20	20
Hydraulic Conductivity (ft/d)	50	5	20
Avg. Linear Velocity (ft/d)	5.5	0.5	40
Transport Distance (ft)	180	700	50
Transport Time (d)	33	1400	1.25

* Calculations assume Darcy flow. Transport time is based upon average linear velocity and does not address hydrodynamic dispersion
Source: Snyder et al. 2004. Copyright 2004 National Ground Water Association. Reprinted with permission.

For binary mixing, the fraction of native groundwater (X_n) in a sample can be calculated provided tracer concentrations in the source waters are known:

$$X_n = \frac{(C_n - C_p)}{(C_n - C_s)} \qquad (11.1)$$

where C_s, C_n, and C_p are the tracer concentrations in the sample, native groundwater, and pond water, respectively. Using equation 11.1 in conjunction with average Cl^- concentrations, native groundwater was found to constitute between ~3% and ~15% of the water in both monitor wells at the WWTF. Similar calculations using TDS provided an estimate of ~15%. This fraction is small compared to the uncertainties involved in hydraulic conductivity values and estimates of transit time between the ponds and the monitor wells.

Transit times between sample stations on the same flow path were assessed using a Darcy's Law calculation of the average linear velocity {Freeze, 1979 #24}:

$$V_{avg} = \frac{K \nabla h}{\varepsilon} \qquad (11.2)$$

where V_{avg} is the average linear velocity, K the hydraulic conductivity, ∇h (grad-h) the water table gradient, and ε the effective porosity. Transit time was obtained simply by dividing the distance between stations by V_{avg}. This methodology does not account for hydrodynamic dispersion. However, all of the sites have operated for at least several years under approximately the same conditions, and likely have approached steady-state. Moreover, the magnitude of uncertainty in hydraulic conductivity exceeds typical field dispersivity values reported by {Neumann, 1990 #26} for the distance scales applicable at the study sites. Table 11.9 lists the hydraulic parameters and calculated transport times between the source water and a characteristic sample point at each site. These times are on the order of hundreds of days at Wildhorse, tens of days at Desert Rose, and one day at the WWTF. It should be emphasized that in the presence of steep variable gradients, and in an unconfined potentially heterogeneous porous medium of uncertain hydraulic conductivity, travel time calculations based on equation 11.2 are approximations.

Transit time calculations indicate that solute concentration increases due to evaporation are negligible at the WWTF. At the golf courses, ET is significant when applied water passes through the vadose zone (Tables 11.6 & 11.7). For instance, Table 11.7 demonstrates a 5-fold increase in Cl^- content as water moves from the near surface to ~3 feet (105 cm) below the surface. Because the water table is shallow, it is likely that this Cl^- concentration typifies those in the saturated zone beneath the Wildhorse golf course. Though a detailed analysis was beyond the scope of this investigation, the limited data presented in this chapter suggest biological attenuation of most target compounds during passage through the soil. With the exception of triclosan, none of these compounds show proportionally increased concentrations and some are removed to less than the method report limits.

At Wildhorse, none of the compounds detected in the pond water, which is assumed a representative source, were detectable in water from down gradient monitor wells. This apparent removal may result from dilution because Cl^- concentrations in all of the wells are less than that at the 3.4-foot (105 cm) level in the soil at the vadose zone sample site. All wells were in residential areas irrigated by potable water, not reuse water. If it is conservatively assumed that dilution is from potable water (78 ppm Cl^-), then using equation 11.1 only ~8% to ~25% of the water in the wells originates from the reuse water applied on the golf course. Because most of these compounds are not present in the drinking water system, the introduction of potable water causes dilution of EDCs and PPCPs at the vadose zone site to below detection levels, with the exception of triclosan and oxybenzone. If the diluent waters were residential leaching fractions typified by water in well MW8-GV, then the respective proportions of reuse water would range from ~20% to <1%. In either case, triclosan and oxybenzone would be expected in well FR4-GV, where none were detected. This suggests that these compounds, and potentially other EDCs and PPCPs, attenuate beyond the extent accounted for through simple dilution. The distance in which attenuation takes place is thus less than ~700 feet (~214 m), which is the distance between the edge of the golf course and well FR4-GV. Using this distance and a reasonable average linear velocity gives a residence time of ~1400 days maximum (Table 11.9).

Desert Rose samples suggest greater attenuation rates and shorter attenuation distances. Transit time is on the order of one month between the irrigation reservoir and well DROSE, which is ~180 feet (~55 m) distant. Along that flow path, levels of all detected compounds except oxybenzone dropped. Note that target compound concentrations from all the wells are lower than in the feed. This suggests the possibility of attenuation within vadose zone soils as opposed to within sediment below the water table. This possibility was tested using equation 11.1 and the Wildhorse soil data. At Desert Rose, steady-state flow can be presumed between the pond and well DROSE. Water migrating along this path encounters the downward-migrating leaching fraction from overlying irrigated turf. If ET induced a 5-fold Cl^- increase within the leaching fraction, as suggested at Wildhorse, then water at well DROSE consisted of ~70% pond water and ~30% leaching fraction. Even if soil processes were to have removed virtually all EDCs and PPCPs, a 30% dilution would be insufficient to account for the observed concentrations at well DROSE. As indicated in Table 11.6, meprobamate, dilantin, carbamazepine, and DEET concentrations were between ~6% and ~57% of those in the feed. Fluoxetine levels may have been similarly reduced. It was present at only twice the detection level in the feed, so no concentration reduction beyond ~50% is discernable. Thus, even if target compound dilution takes place, these figures strongly suggest attenuation over distances of hundreds of feet and timeframes of tens of days.

At the WWTF, transit times between the ponds and the wells may be on the order of one day (Table 11.9). Because of the small component of native groundwater in the wells (~3% to ~15%), between 85% and 97% of the reduction in certain target compound levels seen at the wells is therefore attributable to attenuation. If residence times in the ponds are sufficiently long that evapoconcentration is significant, then attenuation accounts for an even greater proportion of apparent target compound losses. Compounds that are attenuated include erythromycin, meprobamate, progesterone, and triclosan. The percentages of attenuation range from ~5% to ~40%. Dilantin levels are relatively unchanged, while both carbamazepine and DEET show unexpected increases. If these increases result from reversible processes (i.e., deconjugation of metabolites) that take place within the hyporrheic zones beneath the ponds, then the observed attenuation of other compounds may likewise be reversible. However, if such hypothetical processes minimally impact the other compounds, then EDC and PPCP attenuation begins almost immediately and proceeds rapidly. Further investigation at down gradient well sites and within the ponds is necessary to better characterize EDC and PPCP compound behavior at these and other sites.

RESULT SUMMARY AND DISCUSSION

BDOC and simulated ASR experiments demonstrated that some EDC and PPCP compounds can rapidly biodegrade while others are much more resilient to degradation. Although attempts were made to distinguish between adsorption and biological degradation, more involved experimentation would be necessary to thoroughly separate these removal processes. In general, BDOC data indicated less rapid contaminant degradation than was observed in actual ASR monitoring (Table 11.10). Differences in substrate availability and contact time may likely explain the differences observed. For some compounds (marked as ambiguous in Table 11.10), the spiked compound concentration was too close to the method reporting limits, which restricted interpretation of compound loss. However, for many compounds, 1 to 2-log removal was observable.

A generalized comparison of the field data to the laboratory and QSAR data (from BIOWIN software – Chapter 12) are presented in Table 11.10. Field observations were limited to those target compounds present at detectable concentrations in the feed water. For some compounds, concentrations in the feed/source waters were close to the limit of detection, hindering interpretation of disappearance. Conversely, some compounds were not detected in feed/source waters, yet appeared at levels just above the detection limits in monitoring wells or soils. This is likely the result of feed/source water and analytical variability. Lower detection limits would undoubtedly aid in the interpretation of these data. While all laboratory and QSAR data are based on aerobic degradation, it is possible that anaerobic degradation may play a role for the water infiltrating the bottom of the WWTF ponds.

Table 11.10
Summary data for laboratory and field experiments*

	Batch	ASR	Field	QSAR
Acetaminophen	R	R	AMB	R
Androstenedione	AMB	AMB	ND	NP
Atrazine	S	M	ND	M
Caffeine	R	R	AMB	NP
Carbamazepine	S	M	M	M
DEET	M	M	M	R
Diazepam	S	R	ND	R
Diclofenac	S	AMB	ND	S
Dilantin	S	M	M	R
Erythromycin	S	AMB	S	S
Estradiol	R	R	ND	R
Estriol	M	R	AMB	R
Estrone	R	R	ND	M
Ethynylestradiol	M	R	ND	M
Fluoxetine	S	R	M	M
Gemfibrozil	R	R	R	R
Hydrocodone	M	R	ND	NP
Ibuprofen	R	R	AMB	R
Iopromide	S	R	AMB	S
Meprobamate	M	M	R	M
Naproxen	M	R	ND	NP
Oxybenzone	M	R	AMB	R
Pentoxifylline	S	R	ND	NP
Progesterone	AMB	AMB	S	NP
Sulfamethoxazole	S	S	AMB	M
Testosterone	M	R	ND	M
Triclosan	AMB	AMB	M	S
Trimethoprim	M	R	M	M

* R = rapidly degraded; M = moderately degraded; S = slowly degraded; ND = not detected; NP = not predicted; AMB = ambiguous

Source: Snyder et al. 2004. Copyright 2004 National Ground Water Association. Reprinted with permission.

REFERENCES

Akkanen, J., S. Penttinen, M. Haitzer, and J. V. K. Kukkonen. 2001. Bioavailability of atrazine, pyrene and benzo[a]pyrene in European river waters. *Chemosphere* 45(4-5): 453-462.

Carlsen, C. L., R. C. Lunis, and D. E. Prudic. 1991. Changes In Water Levels And Water Quality In Shallow Ground Water, Pittman - Henderson Area, Clark County, Nevada, Resulting From Diversion Of Industrial Cooling Water From Ditch To Pipeline In 1985. Presented on the Las Vegas Wash Coordination Committee website: www.lvwash.org. United States Geological Survey data.

Drewes, J., T. Heberer, and K. Reddersen. 2001. Removal of pharmaceuticals during conventional wastewater treatment, advanced membrane treatment and soil-aquifer treatment. In *2nd International Conference on Pharmaceuticals and Endocrine Disrupting Chemicals in Water*. Minneapolis, Minnesota.

Drewes, J. E., P. Fox, and M. Jekel. 2001. Occurrence of iodinated X-ray contrast media in domestic effluents and their fate during indirect potable reuse. *J. Environ. Sci. Health Part A-Toxic/Hazard. Subst. Environ. Eng.*, 36(9): 1633-1645.

Freeze, R. A., and J. A. Cherry. 1979. Groundwater. Prentice-Hall: Englewood Cliffs, New Jersey.

Neumann, S. P. 1990. Universal scaling of hydraulic conductivities and dispersivities in geologic media. *Water Resources Research* 26(8): 1749-1758.

Snyder, S. A., J. Leising, P. Westerhoff, Y. Yoon, H. Mash, and B. J. Vanderford. 2004. Biological attenuation of EDCs and PPCPs: Implications for water reuse. *Ground Water Monit. Rem.* , 24(2): 108-118.

CHAPTER 12
COMPUTER MODELING

INTRODUCTION

EDCs and PPCPs represent a spectrum of structures with dramatically diversified features and quite different physico-chemical properties. This may result in different fate and removal by various water treatment processes. It is unrealistic to perform actual treatment experiments for every possible compound from the large candidate pool. As a result, Quantitative Structure-Property Relationship (QSPR) modeling is explored as a potential tool to prescreen compounds and help focus research on more persistent compounds during typical drinking water disinfection processes. In this work QSPR has been investigated for the removal of target EDCs and PPCPs. Two partitioning processes (e.g., air-water and water-carbon) and one membrane filtration process (ultrafiltration UF) have been investigated as nondestructive drinking water removal mechanisms; while biodegradation, hydrolysis, and oxidation (chlorine, ozone) have been investigated as destructive removal mechanisms. The QSPR approach utilized for predicting fate properties for EDCs and PPCPs in water treatment processes are summarized in Table 12.1.

For partitioning processes, the central modeling parameter is the activity coefficient of the chemical solute in water. EDCs and PPCPs typically occur as dilute solutions in contaminated drinking waters. A model of the dilute solution activity coefficient can be used with a few simple thermodynamic relationships to predict environmental quantities such as Henry's constant, aqueous solubility, and the octanol-water partition coefficient (log K_{ow}). Basic thermodynamic relationships are used to directly estimate liquid-liquid miscibility for solubility calculations. Liquid-phase adsorption onto microporous activated carbon can be predicted using the Polanyi universal isotherm theory based on pore-filling (Polanyi 1920; Ashworth et al. 1988) with reference to a UNIFAC-generated estimate of the solute's adsorption potential in saturated water. UNIFAC (UNIQUAC Functional-group Activity Coefficient) is a QSPR that has been used successfully to model the activity coefficients of hydrophobic chemicals in dilute solutions (Fredenslund, Jones, and Praunitz 1975). A special set of UNIFAC functional group interaction parameters has been developed specifically for dilute aqueous mixtures (Rogers 1994). For stripping processes, the air-water partition coefficient (i.e., Henry's law constant) can be obtained directly from the structure-based UNIFAC model by multiplying the calculated infinite dilution activity coefficient by the solute's vapor pressure.

Accepted procedures from the literature were examined for biodegradation tendency (Howard-Boethling's method) and elevated-pH hydrolysis rate (Taft's method). Biodegradation potential is predicted based on fragmentation approach which provides qualitative information on the tendency of the target compound to go through biodegradation reactions. For hydrolysis, various semi-empirical approaches have been discussed in terms of their applicability, advantages and disadvantages.

Table 12.1
Overview of predictive methods for fate and transport properties

Fate and Transport Property	Chemical Information	Approach	Data Input
Biodegradation rate (or tendency)	Substructure fragments	Howard-Boethling group contribution	Functional group correlation parameters
Hydrolysis rate constant at basic pH Electronic and steric model (Taft equation)	Steric (*Es*) and electronic (σ^*) parameters of tabulated chemicals	Multi-variable linear Regression (MLR) analysis of tabulated data	Correlation parameters (group contribution) optimized
Percent removal by: Ozone, chlorine, UF	Adsorption, distribution and metabolism data	Monte Carlo (MC) statistical mechanics simulations and MLR analysis	2D molecular structures
Air-water partitioning (Henry's law constant)	Structure fragments	Activity coefficient model (ie., UNIFAC) with thermodynamic relationships, or Hine-Mookerjee as 2nd option	Vapor pressure and BIP
Adsorption onto activated carbon	Substructure fragments	Liquid-phase Polanyi isotherm model	Aqueous solubility limit, molar volume at Tb, liquid density

The QSPR investigation of ozonation, chlorination, and ultrafiltration (UF) processes use a molecular modeling approach. It is different from the majority of previous QSPR research in that all descriptors, if not experimentally available, were calculated based on 3D molecular structures. This approach allows more accurate prediction of physicochemical properties and the inclusion of molecular descriptors. The research described here utilized Monte Carlo (MC) statistical mechanics simulations provided by commercial software (Schrödinger) to generate accurate 3D molecular descriptors and physicochemical properties for QSPR development. These descriptors are then used in multiple linear regressions to determine their relevance to ozonation, chlorination, and UF removal and for the further development of 3D QSPR models.

ADSORPTION ONTO ACTIVATED CARBON

Basic Theory

The Freundlich equation was used to describe the equilibrium relationship of sorbate *i* between its solution phase concentration C_i and the solid phase concentration q_i.

$$q_i = K_i C_i^{1/n} \qquad (12.1)$$

where K_i and 1/n are empirical constants, with the former related to adsorbent capacity and the latter related to adsorbent intensity. These two constants can be obtained either via bench-scale isotherm experiments or via QSPR models. Two models were considered as predictive tools, with their underlying theories and prediction approaches briefly explained here.

Polanyi Theory

Isotherms derived from Polanyi potential theory are useful for describing adsorption by pore filling of micro porous adsorbents (e.g., GAC). This theory assumes a thin layer of fixed volume, refined by equi-potential contours above the surface, exists on the surface of the adsorbent and can accommodate the adsorbed molecules via weak interactions with the surface (Polanyi 1920). The empirical relationship between the potential field and the adsorbed volume above the surface is called the Polanyi universal isotherm. Polanyi adsorption potential theory defines a single characteristic curve for a specific GAC and single solute isotherm data. The Polanyi uniform adsorbate isotherm takes the following general form:

$$W = W_0 \exp[-B(\varepsilon/V_m)^G] \qquad (12.2)$$

where W is the volume adsorbed (ml/g), W_0 is the maximum pore volume for adsorption (ml/g), ε is the adsorption potential defined as the difference between the chemical potentials of the sorbate in freely dissolved phase and in adsorbed state, V_m is the adsorbate liquid molar density at boiling point T_b. B $(ml/cal)^G$, G, and W_0 are fitted Polanyi parameters.

Some important features of this theory are:

- Considers the interactions between water molecules, water molecules and the adsorbate, and water molecules and the adsorbent
- Can describe nonlinear isotherm behavior (favorable and unfavorable)
- Parameters are available for several commercial activated carbons and synthetic resins

Researchers previously have used this theory to correlate adsorption isotherms of gases, mixed vapors, and organic compounds. Table 12.2 presents some representative experimental and predicted Freundlich isotherms for F-400 carbon (Sontheimer, Crittenden, and Summers 1988). The comparison between the experimental and predicted values illustrates the application potential of Polanyi theory in chemical adsorption predictions.

Table 12.2
Experimental and predicted Freundlich parameters for F-400 carbon

Compound	Freundlich K		Freundlich 1/n	
	Experimental	Predicted	Experimental	Predicted
Chloroform	30.4	40.9	0.5325	0.5756
Trichloroethane	191.7	186.1	0.4328	0.4857
Ethylbenzene	714.4	706.2	0.2953	0.3211
o,p-Xylene	894.6	862.1	0.2587	0.3026
cis1,2-Dichloroethene	46.9	48.4	0.5562	0.6047

Prediction of EDC/PPCP adsorption on Calgon F-400 liquid phase powdered active carbon was conducted using this theory. Using the Polanyi isotherms, Freundlich parameters for different target contaminants were obtained.

Approaches for Isotherm Development

The Polanyi model is not applicable to chemisorption or whenever ion exchange is important. Neither the displacement effect nor competition from NOM on the solute adsorption onto the surface of the adsorbent is considered.

Adsorption isotherms for selected EDC/PPCPs were generated based on Equation 12.2 following the procedures outlined below:

1. Generate adsorption potentials for the adsorbate over the bulk concentration range of interest according to equation 12.3.

$$\varepsilon = RT\ln(\frac{C_i^S}{C_i}) \qquad (12.3)$$

This requires the aqueous solubility limit, experimental or predicted. The compounds were assumed neutral for this step.

2. Obtain the molar volume at T_b of the adsorbing chemical. Use data or the group contribution method of Shroeder (Lyman, Reehl, and Rosenblantt 1982)
3. Correct the adsorption potentials by dividing them by the molar volume normalizing parameter.
4. Use a Polanyi isotherm to generate a series of adsorption capacities (adsorbed volume) versus bulk concentration.
5. Convert to mass loading by multiplying the adsorbed volumes by the adsorbate's liquid density.
6. Fit the equilibrium mass loadings by linear regression to the Freundlich isotherm equation.

The Polanyi parameters (Table 12.3) for WPM (a powdered form of F-400 carbon), required for step 4 calculations, were determined from adsorption isotherms (Figure 12.1) provided by Calgon Corporation using non-linear regression methods.

For this evaluation, bench-scale experiments were performed for five selected EDCs to obtain their adsorption isotherms (Yoon et al. 2003). Selected contaminants were adsorbed on WPM for 2 hours before the concentrations were measured. The adsorption isotherm for each compound was determined using at least six concentrations. The isotherms were fitted to derive the corresponding Freundlich parameters K and 1/n, and compared with predictions from the Polanyi model following the procedures outline above.

Table 12.3
WPM Polanyi parameters obtained by non-linear regression

Wo	1.6645
B	0.0868
G	1.2058

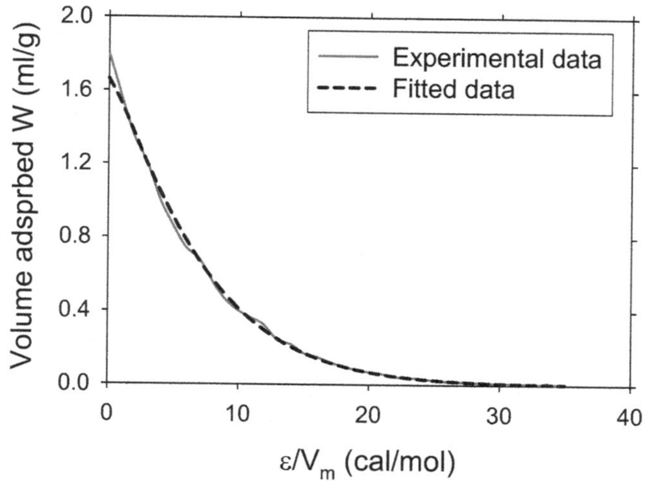

Figure 12.1 Polanyi universal isotherm for WPM activated carbon. Solid line is the experimental data obtained at 25°C. Dashed line is the regression analysis.

Results from QSAR/QSPR

Solubilities of the five selected target compounds were first estimated using a method developed by Meylan et al. and are listed in Table 12.4 (Meylan, Howard, and Boethling 1996). Based on the solubility, molar volume, and bulk concentration of each compound, linear least square regressions were conducted to obtain the equilibrium Freundlich parameters and they are presented in Table 12.5. The compounds listed in Table 12.4 were all neutral at ambient pH.

Assuming the Freundlich isotherm (equation 12.1) is valid for the target analyte, bench-scale adsorption results are presented in Figure 12.2 by plotting solid phase concentration versus bulk liquid phase concentration in logarithmic scales and compared with predictions from Polanyi model. Figure 12.2 demonstrates that the values calculated using Polanyi theory and those obtained experimentally in distilled water follow similar trends, but the calculated capacity values are higher than those obtained experimentally. This can be explained in terms of equilibrium and non-equilibrium behavior. The calculated values are from the Polanyi isotherm, which was obtained at an equilibrium adsorption condition (24 h), whereas the experimental adsorption was done at non-equilibrium conditions (2 h). Also, the difference between the equilibrium and non-equilibrium values of a contaminant depends on the solubility of that particular compound. Contaminants with greater water solubility show less difference between equilibrium and non-equilibrium values. Since higher solubility contaminants will have a higher driving force, they will more closely approach equilibrium in a 2-hour timeframe than compounds with lower water solubility. Batch adsorption experiments performed for bisphenol A and ethynyl estradiol under equilibrium conditions (24 hours) have confirmed the non-equilibrium behavior observed for 2-hour experiments. The experimental adsorptions for these two chemicals obtained at 24 hours are predicted well by Polanyi theory.

Table 12.4
Estimated solubility for target compounds

Compound	Water Solubility
Bisphenol A	120 mg/L
17β-estradiol (E2)	3.6 mg/L
17α-ethynyl estradiol (EE2)	11.29 mg/L
Fluoranthene	0.26 mg/L
Galaxolide	0.089 mg/L

Table 12.5
Comparison of Freundlich parameters for predicted equilibrium and bench-scale non-equilibrium conditions

Chemical	Experiment		Model	
	K	1/n	K	1/n
Bisphenol A (BPA)	0.303	0.30	8.649	0.43
17β-estradiol (E2)	0.968	0.35	10.482	0.33
17α-ethynyl estradiol (EE2)	0.240	0.27	9.301	0.23
Fluoranthene	2.054	0.34	13.938	0.30
Galaxolide	4.630	0.48	11.586	0.27

Based on the above observations, Polanyi theory was then used to predict Freundlich parameters for other target compounds for which no experimental data are available (see Appendix Table F.1). By examining the predicted values against some commonly used physico-chemical properties such as solubility and log K_{ow}, no apparent dependence was observed. However, those with a solubility of above 10^4 mg/L tend to have a low Freundlich constant (K) while the rest do not follow any simple pattern. In general, compounds with a higher K_{ow} tend to have higher adsorption on GAC with quite a number of exceptions. Based on the predicted adsorption, most compounds should have at least a moderate adsorption capacity on GAC, which makes GAC a good option for the removal of these microcontaminants. The experimental data presented in Chapter 5 confirm this observation, showing satisfactory removal for most target analytes (Table 5.1).

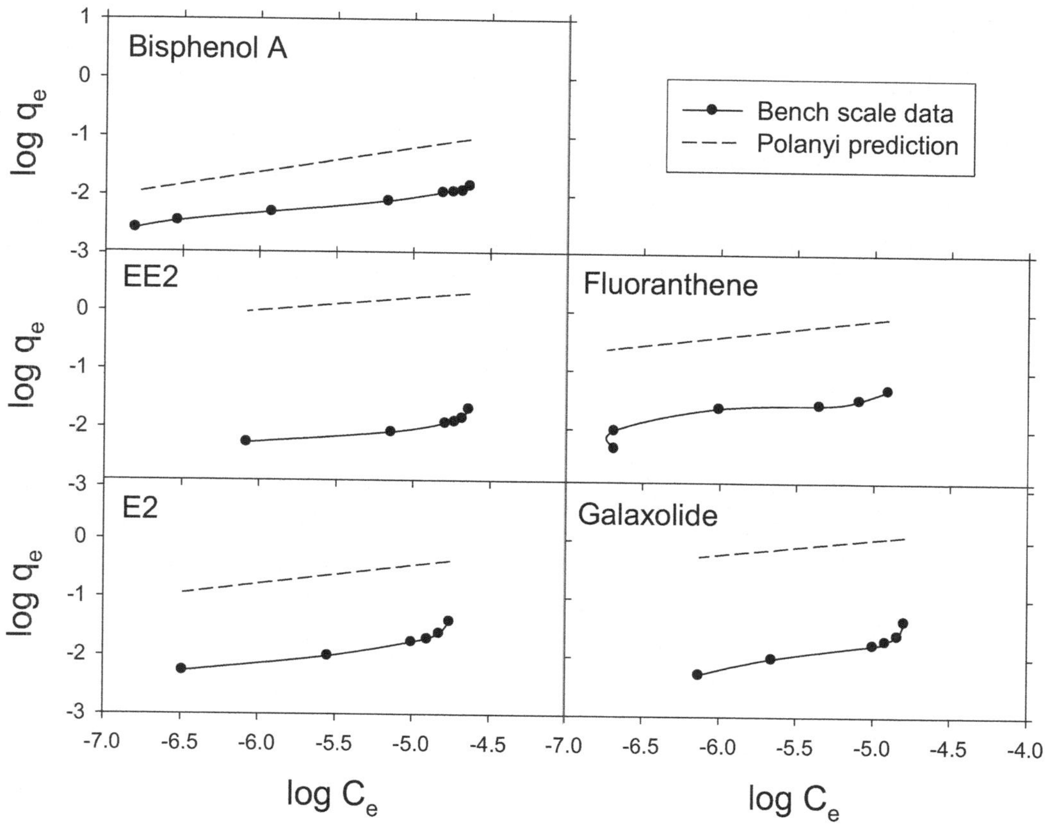

Figure 12.2 Comparison of equilibrium Polanyi predictions and bench-scale data on WPM PAC

AIR STRIPPING

Basic Theory

Henry's law is an important factor in defining partitioning of partially soluble species between air and water phases. Henry's constant is a measure of the equilibrium distribution of a chemical in air relative to that dissolved in aqueous solution. It governs the condition in which a very dilute species reaches partitioning equilibrium between air and water, and is strongly temperature dependent. Its value generally increases with increasing temperature and the consequence of this behavior is that the solubilities of gases often decrease as the temperature rises. Knowledge of Henry's constant values is useful in evaluating drinking water systems for the removal of chemicals by volatilization.

Henry's law constant (H_c) can be obtained by applying the Lewis and Randall rule and multiplying the infinite dilution activity coefficient (γ_i^∞) by the vapor pressure (P_i^S) of the pure solute:

$$H_C = \gamma_i^\infty \cdot P_i^S \qquad (12.4)$$

The quantity γ_i^∞ is obtained from an activity coefficient model, and the pure component vapor pressure of the organic solute comes from experimental data or a temperature-dependent correlation (e.g., Antoine's equation).

Choice of Prediction Method

The estimation of Henry's constant can be done using the following two methods:

1. *UNIFAC (UNIQUAC Functional-group Activity Coefficient) Model*: UNIFAC is a group contribution method for infinite dilution activity coefficient (γ_i^{∞}) calculations. It assumes γ_i to be the sum of two contributions: a combinatorial part due to differences in size and shape of group fragments, and a residual part due to energy interactions between groups:

$$\ln \gamma_i = \ln \gamma_i^{Combinatorial} + \ln \gamma_i^{Residual} \qquad (12.5)$$

where: $\ln \gamma_i^{Residual} = f$ (functional group binary interaction parameter)

2. *Method of Hine & Mookerjee* (Hine and Mookerjee 1975): This method uses the Log of the Water-Air Partition Coefficient (LWAPC) group contribution approach to predict the Henry's constant value based on two equations:

$$\mathrm{LWAPC} = \log(\frac{1}{Hc}), \quad Hc \text{ is unit-less} \qquad (12.6)$$

$$\mathrm{LWAPC} = \sum_k \gamma_k^{(i)} h_k \qquad (12.7)$$

where: h_k = contribution of group "k"(unit-less),

$\gamma_k^{(i)}$ = number of group "k" in molecule "i"

This method uses a diverse functional group library, and predicts the Henry's constant value directly without requiring any other data inputs. However, the UNIFAC method was used as the predictive method of choice in this study because the *Hine and Mookerjee method* is limited to predictions at 25°C.

Calculations

Air-Water Partitioning Calculation Procedure:

1. Perform activity coefficient calculation at infinite dilution by applying binary interaction parameters (BIPs) of substructure fragments from a tabulated library.
2. Obtain the pure component vapor pressure of the organic solute from experimental data or a temperature-dependent form (e.g., Antoine's equation).
3. Henry's law constant (H_c), or air-water equilibrium relationship, is obtained using Equation 12.4.

Henry's constants were predicted for 40 different EDCs/PPCPs for several sets of binary interaction parameters (VLE, LLE, ENV and GLB; see (Rogers 1994) for details) using the UNIFAC model. The predicted values are compared with literature values obtained from the Syracuse Research Corporation (SRC) physical properties database. Figure 12.3 shows a comparison between the experimental and predicted values for these 40 chemicals. Air-water partion values were calculated because quantities are more readily available from molecular modeling libraries, and they were considered potentially useful for some of the treatment processes.

Observations

The Henry's constant predictions for most of the 40 chemicals in Figure 12.3 agree reasonably well with the literature values. Predictions using the LLE, ENV, and GLB parameters are generally better than those from the VLE parameter set. However, a larger number of compounds can be predicted using the VLE parameters because its available fragment library is larger.

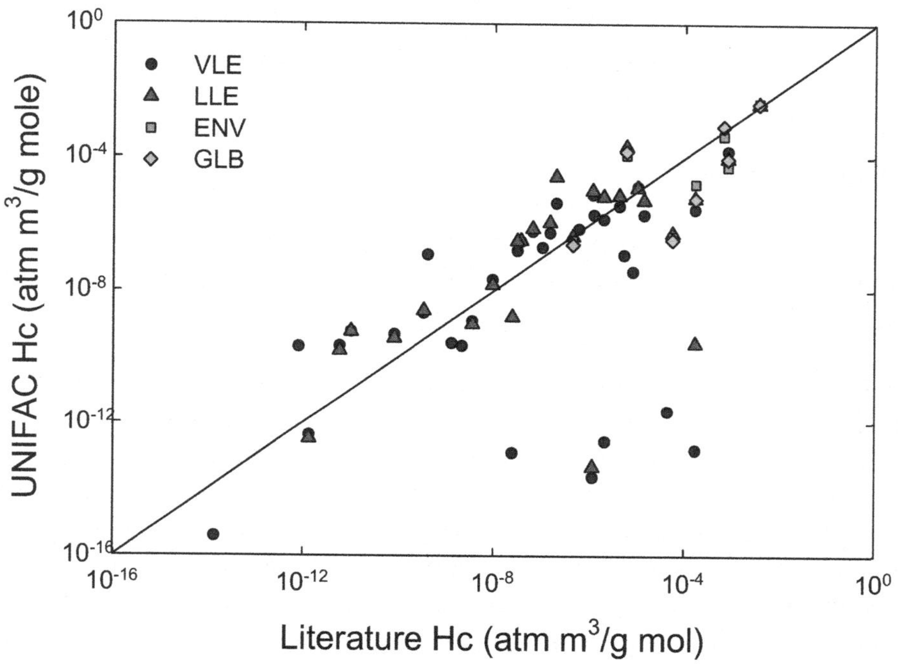

Figure 12.3 Comparison of predicted and experimental Henry's constants

BIODEGRADATION TENDENCY

Basic Theory

Biodegradability is represented by various parameters including half-lives, biodegradation rates and constants, theoretical oxygen demand (ThOD), biological oxygen demand (BOD), and others. Biodegradation of an organic chemical in a natural system may be classified as:

1. Primary biodegradation (alteration of molecular integrity)
2. Ultimate biodegradation (complete mineralization)
3. Acceptable biodegradation (toxicity ameliorated)

The linear and non-linear probability (Yi) of the primary biodegradability of an unknown compound can be determined by:

Linear model: $Y_i = a_0 + a_1 f_{11} + a_2 f_{12} + \ldots + a_i f_{ij} + e_j$ (12.8)

Non-linear model: $Y_i = \frac{e^{(a_0 + a_1 f_{11} + a_2 f_{12} + \ldots + a_i f_{ij})}}{1 + e^{(a_0 + a_1 f_{11} + a_2 f_{12} + \ldots + a_i f_{ij})}}$ (12.9)

where, Y_i = binary indicator for probability of biodegradation: 0 defined as slow biodegradation, 1 defined as fast biodegradation, and 0.5 serves as the point separating fast and slow biodegradation

f_{ij} = number of *i*th substructures in *j*th chemical

a_0 = equation intercept (or equation constant)

a_i = regression coefficient for *i*th substructure

e_j = error term

Prediction Methods

Two different methods for primary biodegradation are used in this study to calculate the probability that a chemical under aerobic conditions with mixed cultures of microorganisms will biodegrade rapidly or slowly:

1. *Syracuse Research Corporation (SRC) methods*: Implemented commercially as BIOWIN software, SRC uses fragment constants developed using multiple linear and non-linear regressions and data from SRC's database for biodegradation rates. Based on the fragment and molecular weight of the contaminant it can: a) estimate the probability of rapid biodegradation or display the experimental rate data, and b) estimate the primary and ultimate biodegradation times for complete degradation (days, weeks, months or longer) using the evaluations of 200 chemicals by 17 biodegradation experts (Boethling et al. 1994). The BIOWIN program also has two new predictive models for assessing a chemical's biodegradability in the Japanese MITI (Ministry of International Trade and Industry) biodegradation test.
2. *Howard-Boethling method*: This method involves calculating an index (Y_i) that characterizes aerobic biodegradation rate in an ambient environment (Howard et al. 1992; Boethling et al. 1994).

Both methods were selected in this study for biodegradation predictions with a critical assessment of each.

Calculations

Both methods were used to calculate the biodegradation probabilities of 40 candidate EDCs/PPCPs representing a variety of chemical types. The calculations are shown in Appendix Table F.2. Manual validation of the calculations has been performed for five representative EDCs.

In Table 12.7, the predicted probabilities for 10 chemicals have also been compared with the probabilities based on the experimental half-lives for the biodegradation of these chemicals reported in the Handbook of Environmental Biodegradation Rates (Hammett 1937). Classification of a chemical's biodegradation probability as fast, moderate, slow, and resistant, based on its biodegradation half-life is shown in Table 12.6.

Observations

The predictions of contaminant biodegradation from both methods are in good agreement. The values predicted by the Howard-Boethling method are close to values predicted by the BIOWIN program. However, the probability of biodegradation being fast or slow (i.e. whether the probability value is greater than or less than 0.5) is in reasonable agreement for both methods for 33 of the 40 test chemicals.

Chemicals that have halogens in their structure often have slow rates of biodegradation. In general, 23 of the test EDCs/PPCPs showed fast predicted biodegradation rates. A comparison between the biodegradation probabilities from prediction and experimental half-lives for 10 chemicals is shown in Table 12.7. As evident from Table 12.7, there is a good agreement between the biodegradation probability based on experimental biodegradation half-lives and the predicted values from the BIOWIN and Howard-Boethling methods.

Table 12.6
Classification of a chemical's biodegradation probability

Class	Low $t_{1/2}$	High $t_{1/2}$
Fast	1 day	7 days
Moderately fast	7 days	4 weeks
Slow	4 weeks	6 months
Resistant	6 months	1 year

Table 12.7
Comparison of biodegradation probabilities

Chemical Name	Biodegradation rate probability		
	Experimental	BIOWIN	Howard-Boethling
Trichlorofon	Fast to slow	Slow	Slow
Chlordane	Slow	Slow	Slow
Dimethoate	Medium fast	Fast	Fast
Dieldrin	Slow	Fast to slow	Slow
Amitrole	Slow	Slow	Slow
Carbyl	Fast to moderately fast	Fast	Fast
Chloroform	Slow	Slow	Slow
Methoxychlor	Slow	Slow	Slow
DDE	Slow	Slow	Slow
Heptachlor	Slow	Slow	Slow

Background

Hydrolysis Rate Constant at Basic pH

Hydrolysis is the reaction of organic compounds with water in which the water molecule (or hydroxide ion) displaces an atom or group of atoms in the compound. Hydrolysis is possibly the most important reaction of organic substances with water in aqueous environments, and is a significant process in describing the environmental fate of many organic chemicals. Hydrolysis of organic chemicals is typically a pseudo first-order reaction in terms of the concentration of the organic species.

The hydrolysis rate constant includes contributions from acid and base catalyzed hydrolysis and neutral processes. At elevated pH where basic hydrolysis becomes dominant, the hydrolysis rate constant can be defined as:

$$k_T = k_0 + k_{OH}[OH^-] \tag{12.10}$$

where: k_0 = rate constant for neutral hydrolysis

k_{OH} = rate constant for specific base-catalyzed hydrolysis

Estimating the values of k_0 and k_{OH} is the main problem associated with understanding hydrolysis and performing predictions.

Choice of Predictive Method

Semi-empirical correlations are typically used to predict the hydrolysis rate constant. The features of this approach are summarized below:

- Hydrolysis rate constants, which can be used to predict the hydrolysis rate of organic chemicals in water, are estimated based on linear free energy relationships (LFER).
- LFER is an empirical correlation between the standard free energies of reaction (ΔF°) or activation energy (ΔF) for a series of reactions, all of which subject to similar reactant structures and reaction conditions.
- LFER relates observations for one system to observations for other similar systems.
- Various estimation methods for hydrolysis rate constants, all based upon the LFER methodology, are available.

Three different methods can be used for predicting the hydrolysis rate constant:

1. *Hammett LFER Correlation* (Hammett 1938)**:** This model relates reactivity to molecular structure. Usually, it is expressed in the following form:

$$\underbrace{\log k_j}_{Derived} = \underbrace{\log k^\circ}_{Parent} + \rho\sigma \tag{12.11}$$

where: $k_j \equiv k_{OH}, k_H, or\ k_0$

Of the two empirical constants, σ is characteristic of the substituent functional group and independent of reaction, whereas ρ is determined by the reaction and its conditions and is independent of the substituent. Both parameters are unit-less correlation constants.

2. *Taft Correlation* (Taft 1956)**:** This equation considers polar, steric, and resonance effects of a substituent group on the hydrolysis rate. The Taft equation, which is restricted at present to carboxylic esters, is represented as:

$$\underbrace{\log k_j}_{Derived} = \underbrace{\log k^0}_{Parent} + \rho^* \sigma^* + \delta E_s \qquad (12.12)$$

where: $k_j \equiv k_{OH}, k_H, or\ k_0$

In this equation, σ^* is a measure of the polar effects and E_S is a measure of the steric effects of the substituent. The two constants are unit-less.

3. *Correlation Based on pKa* (Wolfe, Zepp, and Paris 1978; Wolfe, Steen, and Burns 1980)**:** In this method, log k_{OH} is correlated with pKa (the negative log of the acid dissociation constant) of the conjugated acid of the leaving group. The general correlating equation is:

$$\log k_{OH} = a \cdot pK_a + b \qquad (12.13)$$

where *a* and *b* are empirical constants.

The Hammett correlation and Taft correlation are basically substituent-effect approaches. Substituent changes in the hydrolysable molecule (RX) may occur either in the central portion of the molecule (R) or in the leaving group (X). The general principles of choosing a correlation method are:

1. When substituent changes are made in R, and R is aromatic, the hydrolysis rate constant(s) may be correlated with a Hammett correlation.
2. If RX is aliphatic and substituent changes are made in the R group, the Taft correlation is used. Correlations can be developed beyond carboxylic esters.
3. If substituent changes are made in the leaving group, X, the Hammett and Taft correlations are both potentially applicable to aromatic and aliphatic moieties, respectively. Alternatively, a correlation between the rate constants and the pKa of the leaving group could be applied.

All three methods are commonly used for estimation of k_0 and k_{OH} depending on the chemical's structure. Table 12.8 has summarized the chemical range of LFER correlations for hydrolysis. Note that some of the correlations were developed based on data from organic/aqueous mixtures.

Calculations

The general procedure for estimating the rate of hydrolysis of an organic chemical under basic conditions is:

- Categorize the organic chemical in terms of hydrolysable functional groups present.
- Estimate k0 from Hammett correlation.
- Estimate kOH from Hammett correlation, Taft correlation, or pKa correlation based on the functional groups.

1. If k0 and/or kOH are needed at a temperature, T2, other than 298K, convert to 298K with the empirical relationship (Lyman, Reehl, and Rosenblantt 1982):

$$\log k_{25°C} = \log k_{T_2} - 3830\left(\frac{T_2 - 298}{298T_2}\right) \qquad (12.14)$$

2. Calculate kT for the pH of interest according to Equation (12.15):

$$k_T = k_0 + \left(k_{OH} \times 10^{pH-14}\right) \qquad (12.15)$$

Table 12.8
Chemical range of literature hydrolysis LFER correlations

Estimate	Basis	Chemical Classes Covered
k_0	Hammett Correlation	Benzyl halides; dimethyl benzyl halides; benzyl tosylates (all in mixed organic/aqueous solvents.)
k_{OH}	Hammett Correlation	Benzene ring-substituted compounds based on $ArCOOCH_3$, $ArCOOCH_2CH_3$, $ArCH_2COOCH_2CH_3$, $ArCH{=}CHCOOCH_2CH_3$, $ArCONH_2$, $ArOCOCH_3$, $ArCH_2OCOCH_3$, $ArCON(CH_3)_2$, $ArCONHCH_3$, $ArCH_2Cl$, and $ArOSi(CH_2CH_3)_3$ (all in mixed organic/aqueous solvents).
k_{OH}	Taft Correlation	Dialkyl phthalate esters
k_{OH}	Correlation with pKa of leaving group	Aryl esters of methylphosphonic acid ($(CH_3)_2CHOP(O)(CH_3)OAr$); carbamates of the form: (1) $(C_6H_5)NHCOOAr$; (2) $CH_3N(C_6H_5)COOAr$; (3) $CH_3NHCOOAr$; (4) $(CH_3)_2NCOOAr$.

Table 12.9 has summarized the recommended correlation methods for a number of contaminants.

Observations

Nearly half of the test chemicals in Table 12.9 do not have hydrolyzable groups and thus will not hydrolyze under elevated pH conditions. Most of the remaining compounds do not belong to the chemical families in Table 12.8, so no recommendation is given for those chemicals. The hydrolysis rate constants of the remaining chemicals could potentially be predicted by either the Hammett correlation or the Taft correlation. The shortage of reference chemical data and/or Hammett and Taft parameters rendered predictions for these test chemicals impossible. To illustrate the method, the hydrolysis rate constant for diethyl phthalate was estimated from available reference data and Taft parameters:

1. The following Taft equation was developed by Wolfe (Wolfe, Steen, and Burns 1980) to estimate hydrolysis rates of phthalate esters at 30°C

$$\log k_{OH} = 4.59\sigma^* + 1.52E_S - 1.02 \quad (12.16)$$

2. From literature, $\sigma^* = -0.10; E_S = -0.07$ for substitute –CH3 (Lyman, Reehl, and Rosenblantt 1982) .

3. Substituting the values of σ* and Es in Eqn. (12.16), then $\log k_{OH} = -1.585$ at 30°C; and $k_{OH} = 0.026M^{-1}s^{-1}$

4. According to Eqn. (12.14), $\log k_{OH} = -1.797$ at 25°C; and $k_{OH} = 0.016M^{-1}s^{-1}$

Howard recommends $k_{OH} = 0.025M^{-1}s^{-1}$ at 30°C (Boethling et al. 1994). It is not clear whether Howard's value is experimental or estimated.

Table 12.9
Hydrolysis rate constant estimation method recommendation

Category	Chemical Name	CAS#	Estimation Method	Parameter availability
I	DCPA solution	1861321	Hammett	NO
	Acetylsalicylic Acid	50782	Hammett	NO
	DDT	50293	Hammett	NO
	Fluoxetine	54910893	Hammett	NO
	Methoxychlor	72435	Hammett	NO
	Mirex	2385855	Hammett	NO
II	Acetophenone	98862	N/A	Not needed
	Androst-4-ene-3,17-dione	63058	N/A	Not needed
	Anthracene	120127	N/A	Not needed
	Benzo[a]pyrene	50328	N/A	Not needed
	Bisphenol A	80057	N/A	Not needed
	Cholesterol	57885	N/A	Not needed
	Estriol	50271	N/A	Not needed
	Guaiacol	90051	N/A	Not needed
	Ibuprofen	15687271	N/A	Not needed
	Nabumetone	42924538	N/A	Not needed
	Naproxen	22204531	N/A	Not needed
	Nonylphenol	25154523	N/A	Not needed
	Para-Dodecylphenol	104438	N/A	Not needed
	Pentachloronitrobenzene	87865	N/A	Not needed
	Phenol, 4-tert-Butyl	98544	N/A	Not needed
	Progesterone	57830	N/A	Not needed
	Resorcinol	108463	N/A	Not needed
	Testosterone	58220	N/A	Not needed
	Vanillin	121335	N/A	Not needed
	Coprostanol	360689	N/A	Not needed
	Diethylstilbestrol	56531	N/A	Not needed
	Aniline	62533	N/A	Not needed
	Benzenehexachloride (BHC)	118741	N/A	Not needed
	Musk Ketone	81141	N/A	Not needed
III	Toxaphene	8001352	No published corr.	–
	Nonachlor, trans-	39765805	No published corr.	–
	Heptachlor	76448	No published corr.	–
	Nadolol	42200339	No published corr.	–
	Propranol	525666	No published corr.	–
	Aldrin	309002	No published corr.	–
	Allethrin, d-trans	584792	No published corr.	–
	Bromacetic acid	79083	No published corr.	–
	1,2-dibromoethane	106934	No published corr.	–
	Chloroacetic acid	79118	No published corr.	–
	o-Chloroaniline	95512	No published corr.	–
	Chlorpheniramine	132229	No published corr.	–
	Chloroform	67663	No published corr.	–
	Pentachloronitrobenzene	82688	No published corr.	–
IV	Diethyl Phthalate (DEP)	84662	Taft	YES
	Butyl benzyl phthalate (BBP)	85687	Taft	NO

Class I Hammett correlation applies (limited parameters)
Class II No hydrolyzable groups (negligible hydrolysis)
Class III No correlation for chemical class and/or difficult to classify (e.g. polyfunctional)
Class IV Taft correlation applies (limited parameters)

OXIDATIVE REMOVAL

Basic Theory

Various molecular descriptors have been used in ozone oxidation QSPR models. Electron donating substitutes favor ozone oxidation reactions while electron-withdrawing groups tend to decrease the degradation rate of organic contaminants. Their effect was incorporated in the predictive model as the energy of the highest occupied molecular orbital (HOMO) and energy of the lowest unoccupied molecular orbital (LOMO). The correlation between HOMO and ozone kinetics rate constants suggests a nucleophilic mechanism during oxidation of amine herbicides, substituted alkanes, alkenes, phenols, phenols and benzene. In one case, the reactivity of substituted benzenes to ozone was related to Hammett's constant. Other molecular descriptors pertaining to oxidation reactions, such as polarity, oxidation potential, and hydrophobicity have been used for oxidation QSPR models. Most of these models are applicable only to a group of compounds with similar structures differing only in type or number of substitutes, thus they are limited in their applications (Hoigne and Bader 1983a; Hoigne and Bader 1983b; Cooper et al. 1992; Percival, Marston, and Wayne 1995; Tang 2004).

Chlorination reactions are a mix of three types of reactions: oxidation, substitution, and addition. Information on free chlorination removal mechanisms is limited. Removal of contaminants by chlorine is, in general, lower than more powerful oxidative processes, such as ozonation, and UV-H_2O_2.

This work represents initial efforts in developing 3D QSAR models as tools to predict the fate of microcontaminants in various drinking water treatment processes from the limited data that are presently available. The differences between this approach and previous efforts are mainly the type of molecular descriptors. The majority of previous QSPR research was based on empirical models, in which predictions were based on 2D molecular structures and limited information on molecular properties. The applicability of some empirical QSPR research was further limited due to the fact that only data obtained under strictly controlled lab conditions were utilized and compounds of similar structural features were evaluated. In contrast to traditional empirical models, the research described here utilized Monte Carlo (MC) statistical mechanics simulations provided by commercial software to generate more accurate 3D molecular descriptors and physico-chemical properties for QSAR development. These descriptors were then used in multiple linear regressions to determine their relevance to ozonation and chlorination removals, and for the further development of 3D QSPR models.

Model Development

The removal data for selected EDCs and PPCPs by ozone and free chlorine are adopted from previous bench-scale (Westerhoff et al. 2005) and pilot-scale work. 2D molecular structures were obtained from National Library of Medicine (NLM) online resource (http://chem.sis.nlm.nih.gov/chemidplus/). LigPrep, QikProp, and Strike from Schrödinger were used as the main tools for model development. LigPrep utlized the 2D molecular structures and generated 3D molecular structures with minimized molecular energy and various possibilities for molecular chirality. Information obtained from NLM was used to select the unique structure with correct chiral carbons. A total number of 32 descriptors, including molecular descriptors and physico-chemical properties, were generated by QikProp in fast mode for each of the target compounds based on the 3D molecular files generated by LigPrep reflecting Monte Carlo simulation studies (Duffy and Jorgensen 2000; Jorgensen and Duffy 2000). The seven non-relevant medicinal chemistry properties were eliminated from further consideration and this left 25 properties as the possible candidates for QSPR model construction (Table 12.10). The detailed calculation results are included in the Appendix Table F.3. It is assumed that the removal of the selected compounds is a function of QikProp generated descriptors following the general format of:

$$\%removal = f\begin{pmatrix} \text{geometrical properties (volume, SASA, PISA, FOSA}\cdots) \\ \text{physicochemical properties (}K_{ow}, \text{S}, H_c, \text{polarizability}\cdots) \\ \text{structural characteristics (\# of groups, hydrogen bonding}\cdots) \\ \text{previously reported properties} \end{pmatrix} \quad (12.17)$$

The multiple linear regressions from Strike and literature information helped identify the most relevant parameters.

Table 12.10
Descriptors and properties generated by QikProp used for QSPR model development

Descriptor/Property	Explanation
Molecular descriptor	
MW, volume	Molecular weight and volume
SASA, FOSA, FISA, PISA, WPSA	Molecular surface area, hydrophobic, hydrophilic, π and weakly polar component of surface area
#amine, #amidine, #acid, #amide	Number of non-conjugated functional groups for each type
#rotor	Number of non-trivial (not CX_3), non-hindered (not alkene, amide, small ring) rotatable bonds
#rctvFG, #metabol	Number of reactive functional groups, number of metabolites
donorHB, accptHB	Number of hydrogen bonding donated or accepted by the solute from water
dipole	Dipole moment
QPpolrz	Polarizability
IP, EA	Ionization potential and electron affinity
Physicochemical properties	
S	Solubility
QplogPC16	Hexadecane/gas partition coefficient
QPlogPoct	Octanol/gas partition coefficient
QPlogPw	Water/gas partition coefficient
QplogPo/w	Octanol/water partition coefficient

Observations with Ozonation

Recognizing that the removal of the same compounds for different waters will have some variability, only average removals were considered in order to obtain a generalized mechanism. A consistency in ozone removal for most compounds was observed among bench, pilot, and full-scale experiments presumably due to the fundamental simplicity of ozone oxidation reactions (Chapter 7).

Only ten compounds of the 62 evaluated showed less than 20% removal by ozonation, indicating very strong reactivity of ozone toward most compounds. Of the 11 PAHs, benzo(a)pyrene was removed by 75% and the remaining 10 had over 80% removal. The structures of eight selected PAHs show a very high degree of unsaturation, which results in their higher removal because ozone is highly reactive toward double bonds. The six steroids exhibited larger than 90% removals in all waters due to several reasons. Steroids have benzene ring structures or double/triple bonds, which are highly reactive with ozone. Additionally, some steroids contain either phenolic or aliphatic hydroxide groups, which can be easily oxidized by ozone. These observations suggest that π surface area and descriptors relating to functional groups easily attacked by ozone are likely the most important parameters to be included in ozone removal models. For instance, sulfide groups are subject to electrophilic attack and result in the formation of sulfoxide derivatives (Spry 1972; Andreozzi et al. 2005).

QSPR model for ozonation process

Four parameters were found to be most relevant to ozone removal after a multiple linear regression analysis was performed with Strike and the model follows the simple format of:

$$\%(ozone\ removal) = 67.3 + 0.0506 \cdot (PISA) + 5.20 \cdot (\#metab) - 4.34 \cdot (\#rtvFG) - 0.114 \cdot (WPSA)$$

where *PISA* is the π component of the total solvent accessible surface using a probe with a 1.4 Å radius, *WPSA* is the weakly polar surface area, *#metab* is the number of likely metabolic reactions, which corresponds to bonds amenable to electrophilic attack, and *#rtvFG* is the number of reactive functional groups, which is indicative of bonds unstable to nucleophilic attack and will partly correct for over-counted *PISA*. For instance, an acceptor carbonyl double bond, contributing to *PISA*, is actually not reactive towards ozone and is subtracted from *#rtvFG*. Details on these two descriptors are provided in the Schrödinger references. Among the four descriptors, WPSA seems most critical in predicting ozone oxidative removal, and it alone can result in an R^2 of 0.7 in a single dependent variable QSPR model.

The results are presented in Figure 12.4. The R^2 from the linear regression comparing predicted to actual contaminant removal was 0.837. The consistency between model calculations and experimental data indicates that the four parameters used in the QSPR model are appropriate. The 55 compounds in the training set, randomly selected from the bench-scale results, can generally be described well by the QSPR model, with musk ketone as the worst outlier. This may be due to some unique properties from charges on the two nitro groups. The triangles in the figure, half from the remaining chemicals in bench-scale and half from pilot-scale results, are predictions calculated from the QSPR model and are in reasonably good agreement with experimental data. Experimental variability and/or inaccuracy of the model are possible reasons for the observed deviations.

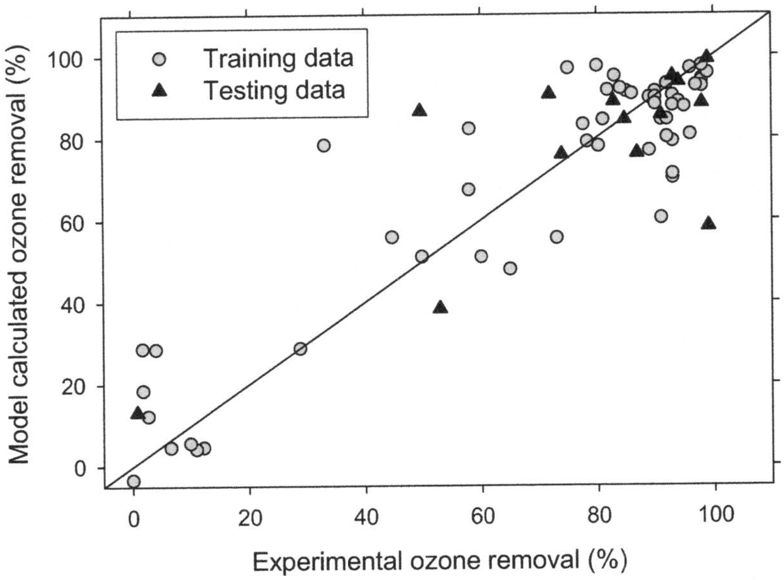

Figure 12.4 Comparison between the experimental obtained ozone removal of selected EDCs and PPCPs and model calculated results

Chlorination

Only data from suppressed pH (5.5) chlorination experiments were used to minimize impact from dissociation of acidic or basic functional groups and impact of pH on reaction kinetics. Using the same logics as stated for ozonation QSPR model development, chlorine removal was found related to five descriptors:

$$\%(\textit{chlorine removal}) = 106.8 + 0.791 \cdot \%(\textit{ozone removal}) + 7.89 \cdot (\#rtvFG) + 4.80 \cdot (QP \log Pow) + 0.175 \cdot (FISA) - 15.0 \cdot (IP)$$

where #rtvFG is the same descriptor used in the ozone model, with the remaining terms in the above model corresponding to: *QP*log*Pow* — experimental or predicted (if not experimentally available) octanol/water partition coefficient, *FISA* — hydrophilic surface area, and *IP* — ionization potential in eV. The lower R^2 of 0.705 compared with ozone removal is expected because free chlorine reactions involve several reaction mechanisms. Free chlorine can oxidize primary alcohols to carboxylic acids and form esters with unreacted alcohols (Nwaukwa and Keehn 1982). Other species, such as secondary alcohols and ethers, also can be oxidized by free chlorine to form various reaction products. Free chlorine also can initiate substitution and addition reactions. Among the dependent variables, % (ozone removal) and IP are statistically the most important as determined by the T value, a parameter generated during multiple linear regression analysis.

ULTRAFILTRATION REMOVAL

Model Development

The same statistical analysis method employed for ozonation and chlorination was used for this process. Once again, the general format of the QSPR model (eq 12.17) was initially used to build the model with an emphasis on those parameters reflecting molecular size and potential interactions with surrounding media and other components in the testing solutions.

The experimental data used for QSPR model development were reported in Chapter 9 using pilot-scale ultrafiltration process on secondary effluent. In this case, the membrane had been in operation for several weeks before testing to minimize potential artifacts from the virgin membrane material. In this evaluation, secondary effluent in an equalization tank was spiked with a mixture of the target compounds. After thorough mixing, the spiked secondary effluent was pumped through the UF membrane. The contaminant concentrations before and after UF treatment for all spiked compounds were monitored and the corresponding removal efficiencies reported.

Observation with UF process

Previous studies have discussed the mechanisms for contaminant removal by membrane processes. Size/steric exclusion, hydrophobic adsorption, electrostatic repulsion and diffusivity have been identified for their key roles in governing the rejection of trace contaminants (Bellona et al. 2004). A weak dependence of retention on log K_{ow} was reported; however, water chemistry and membrane properties also are relevant (Shetty, Malki, and Chellam 2003; Yoon et al. 2006). UF membranes, with relatively higher MW cut-offs, are not as efficient as NF and RO for contaminant rejection. Some rejection has been reported for diclofenac and ibuprofen, likely due to the interaction of these molecules with macromolecular humic substances (Burba, Geltenpoth, and Nolte 2005).

The natural organic matter contained in the secondary effluent used as feed water for the UF membrane in this research was expected to trigger similar types of interactions and hence removal of some of the selected target compounds was expected. Consistent with this assumption, 23 out of the 60 target contaminants were removed by more than 40% by the UF process. The removal of these contaminants does not seem to correlate directly with any single molecular descriptor or physico-chemical property. For instance, log K_{ow}, a commonly used descriptor for membrane processes, resulted in an R^2 of 0.40 when related to percent removal. Molecular size did not appear to be a determining factor, as larger molecules did show appreciably higher removal than did smaller molecules, while noting that the relative range of molecular size is only 600 Daltons. For instance, erythromycin is a relatively large molecule with a molecular weight of 734 Daltons and molecular volume of approximately 200 $Å^3$. The apparent removal of erythromycin was 69% through the UF membrane, while the significantly smaller phenanthrene (178 Daltons) was removed by 92%. These observations were evaluated by plotting percent removal by UF against MW, and no significant trend found.

The 10 PAHs are grouped together and their retention seems correlated with several factors. When listed in increasing dependence, these factors are log S, log K_{ow}, electron affinity, and π surface area. The corresponding R^2 relationship of removal by UF was 0.458, 0.506, 0.527 and 0.790, respectively. The dependence on π surface area suggests that the hydrophobic interaction between these contaminants and humic substances plays a key role in the removal mechanism during UF treatment (Burba, Geltenpoth, and Nolte 2005).

QSPR model for UF

The statistical approach adopted in this research was initiated by including all molecular descriptors and properties listed in Table 12.11 except #rctvFG, #metabol, and three partitioning coefficients (QplogPC16, QplogPoct and QPlogPw). The two parameters relevant to reactivity from ozonation and chlorination were #rctvFG and #metabol. The three partitioning coefficients were not considered because gas phase is not relevant to membrane filtration.

Multiple linear regression analyses were performed on the 60 target compounds with available removal data. As expected, the R^2 representing the linearity of the regression analysis as well as the extent of consistency between the model predictions and experimental data improves dramatically as more dependent parameters are included in the QSAR model (Figure 12.5). However, beyond 4 descriptors, the inclusion of more descriptors does not improve QSAR performance. Table 12.11 shows the relevant parameters during the statistical analysis and their relative significance represented by T values. When six parameters were included in the model, the removal of EDCs and PPCPs seemed to depend on the number of acidic groups, hydrogen bonding, polarizability, the number of amide groups, ionization potential or the highest occupied molecular orbital (HOMO) and solubility. The T values for each parameter decreases along this list, indicating the decreasing significance of the corresponding parameters. Since solubility was the least relevant, it was eliminated during the next QSAR evaluation when only five parameters were allowed in the model. The most important parameter in the QSAR model was the number of acidic groups, which remained throughout all statistical analysis regardless of how many parameters were allowed. Hydrogen bonding between the solute and water molecules and the polarizability are the next two most important descriptors in the model, both of which have been identified as key factors in previous research during RO membrane filtration (Rodriguez et al. 2004). The results summarized in Table 12.11 indicate that a four-parameter QSAR model is most appropriate to describe the UF process for the selected group of contaminants considering the compromise between the number of dependent variables and accuracy of the model.

$$\%(UF\ removal) = 5.7 + 2.87 \cdot (QPpolrz) - 47.2 \cdot (\#acid) - 39.2 \cdot (\#amide) - 8.32 \cdot (hb\ accptHB)$$

A comparison between the QSPR model calculation and the experimental results is presented in Figure 12.6. A generally good agreement between the model and empirical data was observed, except for heptachlor for which the model indicates an 80% removal versus the experimental removal of <1%. The reason for this large discrepancy is still unknown.

The importance of K_{ow} during membrane processes has been discussed extensively in the literature. The rejection of mono-substituted benzenes was correlated almost linearly with log K_{ow} (Kiso et al. 2001). Interestingly, log K_{ow} did not show relevance to UF removal in this study until only one and two parameters were allowed in the model, by which stage the R^2 values were much lower with values of 0.395 and 0.562, respectively. The weak dependence on log K_{ow} is presumably due to the fact that multiple removal mechanisms are involved in membrane process as recognized by previous research.

Steric hindrance is one of the factors that controls the rejection of organic molecules by NF and RO membranes (Kiso et al. 2001; Bellona et al. 2004). However, descriptors representative of this effect, such as molecular volume or other similar molecular geometrical parameters, failed to show significant correlations with UF removal. One possible explanation is that the molecular weight cut off of UF membranes (around 1K Dalton) and training molecular volumes fall within a relatively narrow range, hence there is no apparent affect on UF removal.

Table 12.11
Evolvement of significant dependent variables in QSPR modeling for the UF membrane

# of descriptors	#acid	H bond	QPpolrz	#amide	IP	QPlogS	Log Kow	R^2
6	√	√	√	√	√	√	×	0.645
5	√	√	√	√	√	×	×	0.643
4	√	√	√	√	×	×	×	0.638
3	√	√	√	×	×	×	×	0.615
2	√	×	×	×	×	×	√	0.562
1	×	×	×	×	×	×	√	0.396

T value decreasing →

Note: √ indicates a parameter found significant in the model and x indicates a parameter excluded from the model

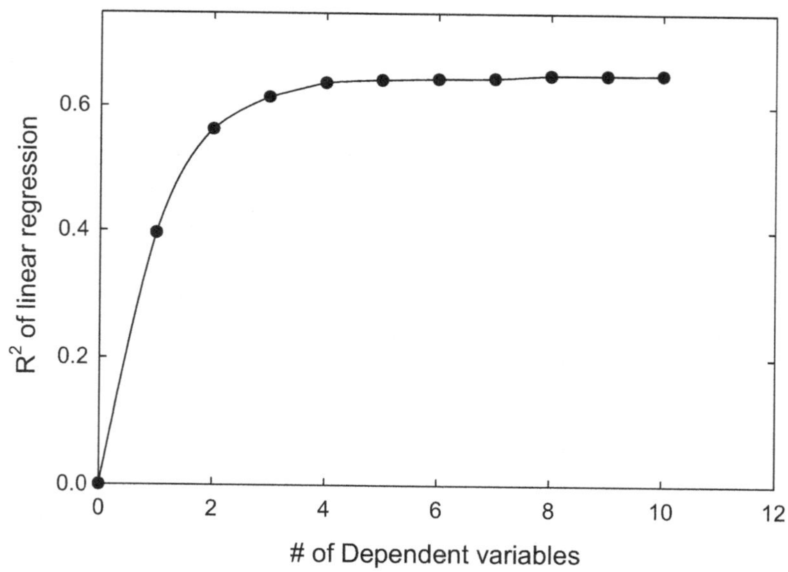

Figure 12.5 Progress in R^2 as the number of dependent variables increases

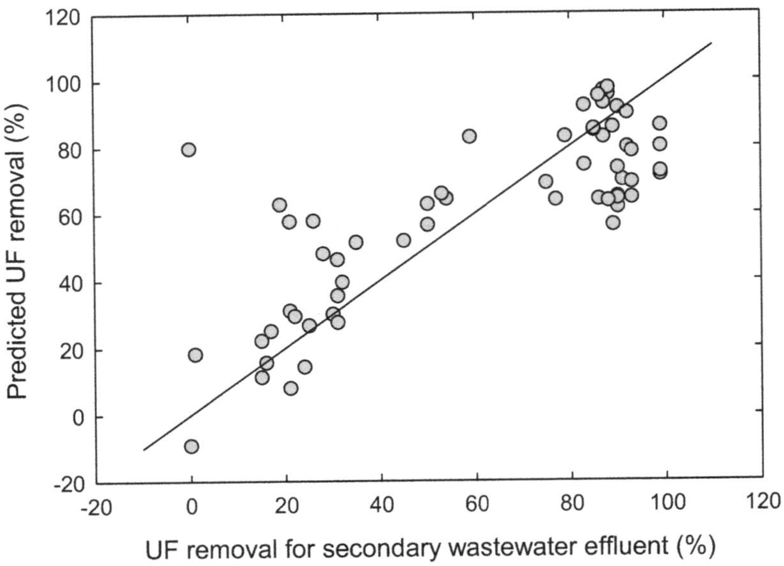

Figure 12.6 Comparison between the experimental obtained UF removal of selected EDCs and PPCPs and model calculated results

The PVDF (polyvinylidene fluoride) membrane investigated in this research is well known for its chemical resistance and carries no charge. Consequently, strong interactions between membrane materials and contaminants are not expected. UF itself is not efficient for the removal of microcontaminants; however, in the presence of surfactants (Bielska and Szymanowski 2004) and humic substances (Kim et al. 2005), improved removal of metal species and organic contaminants can be achieved due to interaction of contaminants with complexing agents and surfactants. This micelle or humic substance enhanced UF was reported for a number of heavy metal ions, such as cobalt, cadmium, copper, lead, nickel and zinc (Ahmadi et al. 1994) and for some organic contaminants (Kim et al. 1998; Baek and Yang 2004; Lee et al. 2005; Purkait, DasGupta, and De 2005). The removal distribution for selected compounds by UF membrane simply reflects the same removal mechanism (i.e. via the interaction between natural organic matter and micro-contaminants) for which hydrophobic interaction is believed to be main driving force.

RESULT SUMMARY AND DISCUSSION

In this chapter, seven processes relevant to water treatment have been discussed in regard to either the selection of currently available models or the development of new QSPR models. In spite of the limitations of computer models, they can be powerful tools to aid in emerging contaminant research.

For activated carbon, Polanyi theory has been demonstrated efficient in predicting the adsorption behavior of select compounds. Based on the predicted data and observations discussed above, the following conclusions are drawn:

- A Polanyi isotherm for WPM powdered carbon has been used for making EDC and PPCP adsorption predictions. The successful use of this theory by previous researchers with other organic chemicals has been shown in Table 12.2 for GAC. Comparing the trends of experimental data to Polanyi predictions, it is concluded that this theory was appropriate liquid phase adsorption of target compounds on powdered activated carbon.

- Polanyi's approach accounts for physiorption by van der Waals forces. However, this work has extended its use to liquid phase systems where solvent-solute and solvent-surface interactions are present. Chemisorption and ion exchange are not considered.
- The advantage of using Freundlich parameters is that a direct method to predict adsorption loading at different concentrations is possible. The parameter K describes the extent of adsorption, and the n value accounts for the non-linearity of the adsorption isotherm. Also, Freundlich parameters for various chemicals can be predicted at any temperature from the same Polanyi universal isotherm.
- The use of the Polanyi/Freundlich equations is limited to equilibrium adsorption. Input parameters to the method are solubility and molar volume at the normal boiling point. Uncertainties in these values can propagate errors in the predicted adsorption capacity.
- Examination of more structurally diverse adsorbates is needed to determine what EDC/PPCP types are not well represented by the liquid-phase Polanyi approach. The limiting factor in this work was obtaining the secondary property inputs (i.e. solubility and molar volume). Also, equilibrium isotherm data for many chemicals have not been published.

For air stripping, the focus was on the capability of predicting Henry's law constant since this is the main factor for this physical removal process. The conclusions are drawn on the basis of predicted Henry's constant values from different methods:

- The UNIFAC method is predictive, requiring chemical structure information and a choice of group interaction parameters regressed from known systems. It can represent the effect of temperature, but it is limited to adsorbates that can be built from UNIFAC's fragment library and for which the pure component vapor pressure is available.
- Existing published binary interaction parameter (BIP) sets (VLE, LLE, ENV, GLB) need to be further evaluated to see if a new, specialized set is warranted for EDCs/PPCPs. The use of the UNIFAC model entails reduction of experimental data to obtain BIPs characteristic of certain structural fragments, which can then be used in other systems containing the same structural fragments.
- Parameters in the UNIFAC library should be expanded for non-organic functional groups to incorporate a larger number of structures.
- The Hine and Mookerjee group contribution method is an alternative to UNIFAC that predicts $1/H_C$ (i.e. LWAPC) directly using only additive group contributions (limited to 25°C, as published). This temperature limitation led to the choice of UNIFAC in this effort.

For biodegradation removal of EDCs and PPCPs, several successful models are available and all are built based on the group contribution method. The conclusions drawn from the predicted values are:

- The BIOWIN program was used to predict the timeframe for both primary and ultimate biodegradation of EDCs. The Howard-Boethling method was used only for its linear and non-linear predictions.

- The predictions from BIOWIN and Howard-Boethling using the linear and non-linear approaches are comparable. However, the MITI biodegradation test predictions for most target compounds are not in agreement with any of the methods' predictions. This work has not explored this discrepancy.

For base-assisted hydrolysis reaction, methodologies have been discussed and developed; yet challenges remain due to significant lack of information. The following conclusions can be drawn from this critical analysis:

- Hammett and Taft correlations are common correlation methods used to predict hydrolysis rate constants. Correlation with pKa is an alternative method if the pKa of the conjugate acid of the leaving group is available.
- The rate constants correlated from systems containing mixtures of organic and aqueous solvents should be regarded as lower limits of the values that might be observed in very dilute aquatic environments.
- There is a severe lack of literature data and correlation parameters for EDC/PPCP chemical types. When data was available, the correlations remain limited to the chemical classes for which each was developed (see Table 12.8).
- To be more applicable to EDC/PPCPs, we recommend a research effort to compile LFER parameters for Hammett and Taft correlations tailored to common EDC/PPCP chemical types. Proper classification of EDC/PPCPs would be a major part of developing and applying such correlations.

Two disinfection processes and a UF membrane process have been investigated with a 3D QSPR modeling approach by utilizing Monte Carlo statistical mechanics simulations. EDCs and PPCPs can be efficiently removed by ozonation and, to a lesser extent, by chlorination. The two QSPR models developed by statistical analysis indicate that the removals of these microcontaminants by the two drinking treatment processes appear to be related. Ozone removal is largely determined by WPSA, followed by PISA and the number of functional groups that can be easily oxidized. IP was found to be most critical in determining chlorine removal. The dependence of both QSPR models on multiple parameters demonstrates the presence of more than one type of reaction mechanism during ozonation and chlorination processes. The statistical analysis of UF for contaminant removal revealed that molecular interactions were important, indicated by the inclusion of functional groups, polarizability and hydrogen bonding.

REFERENCES

Ahmadi, S., L. K. Tseng, B. Batchelor, and S. S. Koseoglu. 1994. Micellar-enhanced ultrafiltration of heavy-metals using lecithin. *Sep. Sci. Technol.*, 29(18): 2435-2450.

Andreozzi, R., M. Canterino, R. Marotta, and N. Paxeus. 2005. Antibiotic removal from wastewaters: The ozonation of amoxicillin. *Journal of Hazardous Materials*, 122(3): 243-250.

Ashworth, R. A., G. B. Howe, M. E. Mullins, and T. N. Rogers. 1988. Air water partitioning coefficients of organics in dilute aqueous-solutions. *J. Hazard. Mater.*, 18(1): 25-36.

Baek, K., and J. W. Yang. 2004. Simultaneous removal of chlorinated aromatic hydrocarbons, nitrate, and chromate using micellar-enhanced ultrafiltration. *Chemosphere*, 57(9): 1091-1097.

Bellona, C., J. E. J. E. Drewes, P. Xu, and G. Amy. 2004. Factors affecting the rejection of organic solutes during NF/RO treatment--a literature review. *Water Res.*, 38(12): 2795-2809.

Bielska, M., and J. Szymanowski. 2004. Micellar enhanced ultrafiltration of nitrobenzene and 4-nitrophenol. *J. Membr. Sci.*, 243(1-2): 273-281.

Boethling, R. S., P. H. Howard, W. Meylan, W. Stiteler, J. Beauman, and N. Tirado. 1994. Group-contribution method for predicting probability and rate of aerobic biodegradation. *Environ. Sci. Technol.*, 28(3): 459-465.

Burba, P., H. Geltenpoth, and J. Nolte. 2005. Ultrafiltration behavior of selected pharmaceuticals on natural and synthetic membranes in the presence of humic-rich hydrocolloids. *Analytical and Bioanalytical Chemistry*, 382(8): 1934-1941.

Cooper, D. L., T. P. Cunningham, N. L. Allan, and A. McCulloch. 1992. Tropospheric lifetimes of potential CFC replacements - Rate coefficients for reaction with the hydroxyl radical. *Atmospheric Environment Part a-General Topics*, 26(7): 1331-1334.

Duffy, E. M., and W. L. Jorgensen. 2000. Prediction of properties from simulations: Free energies of solvation in hexadecane, octanol, and water. *J. Am. Chem. Soc.*, 122(12): 2878-2888.

Fredenslund, A., R. L. Jones, and J. M. Praunitz. 1975. Group-contribution estimation of activity coefficients in nonideal liquid mixtures. *AIChE Journel*, 21: 1086-1099.

Hammett, L. P. 1937. The effect of structure upon the reactions of organic compounds benzene derivatives. *J. Am. Chem. Soc.*, 59: 96-103.

Hammett, L. P. 1938. Linear free energy relationships in rate and equilibrium phenomena. *T. Faraday Soc.*, 34: 156-165.

Hine, J., and P. K. Mookerjee. 1975. The intrinsic hydrophilic character of organic compounds. Correlations in terms of structural contributions. *Journal of Organic Chemistry*, 40(3): 292-298.

Hoigne, J., and H. Bader. 1983a. Rate constants of reactions of ozone with organic and inorganic-compounds in water. 1. Non-dissociating organic-compounds. *Water Res.*, 17(2): 173-183.

Hoigne, J., and H. Bader. 1983b. Rate constants of reactions of ozone with organic and inorganic-compounds in water. 2. Dissociating organic-compounds. *Water Res.*, 17(2): 185-194.

Howard, P. H., R. S. Boethling, W. M. Stiteler, W. M. Meylan, A. E. Hueber, J. A. Beauman, and M. E. Larosche. 1992. Predictive Model for Aerobic Biodegradability Developed from a File of Evaluated Biodegradation Data. *Environmental Toxicology and Chemistry*, 11(5): 593-603.

Jorgensen, W. L., and E. M. Duffy. 2000. Prediction of drug solubility from Monte Carlo simulations. *Bioorganic & Medicinal Chemistry Letters*, 10(11): 1155-1158.

Kim, C. K., S. S. Kim, D. W. Kim, J. C. Lim, and J. J. Kim. 1998. Removal of aromatic compounds in the aqueous solution via micellar enhanced ultrafiltration: Part 1. Behavior of nonionic surfactants. *J. Membr. Sci.*, 147(1): 13-22.

Kim, H. J., K. Baek, B. K. Kim, and J. W. Yang. 2005. Humic substance-enhanced ultrafiltration for removal of cobalt. *Journal of Hazardous Materials*, 122(1-2): 31-36.

Kiso, Y., T. Kon, T. Kitao, and K. Nishimura. 2001. Rejection properties of alkyl phthalates with nanofiltration membranes. *Journal of Membrane Science*, 182(1-2): 205-214.

Lee, J., J. S. Yang, H. J. Kim, K. Baek, and J. W. Yang. 2005. Simultaneous removal of organic and inorganic contaminants by micellar enhanced ultrafiltration with mixed surfactant. *Desalination*, 184(1-3): 395-407.

Lyman, W. J., W. F. Reehl, and D. H. Rosenblantt. 1982. Handbook of Chemical Propertiy Estimation Methods. McGraw Hill Book Company, NY.

Meylan, W. M., P. H. Howard, and R. S. Boethling. 1996. Improved method for estimating water solubility from octanol water partition coefficient. *Environmental Toxicology and Chemistry*, 15(2): 100-106.

Nwaukwa, S. O., and P. M. Keehn. 1982. The oxidation of alcohols and ethers using calcium hypochlorite [$Ca(OCl)_2$]. *Tetrahedron Letters*, 23(1): 35-38.

Percival, C. J., G. Marston, and R. P. Wayne. 1995. Correlations between Rate Parameters and Calculated Molecular-Properties in the Reactions of the Hydroxyl Radical with Hydrofluorocarbons. *Atom. Environ.*, 29(3): 305-311.

Polanyi, M. Z. 1920. Adsorption from solutions of limited soluble substances. *Z. Phys.*, 2: 111-116.

Purkait, M. K., S. DasGupta, and S. De. 2005. Separation of aromatic alcohols using micellar-enhanced ultrafiltration and recovery of surfactant. *J. Membr. Sci.*, 250(1-2): 47-59.

Rodriguez, G., S. Buonora, T. Knoell, J. D. Phipps, and H. Ridgway.2004. Rejection of Pharmaceuticals by Reverse Osmosis Membranes: Quantitative Structure Activity Relationship (QSAR) Analysis. Final Report, NWRI Project 01-EC-002

Rogers, T. N. 1994. Predicting Environmental Physical Properties from Chemical Structure using a Modified UNIFAC Model. Department of Chemical Engineering, Michigan Technological University. PhD.

Shetty, G. R., H. Malki, and S. Chellam. 2003. Predicting contaminant removal during municipal drinking water nanofiltration using artificial neural networks. *J. Membr. Sci.*, 212(1-2): 99-112.

Sontheimer, H., J. C. Crittenden, and R. S. Summers. 1988. Activated Carbon for Water Treatment. DVGW-Forschungsstelle, Karlsruhe.

Spry, D. O. 1972. Oxidation of penicillin and dihydrocephalosporin derivatives with ozone. *J. Org. Chem.*, 37(5): 793-795.

Taft, R. W. 1956. Steric Effects in Organic Chemistry, Chapter 13. Wiley, New York.

Tang, W. Z. 2004. Physicochemical Treatment of Hazardous Wastes. Lewis Publishers: Washington DC.

Westerhoff, P., Y. Yoon, S. Snyder, and E. Wert. 2005. Fate of endocrine-disruptor, pharmaceutical, and personal care product chemicals during simulated drinking water treatment processes *Environmental Science & Technology*, 39(17): 6649-6663.

Wolfe, N. L., W. C. Steen, and L. A. Burns. 1980. Phthalate Ester Hydrolysis - Linear Free-Energy Relationships. *Chemosphere*, 9(7-8): 403-408.

Wolfe, N. L., R. G. Zepp, and D. F. Paris. 1978. Use of structure reactivity relationships to estimate hydrolysis persistence of carbamate pesticides. *Water Research*, 12: 561-563.

Yoon, Y., P. Westerhoff, S. A. Snyder, and M. Esparza. 2003. HPLC-fluorescence detection and adsorption of bisphenol A, 17[beta]-estradiol, and 17[alpha]-ethynyl estradiol on powdered activated carbon. *Water Research*, 37(14): 3530-3537.

Yoon, Y., P. Westerhoff, S. A. Snyder, and E. C. Wert. 2006. Nanofiltration and ultrafiltration of endocrine disrupting compounds, pharmaceuticals and personal care products. *Journal of Membrane Science*, 270(1-2): 88-100.

CHAPTER 13
CONCLUSIONS AND RECOMMENDATIONS

OCCURRANCE OF EDCS/PPCPS IN DRINKING WATER OF THE UNITED STATES

EDC/PPCP occurrence was investigated before and after treatment at 20 drinking water utilities across the United States. Utilities treating source waters with known wastewater influence were selected for their increased potential to detect EDC/PPCP concentrations and to evaluate process removal efficiency. Occurrence results are summarized for the source waters in Table 13.1 and the finished waters in Table 13.2 with the complete data provided in Appendix E. Tables 13.1 and 13.2 show the compounds detected above the method reporting limits (detects), percent frequency of detection (% freq), minimum (min), maximum (max), median, and average (ave). The min, max, median, and ave values were calculated using only EDC/PPCP concentrations above the MRL. Table 13.3 lists the compounds that were not detected in either the source water or finished water at the 20 facilities evaluated.

Results showed DEET had the highest occurrence frequency in the source waters (100%) and in finished waters (90%), indicating that treatment was largely inefficient for removal. The US EPA estimated that approximately one third of the United States population uses products which contain DEET (http://www.epa.gov/pesticides/factsheets/chemicals/deet.htm). The pharmaceuticals carbamazepine and dilantin were both detected in 90% in source waters, and less frequently detected in finished waters (55 and 70%, respectively). Atrazine was, by far, the target analyte detected at the greatest concentrations in drinking waters. Atrazine was detected in 85% of raw waters and 75% of finished waters, with maximum concentrations of 571 and 430 ng/L, respectively. The majority of drinking water treatment processes evaluated were ineffective for the removal of atrazine, with the exception of activated carbon, nanofiltration, and reverse osmosis. Although not shown here, atrazine occurred at significantly greater concentrations in water treatment plants from central and northeast United States compared to western and southwestern regions. While atrazine occurred at the highest concentrations of the contaminants monitored in this study, it is important to consider that the US EPA maximum contaminant limit (MCL) for atrazine in drinking water is 3000 ng/L; therefore, finished drinking water concentrations of atrazine found in this study were an order of magnitude lower than the MCL. The antibiotic sulfamethoxazole occurred in 85% of raw water samples collected, but was only detected in one finished drinking water sample from one utility. This facility had 40 ng/L of sulfamethoxazole in the raw water and 20 ng/L in the finished water, with 50% of this antibiotic removed through the plant. Despite the high reactivity of sulfamethoxazole with the oxidants evaluated during this project, the poor removal at this facility could be explained by the use of chloramine for primary disinfection, which was shown to be less reactive than free chlorine. The pharmaceuticals ibuprofen and meprobamate, and the x-ray contrast agent iopromide were found to be difficult to remove by conventional water treatment processes. These compounds were detected in over 65% of raw and finished drinking waters evaluated, with only small amounts removed during treatment. The antimicrobial triclosan was detected in 30% of the raw water samples at relatively low concentrations, ranging from 1 – 30 ng/L, while it was detected in only one finished water sample at 45 ng/L. The facility that contained the highest level of triclosan in the raw water also contained the only detection of triclosan in the finished water. This utility was also the only location with sulfamethoxazole detected in the finished water. Both triclosan and sulfamethoxazole have rapid removal when in contact with chlorine or chloramine; therefore, the finished water sample clearly had a low amount of contact with chloramine before the oxidant was quenched. The synthetic fragrances musk ketone and galaxolide were detected in only 2 of the 20 raw water samples analyzed, and only musk ketone was detected in one finished water, at a level of 17 ng/L.

Even though steroid hormones seem to garner the greatest amount of concern from the public, these compounds were rarely detected in US drinking waters. Estrone was detected in two raw water samples at 1.0 and 1.4 ng/L and in two finished waters at 1.1 and 2.3 ng/L. These values are near the MRL of 1.0 ng/L. Progesterone and androstenedione were each detected in one raw water, each at less than 2 ng/L, while progesterone was detected twice at 1.1 ng/L in finished water.

These monitoring results show that some pharmaceuticals and personal care products are frequently detectable at ng/L concentrations. Some compounds are not well removed by water treatment processes and will occur in finished drinking water. However, the concentrations of these contaminants are extremely low. These results suggest that the steroid hormone EDCs rarely occur in drinking waters influenced by wastewater effluents. Moreover, the most potent estrogens, the endogenous 17β-estradiol and the birth control pharmaceutical ethynyl estradiol, were not detected at any of the 20 water treatment facilities investigated with an MRL of 1.0 ng/L. Steroid hormones should not be a concern for most full-scale water utilities due not only to lack of occurrence, but also due to relative ease of removal by oxidation. Several target analytes were not detected at any of the facilities evaluated. Efforts to continue monitoring for these contaminants seems futile. If future detection limits are lowered by orders of magnitude, these currently non-detectable contaminants will likely be detected. Determination of toxicological relevance is critical, as detection does not infer health risk and non-detection does not assure safety.

Table 13.1
Summary of EDCs/PPCPs in Raw Drinking Waters (n=20)

	Raw Drinking Water					
	Detects	% Freq.	Min	Max	Median	Ave
DEET	20	100	2.8	28	6.9	9.0
Carbamazepine	18	90	1.2	39	3.1	6.2
Dilantin	18	90	1.1	13	3.2	3.5
Atrazine	17	85	1.3	571	28	101
Sulfamethoxazole	17	85	1.2	44	8.1	14
Ibuprofen	16	80	1.1	24	4.2	6.1
Meprobamate	16	80	1.4	16	5.9	7.0
TCEP	15	75	2	66	13	18
Caffeine	14	70	9.1	87	27	34
Iopromide	14	70	2.2	46	7.6	12
Gemfibrozil	13	65	1.2	11	4.8	5.2
Naproxen	10	50	1.1	16	2.2	5.7
Erythromycin	8	40	1	3.5	2.2	2.2
Acetaminophen	7	35	1.1	9.5	1.6	2.7
Metalochlor	7	35	11	174	15	71
Triclosan	6	30	1	30	1.9	6.4
Oxybenzone	4	20	1.1	7.4	2.9	3.6
Galaxolide	3	15	25	30	28	28
Musk Ketone	3	15	14	17	16	16
Trimethoprim	3	15	1	2.3	2.2	1.8
Estrone	2	10	1	1.4	1.2	1.2
Androstenedione	1	5	1.9	1.9	1.9	1.9
Fluorene	1	5	13	13	13	13
Hydrocodone	1	5	1.9	1.9	1.9	1.9
Progesterone	1	5	1.1	1.1	1.1	1.1

Note: min, median, and ave based only on detectable concentrations

Table 13.2
Summary of EDCs/PPCPs in Finished Drinking Waters (n=20)

	Finished Drinking Water (ng/L)					
	Hits	% Freq	Min	Max	Median	Ave
DEET	18	90	2.1	30	5.1	8.2
Atrazine	15	75	1.4	430	29	74
Meprobamate	15	75	1.6	13	3.8	6.1
Dilantin	14	70	1.1	6.7	2.3	2.7
Ibuprofen	13	65	1	32	3.8	7.9
Iopromide	13	65	1.1	31	6.5	8.5
Caffeine	12	60	2.6	83	23	25
Carbamazepine	11	55	1.1	5.7	2.8	2.8
TCEP	7	35	3	19	5.5	10.1
Gemfibrozil	5	25	1.3	6.5	4.2	3.9
Metalochlor	4	20	14	160	86	86
Estrone	2	10	1.1	2.3	1.7	1.7
Progesterone	2	10	1.1	1.1	1.1	1.1
Erythromycin	1	5	1.3	1.3	1.3	1.3
Musk Ketone	1	5	17	17	17	17
Naproxen	1	5	8	8	8	8.0
Oxybenzone	1	5	1.1	1.1	1.1	1.1
Sulfamethoxazole	1	5	20	20	20	20
Triclosan	1	5	43	43	43	43
Trimethoprim	1	5	1.3	1.3	1.3	1.3

Note: min, median, and ave based only on detectable concentrations

Table 13.3
EDCs/PPCPs not detected in drinking water samples (n=20)

Raw	Finished
	Acetaminophen
	Androstenedione
Benzo[a]pyrene	Benzo[a]pyrene
DDT	DDT
Diazepam	Diazepam
Diclofenac	Diclofenac
Estradiol	Estradiol
Estriol	Estriol
Ethynylestradiol	Ethynylestradiol
	Fluorene
Fluoxetine	Fluoxetine
	Galaxolide
	Hydrocodone
Lindane	Lindane
Pentoxifylline	Pentoxifylline
Testosterone	Testosterone

PROJECT CONCLUSIONS

Conventional and advanced treatment processes were evaluated to determine their efficacy to remove a diverse group of pharmaceuticals and potential EDCs. Bench-scale testing with synthetic and natural waters provided an initial screening of all treatment technologies. Results are summarized in Table 13.4 according to treatment process and corresponding percent removal. Atrazine, iopromide, lindane, meprobamate, and TCEP proved to be the more recalcitrant of the compounds evaluated.

Observed trends from the bench-scale studies were verified by pilot-scale and full-scale evaluations. Since full-scale facilities could not be spiked with contaminants, treatment performance was evaluated using ambient compound concentrations. Results showed that full-scale removal was generally well-predicted by bench-scale and pilot-scale experiments. As expected, results consistently showed that advanced treatment technologies such as ozone, UV/H_2O_2, reverse osmosis, and activated carbon provided superior removal to conventional technologies such as coagulation, chlorination, and chloramination. However, no single treatment process provided complete removal of all target compounds investigated.

Coagulation with either alum or ferric chloride was not efficient at removing target analytes, with only a few compounds exhibiting greater than 30% removal. A weak correlation between the removal of these compounds by coagulation and log K_{ow} was observed, suggesting that particle binding and hydrophobic interactions may play a role in observed removal. Many emerging EDCs/PPCPs are ionized at ambient pH (6-8), whereas log K_{ow} relationships are most important for neutral (uncharged) molecules. Full-scale observations further confirmed coagulation to be ineffective at removing the selected target compounds.

Activated carbon, in both PAC and GAC forms, was very effective for the removal of select contaminants. Adsorption increased with contact time and activated carbon dosage, and was affected by the type of activated carbon and the NOM composition of the water. Compounds showing lower removal in bench-scale PAC testing tended to have earlier breakthrough during bench-scale RSSCTs for GAC. These compounds include charged species (ibuprofen and diclofenac) and large molecules (iopromide). For neutral compounds, log K_{ow} can be used as a preliminary indicator for compound removal by activated carbon. The presence of NOM decreases the adsorption capacity by competing with these contaminants for adsorption sites and by blocking pores in activated carbon. Due to the potential impacts from these factors, differing adsorption rates can be expected and was observed in full-scale GAC data.

Conventional disinfection processes using free chlorine, chloramine, ozone, and UV produced varying degrees of apparent contaminant reduction. Advanced oxidation processes using ozone with the addition of H_2O_2 to enhance hydroxyl radical formation resulted in a marginal increase in removal by ozone alone. Conversely, UV combined with H_2O_2 provided significantly greater removal than UV alone. The following conclusions were drawn for these oxidation processes:

- The efficiency of conventional disinfection processes is as follows: UV (40 mJ/cm^2) $<$ chloramine $\ll$ free chlorine $\ll$ ozone. Consistent results were obtained for bench, pilot, and full-scale experiments, indicating removal of target analytes can generally be predicted at bench-scale.
- For all these processes, the removal of more recalcitrant target compounds increased with larger disinfectant dosage and longer contact times.
- Lowering pH during chlorination increased the removal of target compounds due to the equilibrium shift from hypochlorite to hypochlorous acid.
- Chloramine provided significantly less removal as compared to free chlorine. Longer contact times and higher dosages did not compensate for the inferior oxidative power of chloramine (as compared to free chlorine).
- Ozone was the most powerful disinfectant for removal of target contaminants, with more than 50% of the selected compounds removed to below detection by typical disinfection doses.
- The addition of hydrogen peroxide in combination with ozone provided only a marginal increase in contaminant removal.
- Disinfection doses of UV (≈40 mJ/cm^2) provided the least EDC/PPCP removal of the conventional disinfection processes evaluated. However, the application of higher UV doses (i.e., 1000 mJ/cm^2) greatly increased EDC/PPCP removal. UV with H_2O_2 provided increased contaminant removal, with efficiency essentially identical to that of ozone and ozone AOP.
- The fire retardant TCEP was not effectively removed by any of the oxidation processes.
- The structure of target compounds will define reactivity with disinfectants and subsequent removal/transformation. Electron rich functional groups on contaminants play a vital role in removal/transformation during oxidation processes.

MF and UF membranes were not highly effective for the removal of the majority of target compounds. UF membranes, when applied in an MBR application, produced greater compound removal than the membrane alone due to biodegradation. NF and RO membranes were highly effective for compound removal. Although high rejection of target compounds was observed, a select few breached the RO membrane, due to either membrane imperfections or compound diffusion across the membrane. In NF and RO applications, brine concentrates will contain high levels of EDCs/PPCPs and additional treatment of the waste stream may be necessary.

MIEX provided minimal target compound removal despite the removal of DOC from water. Triclosan and diclofenac were the only two compounds removed by over 90%, while the majority of the compounds showed less than 20% removal. Similar to membrane applications, the brine solution used to regenerate the MIEX may concentrate certain contaminants and require further treatment.

Some EDCs and PPCPs can rapidly biodegrade, while others are quite resistant to biodegradation. In field and pilot studies, it was difficult to differentiate removal through biological degradation from removal through adsorption. However, in some cases good agreement among batch, field, and QSAR predictions was observed. For instance, acetaminophen, caffeine, estradiol, gemfibrozil, and ibuprofen all showed rapid degradation during testing and were predicted to rapidly degrade using the QSAR model. Conversely, carbamazepine, erythromycin, TECP, atrazine, iopromide, and sulfamethoxazole all displayed lower biodegradation potential. During drinking water treatment, biodegradation (i.e., biofiltration) of contaminants should not be considered a highly effective removal process. During water reuse processes, biodegradation (i.e., MBRs and activated sludge) can play a major role in contaminant removal. QSAR modeling provided a rapid evaluation of biological degradability potential; however, the model did not consider adsorption, which may explain the deviations between batch/field observations and model predictions.

Computer models can be quite effective for prediction of emerging contaminant properties and fate. QSPR have been employed to predict the physical-chemical properties of the target contaminants. The models utilized in this project accurately predicted pKa, K_{ow}, surface area, and other properties important in the prediction of contaminant fate during water treatment processes. Predictive models were also used to evaluate the potential for carbon binding, volatilization, biodegradation, and chemical oxidation. Although limited, the data produced indicate that computer modeling provides a useful tool in the evaluation of contaminant removal potential for key treatment processes. The ability to predict chemical properties and fate will provide rapid evaluation of the likelihood that a particular chemical will, or will not, be recalcitrant through a particular unit process.

RECOMMENDATIONS FOR UTILITIES

This report provides guidance for utilities regarding the treatment of a diverse group of EDCs and PPCPs. While the target compounds represent a significant group of contaminants that can occur in source waters, there are many other compounds present that were not considered. The target compounds chosen here represent a diverse range of chemical structure and compound classes. The contaminant structure will largely dictate the fate of the compound during water treatment. This report provides information on the relative removal efficacy of various conventional and advanced processes for a variety of target EDCs and PPCPs.

This report provides methodology on how to identify and quantify these trace contaminants in a variety of water matrices. The research described here has shown that sample preservation and oxidant quenching is critical in order to achieve accurate and precise data. The methodology developed for this project demonstrates the strengths and weaknesses of using a single solid-phase extraction followed by GC-MS/MS and LC-MS/MS analyses. Even with advanced instrumentation, the quantification of trace pharmaceuticals in challenging water matrices (i.e., untreated and primary treated wastewater) can be difficult at ng/L concentrations.

If a utility seeks to evaluate the EDC/PPCP removal efficacy of their particular system, it is recommended that they chose a sub-set of indicator compounds. For instance, the phenolic steroids estrone, ethynyl estradiol, estradiol, and estriol exhibited extremely consistent removal within unit processes. Considering that estrone was the only phenolic steroid detected in the survey of impacted source waters (Table 13.1), estrone can be used as an indicator compound to represent the class of steroid hormones. In other words, monitoring for estrone as opposed to the entire class of phenolic steroids is recommended. The most widely occurring contaminant evaluated was the insect repellant DEET, which was detectable in all 20 of the source waters analyzed. DEET also was one of the most challenging compounds to removal by conventional water treatment processes. For example, DEET could be used as an indicator compound for oxidation process evaluation, with higher levels of DEET removal indicating that those compounds that are more easily oxidized will have even higher levels removal in the process. Atrazine was detectable in 85% of the source waters and was difficult to remove during conventional processes. Since atrazine is detectable at much higher concentrations than most of the pharmaceuticals and steroids, it may serve as a valuable indicator as well, with increased removal of atrazine suggesting efficient removal of steroids and pharmaceuticals. Pharmaceuticals recommended as indicators are sulfamethoxazole, ibuprofen, and meprobamate due to their relatively high occurrence and varying performance during treatment. The x-ray contrast agent iopromide and the flame-retardant TCEP are also suggested as indicator compounds for treatment performance due to their occurrence and resistance to oxidation. As shown in Table 13.3, several compounds were not detected in any of the source waters evaluated and therefore are not recommended for future monitoring programs. By limiting the number of compounds evaluated to those that indicate unit process performance and have the greatest likelihood of occurrence, utilities can reduce analytical costs. The compounds that are present at higher levels generally show less analytical variability as compared to the variability at or near the analytical reporting limits. Future studies should focus on the use of indicators and surrogates (i.e., disinfectant CT and UV transmittance) for treatment performance evaluations, as monitoring for all possible emerging contaminants is not practical nor is it particularly useful in the absence of health relevance data.

No single treatment technology removed all compounds. Coagulation was relatively ineffective for contaminant removal. In the water treatment plants evaluated in the US, disinfection was by far the unit process with the greatest potential for contaminant removal. Without question, activated carbon, RO/NF membranes, and advanced oxidation provided superior removal; however, these processes are relatively rare in typical water treatment plants. Of the disinfectants evaluated, ozone provided the highest degree of contaminant removal. UV disinfection at typical doses ($\approx$40 mJ/cm^2) provided very little removal. Free chlorine provided good removal for approximately half of the target compounds evaluated. Notably, free chlorine was effective in the apparent removal of phenolic steroids (i.e., estrogens). Chloramine also provides a barrier against several EDCs/PPCPs, but is not nearly as strong of an oxidizer as free chlorine. The use of oxidative disinfection will provide apparent removal for several EDCs and PPCPs; however, the formation of DBPs should be considered.

Activated carbon, RO/NF membranes, advanced oxidation, and riverbank filtration showed superior efficiency for contaminant removal. Powder activated carbon offers the advantage of removing contaminants when needed (i.e., seasons of low surface water flows) and provides fresh material on a constant basis. Granular activated carbon is capable of removing all target compounds investigated; however, contaminants with higher water solubilities (such as iopromide) can rapidly break-through GAC, thus requiring frequent regeneration to maintain a high-level of removal. The combination of ozone with hydrogen peroxide provided only a minor increase in contaminant removal compared to ozone alone, and therefore the cost and handling issues of peroxide are likely not worth the small increase in contaminant removal. UV advanced oxidation was found to be excellent for contaminant oxidation and has the significant advantage of lower formation of regulated DBPs as compared to other oxidation processes. Reverse osmosis and nanofiltration offer great promise for contaminant rejection; however, the concentrate waste streams produced during these processes must be considered. The limited evaluation of riverbank filtration shown here indicates a high potential for contaminant removal, and could provide a cost-effective barrier for utilities with suitable source water geography.

It is recommended that utilities consider the costs and benefits before selection of any process or processes for microcontaminant removal. Currently, treatment goals for the majority of emerging contaminants can not be based on human health relevance due to a lack of toxicological data from low-dose exposure. Moreover, analytical technologies will continue to advance and will undoubtedly identify additional microcontaminants and DBPs. Complete removal of any particular contaminant is a function of the analytical reporting limit; therefore, as reporting limits continue to decrease, complete removal will become more difficult to achieve. The use of a particular process today does not insure complete removal tomorrow.

RECOMMENDATIONS FOR FUTURE RESEARCH

Further research is recommended in the following areas:

- Oxidation was found efficient in removing many EDCs and PPCPs; however, the oxidation products were not considered and may be of equal or greater concern than the original contaminant. Research is needed in terms of possible decomposition products and their potential for human health impact.
- Reaction kinetic information is still lacking. Currently, only a limited number of target analytes have reaction rate constants reported in literature. For these compounds, a good correlation of reaction kinetics with removal efficiency was discovered. Therefore, kinetic information can greatly help in the understanding of contaminant fate during treatment.
- While the widespread occurrence of certain EDCs and PPCPs was shown, the toxicological relevance is largely unknown. It is critical that human health relevance is evaluated in order to establish meaningful treatment goals.
- Some target analytes are very persistent to current treatment processes, such as the flame retardant TCEP and the herbicide atrazine. These contaminants were commonly detected in drinking waters and posed the greatest challenges for removal. Future research should focus on key indicator compounds and establish relationships between these more challenging to remove compounds and other emerging contaminants.

- The optimization of the unit processes evaluated was not performed. Future research should consider optimization of various processes in order to maximize contaminant removal while minimizing byproduct formation and energy costs.
- Advanced treatment processes should be evaluated in wastewater applications to evaluate compound removal at their source prior to entering the environment.

Table 13.4
Removal trends summary for various treatment processes under typical conditions

Target Compound	Coagulation		Softening	PAC	Cl_2		NH_2Cl	O_3	O_3/H_2O_2	UV	UV/H_2O_2	Membrane		MIEX
	Alum	Ferric Chloride			pH 5.5	Ambient pH						UF	NF	
Acetaminophen	low	low	low	high	high	high	high	high	high	low	high	low-med	low	low
Androstenedione	low	low	low	med-high	med	low-med	low	high	high	low	med	low	med	low
Atrazine	low	low	low	med-high	low	low	low	med	med-high	low	med	low	med	low
Benzo(a)pyrene	med-high	med-high	med-high	high	high	high	med-high	high	high	low	med-high	high	high	low
Caffeine	low	low	low	med	med	low	low	high	high	low	low	low	low-med	low
Carbamazepine	low	low	low	med-high	high	low	low	high	high	low	low	low	med	low
DDT	low-med	low-med	low-med	med-high	low	low	low	med	med	low	high	high	high	low
DEET	low	low	low	med	low	low	low	high	high	low	low	low	med	low
Diazepam	low	low	low	med	high	low	low	high	high	low	low	low	med	low
Diclofenac	low	low	low	med	high	high	medium	high	high	med	high	low-med	med	high
Dilantin	low	low	low	med	med	low	low	high	high	low	med	low	low	low-med
Erythromycin	low	low-med	low	med	high	high	low	high	high	low	low	med	med-high	
Estradiol	low	low	low	med	high	high	high	high	high	low	high	low	low	low
Estriol	low	low	low	med	high	high	high	high	high	low	high	low	med	low
Estrone	low	low	low	med-high	high	high	high	high	high	low	hgih	low	med	low
Ethynyl Estradiol	low	low	low	med-high	high	high	high	high	high	low	high	low	med-high	low-med
Fluorene	low	low	low	high	low	low	low	high	high	low	low-med	high	high	med
Fluoxetine	low	low	low-med	high	low-med	low	low	high	high	low	high	high	high	low
Galaxolide	low	low	low	med	med	low-med	low	high	high	low	low-med	med-high	med-high	med
Gemfibrozil	low	low	low	low	high	med-high	low	high	high	low	low	low	med	low-med
Hydrocodone	low	low	low	med-high	high	high	med	high	high	low	med	low	med	low
Ibuprofen	low	low	low	low	med	med	low	high	high	low	med	low	med	low-med
Iopromide	low	low	low	low	low	low	low	med	med-high	low	med-high	low	med	low
Lindane	low	low	low	med-high	low	low	low	low	low	low	low	low	med	low
Meprobamate	low	low	low	low	low	low	low	med-high	med-high	low	low	low	low	low
Metolachlor	low	low	low	med	med	low	low	high	high	low	med-high	low-med	med	low
Musk Ketone	low	low	low	med	med	high	low	low-med	low-med	low	low-med	med	med	low-med
Naproxen	low	low	low	med	high	high	low	high	high	low	high	low	low	med
Oxybenzone	low	low	low	high	high	high	high	high	high	low	low	med	high	med
Pentoxifylline	low	low	low	med-high	high	low	low	high	high	low	low	low	low-med	low
Progesterone	low	low	low	high	med	low-med	low	high	high	low	med	med	med	low
Sulfamethoxazole	low	low	low	med	high	high	low	high	high	med	high	low	med	low
TCEP	low	low	low	med	low	low	low	low	low	low	low	low	med	low
Testosterone	low	low	low	med-high	med	med	low	high	high	low	med	low	med	low
Triclosan	low	low	low	high	high	high	high	high	high	med	high	high	high	high
Trimethoprim	low	low	low	med-high	high	high	low	high	high	low	low-med	low-med	med	low

APPENDIX A
LITERATURE REVIEW

Table A.1
Update on method development for potential EDCs and PPCPs investigated in recent years

Compound	Analytical method	Sample volume	Detection limit (ng/L)	Recovery	Reference
Anti-inflammatory/analgesics					
Acetaminophen, Acetylsalicylic acid, Dextropropoxyphene, Diclofenac, Dimethylaminophenazone, Fenoprofen, Hydrocodone, Ibuprofen, Indomethacine, Ketoprofen, Ketorolac, Mefenamic acid, Naproxen, Paracetamol, Phenazone, Piroxicam, Propyphenazone, Tolfenamic acid	All involved preconcentration step either with SPE, SPME or LPME, followed by CE/MS, derivatization GC/MS or LC/MS/MS analysis LLE used occasionally for sample cleanup HPLC- fluorescence detector or DAD used in several cases. Detection limit is about 1000 times higher without SPE and comparable to others with SPE. SPE, enzyme-linked immunosorbent assay, GC/MS was used by Deng et al. (2003) with detection limit at 6 ng/L	0.25 – 1 L, 5-50 mL for SPME or LPME	With SPE, in the order of ng/L	Compound and matrix dependent, generally above 50%	(Ahrer and Buchberger 2001; Deng et al. 2003; González-Barreiro et al. 2003; Vanderford et al. 2003; Cahill et al. 2004; Quintana, Rodil, and Reemtsma 2004; Rodríguez et al. 2004; Soliman, Pedersen, and Suffet 2004; Stolker et al. 2004; Zühlke, Dünnbier, and Heberer 2004; Lin, Chen, and Ding 2005)
Antibiotics					
Azithromycin, Ceftriaxone, Chloramphenicol, Chlortetracycline, Ciprofloxacin, Clarithromycin, Democlocycline, Difloxacin, Doxycycline, Enrofloxacin, Erythromycin, Florfenicol, Lincomycin, Lomefloxacin, Meclocycline, Norfloxacin, Novobiocin, Ofloxacin, Oleandomycin, Oxolinic acid, Oxytetracycline, Pipemidic acid, Roxithromycin, Sarafloxacin, Spiramycin, Sulfachloropyridazine, Sulfadiazine, Sulfadimethoxane, Sulfamerazine, Sulfamethazine, Sulfamethoxazole, Sulfapyridine, Sulfathiazole, Tetracycline, Tilmicosin, Tosufloxacin, Triclosan, Trimethoprim, Tylosin	Mostly by SPE, LC coupled with one stage MS or tandem MS SPE, GC/MS was applied in triclosan analysis by Weigel et al. 2004) Capillary column-switching LC/ESI MS/MS was used to analyze sulfamethoxazole by Pitarch et al. (2004)	40 mL to 1 L, 25 μL for capillary column-switching LC/ESI MS/MS	0.02 - 288	Mostly > 60%, however, 30-47% recovery was reported for trimethoprim by Göbel et al. (2004)	(Zuccato et al. 2000; Hilton and Thomas 2003; Reverté et al. 2003; Vanderford et al. 2003; Göbel et al. 2004; Pitarch et al. 2004; Sørensen and Elbæk 2004; Stolker et al. 2004; Weigel, Kallenborn, and Hühnerfuss 2004; Yang and Carlson 2004; Yang, Cha, and Carlson 2004; Nakata et al. 2005)
β-Blockers					
Atenolol, Betaxolol, Bisoprolol, Metoprolol, Sotalol	Mosly by SPE, LC/MS/MS Capillary column-switching LC/ESI MS/MS was used to analyze betaxolol and metoprolol by Pitarch et al. (2004)	100 mL to 1 L, 25 μL for capillary column-switching LC/ESI MS/MS	Highly depend on compounds, 0.05 - 750	Generally low, falls between 13 to 90%	(Zuccato et al. 2000; Hernando et al. 2004b; Pitarch et al. 2004; Stolker et al. 2004)

(continued)

Table A.1 (Continued)

Compound	Analytical method	Sample volume	Detection limit (ng/L)	Recovery	Reference
Lipid regulator					
Atorvastatin , Bezafibrate, Clofibric acid, Fenofibrate, Gemfibrocil(Gemfibrozil)	All involved preconcentration step either with SPE, LPME, followed by CE/MS, derivatization GC/MS or LC/MS/MS analysis Capillary column-switching LC/ESI MS/MS and HPLC-fluorescence detector reported for bezafibrate LLE used occasionally for sample cleanup	100 mL – 40 L, 25 μL for capillary column-switching LC/ESI MS/MS	Low ng/L to 1.2 μg/L	Generally > 50%, however 28-60% recovery was observed by Hernando et al. (2004) and Stolker et al. (2004) for fenofibrate	(Zuccato et al. 2000; Ahrer and Buchberger 2001; González-Barreiro et al. 2003; Miao and Metcalfe 2003; Hernando et al. 2004b; Macià et al. 2004; Pitarch et al. 2004; Quintana and Reemtsma 2004; Quintana, Rodil, and Reemtsma 2004; Soliman, Pedersen, and Suffet 2004; Stolker et al. 2004; Lin, Chen, and Ding 2005)
Psychiatric drugs					
Carbamazepine (antiepileptic), Citalopram (antidepressant), Diazepam (antianxiety agent/muscle relaxant), Diltiazem (antihypertensive), Enalaprilat (antihypertensive), Fluoxetine (antidepressant), Fluvoxamine (antidepressant), Meprobamate (antianxiety), Meprobamate (antianxiety), Propranolol (antihypertensive), Sertraline (antidepressant), Venlafaxine (antidepressant)	Very similar to Anti-inflammatory/analgesics All involved preconcentration step either with SPE, SPME or LPME, followed by CE/MS, derivatization GC/MS or LC/MS/MS analysis LLE used occasionally for sample cleanup HPLC- fluorescence detector used in several cases. Detection limit is about 1000 times higher without SPE and comparable to others with SPE.	22 mL to 1 L, smaller volume required with capillary column	< 100	Good recovery, mostly between 70-110	(Zuccato et al. 2000; González-Barreiro et al. 2003; Müller et al. 2003; Reddersen and Heberer 2003; Vanderford et al. 2003; Lamas et al. 2004; Pitarch et al. 2004; Soliman, Pedersen, and Suffet 2004; Stolker et al. 2004; van der Ven et al. 2004; Zuehlke, Duennbier, and Heberer 2004; Lin, Chen, and Ding 2005)
Other pharmaceuticals					
Tamoxifen (anti-cancer), Cyclophosphamide (anti-cancer), Warfarin (anticoagulant), Dilantin (anti-convulsant), Albuterol (bronchodilator/antiasthmatic), Salbutamol (bronchodilator/antiasthmatic), 1,7-Dimethylxanthine (CNS stimulant), Furosemide (diuretics), Cimetidine (histamine H2 inhibitor), Ranitidine (histamine H2 inhibitor), Carisoprodol (uuscale relaxant), Iopromide (X-ray contrast agent), Pentoxifylline (Vasodilator)	All by SPE, GC/MS or LC/ESI MS/MS, carisoprodol also analyzed by LLE, GC/EI MS	1 – 40 L	<10	Compound dependent, 21 – 166, ranitidine and cimdetidine are most problematic	(Zuccato et al. 2000; Hilton and Thomas 2003; Vanderford et al. 2003; Cahill et al. 2004; Soliman, Pedersen, and Suffet 2004)

(continued)

Table A.1 (Continued)

Compound	Analytical method	Sample volume	Detection limit (ng/L)	Recovery	Reference
Phenolic xenoestrogens					
tert-Amylphenol, 4,4'-biphenyldiol, Bisphenol A, Bisphenol F, 6-Bromo-2-naphthol, 4-n-Butylphenol, 4-sec-Butylphenol, 4-tert-Butylphenol, 4-Chloro-2-methylphenol, 4-Chloro-3-methylphenol, 4-Cumylphenol, 2,4-Dichlorophenol, 4-n-Heptylphenol, 4-n-Hexylphenol, 3-Hydroxybiphenyl, Nonylphenol, 4-Nonylphenol, 4-n-Nonylphenol, 4-Octylphenol, 4-n-Octylphenol, 4-tert-Octylphenol, Pentachlorophenol, 4-n-Pentylphenol, Phenol	Preconcentration with SPE, SPME, Stir bar sorptive extraction, Polymer-coated hollow fiber microextraction, On-line preconcentration with surface modified polymer, Followed by derivatization GC/MS, LC/MS, or LC/MS/MS, HPLC/UV or electrochemical detector was used in one case	Up to 1 L, smaller volume required for microextrat	Depend on matrix and method, from < 1 to 4000	Generally >60%, with a few exceptions at about 40%	(Peñalver et al. 2002; Braun et al. 2003; Fries and Püttmann 2003; Benijts, Lambert, and De Leenheer 2004; Brossa et al. 2004; Hernando et al. 2004a; Kawaguchi et al. 2004a; Kawaguchi et al. 2004b; Laganà et al. 2004; Liu, Zhou, and Wilding 2004; López-Roldán, de Alda, and Barceló 2004; Rodriguez-Mozaz, de Alda, and Barcéló 2004; Watabe et al. 2004; Yuan et al. 2004; Basheer et al. 2005; Watabe et al. 2005)
Phthalates					
Benzylbutyl phthalate, Bis(2-ethylhexyl) phthalate, Di-n-butyl phthalate	All by On-line SPE, GC/EI MS	15 mL	0.1 - 8	58-97	(Brossa et al. 2002; Brossa et al. 2003)
Steroidal hormones					
Diethylstilbestrol, 17α-Estradiol, 17β-Estradiol, Estriol, Estrone, 17α-Ethinylestradiol, Mestranol	SPE, derivatize GC/EI MS or GC/MS/MS, LC/MS, or LC/MS/MS SPME, HPLC/UV was used for mestranol with detection limit of 700 ng/L	0.1 – 2 L	MS 2, MS/MS 1	63 - 116	(Peñalver et al. 2002; Braun et al. 2003; Benijts, Lambert, and De Leenheer 2004; Brossa et al. 2004; Hernando et al. 2004a; Laganà et al. 2004; Liu, Zhou, and Wilding 2004; Quintana et al. 2004; Rodriguez-Mozaz, de Alda, and Barcéló 2004; Zuehlke et al. 2004)
Phytoestrogens					
Biochanin A, Coumestrol, Daidzein, Genistein	SPE, LC/MS/MS	0.1-1 L	0.2-5.4	> 90%	(Laganà et al. 2004)
Personal care products					
Benzylparaben, Ethylparaben, Methylparaben, Propylparaben	SPE, LC/MS/MS	500 mL	0.2 – 0.5	> 90%	(Benijts, Lambert, and De Leenheer 2004)

Table A.2
Occurrence of most intensively studied EDCs and PPCPs reported in recent years in surface and groundwater

Compound Category	Compounds	Water Source	Location	# of times investigated	Range / impact by wastewater (ng/L)	Reference
Pharmaceuticals						
Analgesics:	acetaminophen	surface	Germany, US	6	n.d. – 10,000 (yes)	(Ternes 1998; Kolpin et al. 2002; **Kolpin** et al. 2004; Stackelberg et al. 2004; Glassmeyer et al. 2005)
	diclofenac	surface and ground	Finland, Germany, Switzerland, United Kingdom	17	n.d. – 568 (yes)	(Heberer et al. 1997; Buser, Poiger, and Muller 1998; Ahrer, Scherwenk, and Buchberger 2001; Farré et al. 2001; Öllers et al. 2001; Sacher et al. 2001; Sedlak and Pinkston 2001; Deng et al. 2003; Hilton and Thomas 2003; Tixier et al. 2003; Ashton, Hilton, and Thomas 2004; Lindqvist, Tuhkanen, and Kronberg 2005; Vieno, Tuhkanen, and Kronberg 2005)
	ibuprofen	surface	Finland, Italy, Switzerland, United Kingdom	15	n.d. – 5044 (yes)	(Heberer et al. 1997; Ternes 1998; Buser, Poiger, and Muller 1999; Stumpf et al. 1999; Farré et al. 2001; Öllers et al. 2001; Calamari et al. 2003; Hilton and Thomas 2003; Tixier et al. 2003; Ashton, Hilton, and Thomas 2004; Kolpin et al. 2004; Lindqvist, Tuhkanen, and Kronberg 2005; Vieno, Tuhkanen, and Kronberg 2005)
	ketoprofen	surface	Finland, Switzerland	4	n.d. – 45 (yes)	(Öllers et al. 2001; Tixier et al. 2003; Lindqvist, Tuhkanen, and Kronberg 2005; Vieno, Tuhkanen, and Kronberg 2005)
	mefenamic acid	surface	United Kingdom	4	n.d. – 366 (yes)	(Hilton and Thomas 2003; Ashton, Hilton, and Thomas 2004)
	naproxen	surface	Finland, Switzerland	4	n.d. – 400 (unknown)	(Öllers et al. 2001; Tixier et al. 2003; Lindqvist, Tuhkanen, and Kronberg 2005; Vieno, Tuhkanen, and Kronberg 2005)
Antibiotics:	ciprofloxacin	surface	Italy, Switzerland	4	n.d. – 20 (yes)	(Golet, Alder, and Giger 2002; Calamari et al. 2003; Kolpin et al. 2004)

(continued)

Table A.2 (Continued)

Compound Category	Compounds	Water Source	Location	# of times investigated	Range / impact by wastewater (ng/L)	Reference
	chlorotetracycline	surface and ground	Germany, US	8	n.d. – 690 (yes)	(Hirsch et al. 1999; Meyer et al. 2000; Lindsey, Meyer, and Thurman 2001; Kolpin et al. 2002; Kolpin et al. 2004)
	sulfamethoxazole	surface and ground	Germany, United Kingdom, US	14	n.d. – 1020 (unknown)	(Holm et al. 1995; Hartig, Storm, and Jekel 1999; Hirsch et al. 1999; Lindsey, Meyer, and Thurman 2001; Sacher et al. 2001; Hilton and Thomas 2003; Ashton, Hilton, and Thomas 2004; Kolpin et al. 2004; Stackelberg et al. 2004)
	sulfamethazine	Surface and ground	Germany, US	4	n.d. – 220 (unknown)	(Lindsey, Meyer, and Thurman 2001)
	erythromycin-H_2O	surface and ground	Germany, Italy, United Kingdom, US	10	n.d. – 1000 (yes)	(Stackelberg et al. 2004)
	trimethoprim	surface and ground	Germany, Italy, United Kingdom	6	n.d. – 249 (yes)	(Hilton and Thomas 2003)
Antiepileptic drugs:	carbamazepine	surface and ground	Switzerland, US	10	n.d. – 1,500 (yest)	(Öllers et al. 2001; Stackelberg et al. 2004)
	primidone	ground	Germany	1	n.d. – 690 (yes)	(Heberer 2002a)
β-blockers:	propranolol	surface	United Kingdom, US	5	n.d. – 215 (yes)	(Hilton and Thomas 2003; Ashton, Hilton, and Thomas 2004; Fono and Sedlak 2005)
Lipid regulators:	clofibric acid	surface and ground	Italy, Switzerland, United Kingdom	11	n.d. – 30 (yes)	(Heberer et al. 1997; Buser, Muller, and Theobald 1998; Ahrer, Scherwenk, and Buchberger 2001; Öllers et al. 2001; Heberer 2002b; Calamari et al. 2003; Hilton and Thomas 2003; Tixier et al. 2003)
	bezafibrate	surface	Finland, Italy	3	n.d. – 57 (yes)	(Calamari et al. 2003; Lindqvist, Tuhkanen, and Kronberg 2005; Vieno, Tuhkanen, and Kronberg 2005)

(continued)

Table A.2 (Continued)

Compound Category	Compounds	Water Source	Location	# of times investigated	Range / impact by wastewater (ng/L)	Reference
Contrast media:	iopromide	surface and ground	Germany	2	-	(Putschew, Schittko, and Jekel 2001)
Muscle Relaxant:	diazepam	surface	Germany, Italy	2	n.d. – 2.13 (yes)	(Ternes 2001; Calamari et al. 2003)
Steroidal hormones						
Estrogens	17α-ethinylestradiol	surface	Germany, Netherlands	6	n.d. –5.1 (yes)	(Belfroid et al. 1999b; Baronti et al. 2000; Huang and Sedlak 2001; Kuch and Ballschmiter 2001; Cargouët et al. 2004)
	17β-estradiol	surface	Germany, Netherlands	6	0.15 - 3.6 (yes)	(Belfroid et al. 1999b; Kuch and Ballschmiter 2001), (Baronti et al. 2000; Huang and Sedlak 2001; Cargouët et al. 2004)
	estrone	surface	Germany, Netherlands	3	0.1-4.1 (yes)	(Belfroid et al. 1999a; Kuch and Ballschmiter 2001)
	17α-estradiol	surface	Germany, Netherlands	3	0.15-2 (yes)	(Belfroid et al. 1999a; Kuch and Ballschmiter 2001)
Personal Care Products						
Surfactants:	nonylphenol;	surface and ground	Germany	6		(Naylor et al. 1992; Crescenzi, Di Corcia, and Samperi 1995; Fendinger et al. 1995; Zoller, Plaut, and Hushan 2004)
	4-nonylphenol	surface	Germany	2	6.7-134 (yes)	(Kuch and Ballschmiter 2001)
	4-tert-octyphenol	surface	Germany	2	n.d. – 54 (yes)	(Kuch and Ballschmiter 2001)
Antimicrobials:	triclosan	surface	US	10	n.d. – 2300 (yes)	(Kolpin et al. 2002; Lindstrom et al. 2002; Singer et al. 2002; Tixier et al. 2002; Sabaliunas et al. 2003; Kolpin et al. 2004; Stackelberg et al. 2004; Glassmeyer et al. 2005; Halden and Paull 2005)
Fragrances:	galaxolide (HHCB)	surface	US	2	n.d. – 100 (yes)	(Mitjans and Ventura 2004; Stackelberg et al. 2004)
	tonalide (AHTN)	surface	US	12	250-700	(Stackelberg et al. 2004)

(continued)

Table A.2 (Continued)

Compound Category	Compounds	Water Source	Location	# of times investigated	Range / impact by wastewater (ng/L)	Reference
Others						
Flame Retardants:	Tris(2-chloroethyl)-phosphate (TCEP)	surface	US	4	75-100 (yes)	(Kolpin et al. 2004; Stackelberg et al. 2004; Glassmeyer et al. 2005)
	Tris(2-chloroisopropyl)-phosphate (TCIPP)	surface	US	12	70-300 (yes)	(Stackelberg et al. 2004)
Stimulant:	caffeine	surface	Germany, US	4	40-250 (yes)	(Ternes, Bonerz, and Schmidt 2001; Buerge et al. 2003; Stackelberg et al. 2004; Glassmeyer et al. 2005)
Plasticizer:	bisphenol A	surface	US	9	n.d. – 12,000 (yes)	(Kuch and Ballschmiter 2001; Fromme et al. 2002; Kolpin et al. 2002; Kolpin et al. 2004; Stackelberg et al. 2004; Suzuki et al. 2004; Glassmeyer et al. 2005)

REFERENCES

Ahrer, W., and W. Buchberger. 2001. Combination of aqueous and non-aqueous capillary electrophoresis with electrospray mass spectrometry for the determination of drug residues in water. *Monatsh. Chem.*, 132(3): 329-337.

Ahrer, W., E. Scherwenk, and W. Buchberger. 2001. Determination of drug residues in water by the combination of liquid chromatography or capillary electrophoresis with electrospray mass spectrometry. *J. Chromatogr. A.* , 910: 69-78.

Ashton, D., M. Hilton, and K. V. Thomas. 2004. Investigating the environmental transport of human pharmaceuticals to streams in the United Kingdom. *Sci. Total Environ.*, 333(1-3): 167-184.

Baronti, C., R. Curini, G. D'Ascenzo, A. Di Corcia, A. Gentili, and R. Samperi. 2000. Monitoring natural and synthetic estrogens at activated sludge sewage treatment plants and in a receiving river water. *Environ. Sci. Technol.*, 34(24): 5059-5066.

Basheer, C., A. Parthiban, A. Jayaraman, H. K. Lee, and S. Valiyaveettil. 2005. Determination of alkylphenols and bisphenol-A: A comparative investigation of functional polymer-coated membrane microextraction and solid-phase microextraction techniques. *J. Chromatogr. A*, 1087(1-2): 274-282.

Belfroid, A. C., A. Van der Horst, A. D. Vethaak, A. J. Schafer, G. B. J. Rijs, J. Wegener, and W. P. Cofino. 1999a. Analysis and occurrence of estrogenic hormones and their glucuronides in surface water and waste water in The Netherlands. *Sci. Total Environ.*, 225: 101-108.

Belfroid, A. C., A. Van der Horst, A. D. Vethaak, A. J. Schafer, G. B. J. Rijs, J. Wegener, and W. P. Cofino. 1999b. Analysis and occurrence of estrogenic hormones and their glucuronides in surface water and waste water in The Netherlands. *Sci. Total Environ.*, 225(1-2): 101-108.

Benijts, T., W. Lambert, and A. De Leenheer. 2004. Analysis of Multiple Endocrine Disruptors in Environmental Waters via Wide-Spectrum Solid-Phase Extraction and Dual-Polarity Ionization LC-Ion Trap-MS/MS. *Anal. Chem.*, 76(3): 704-711.

Braun, P., M. Moeder, S. Schrader, P. Popp, P. Kuschk, and W. Engewald. 2003. Trace analysis of technical nonylphenol, bisphenol A and 17α-ehinylestradiol in wastewater using solid-phase microextraction and gas chromatography-mass spectrometry. *J. Chromatogr. A*, 988: 41-51.

Brossa, L., R. M. Marcé, F. Borrull, and E. Pocurull. 2002. Application of on-line solid-phase extraction-gas chromatography-mass spectrometry to the determination of endocrine disruptors in water samples. *J. Chromatogr. A*, 963(1-2): 287-294.

Brossa, L., R. M. Marcé, F. Borrull, and E. Pocurull. 2003. Determination of endocrine-disrupting compounds in water samples by on-line solid-phase extraction-programmed-temperature vaporisation-gas chromatography-mass spectrometry. *J. Chromatogr. A*, 998: 41-50.

Brossa, L., E. Pocurull, F. Borrull, and R. M. Marcé. 2004. Solid-phase extraction/high-performance liquid chromatography-electrospray mass spectrometry to determine endocrine disruptors in water samples. *Chromatographia*, 59(7-8): 419-423.

Buerge, I. J., T. Poiger, M. D. Muller, and H. R. Buser. 2003. Caffeine, an anthropogenic marker for wastewater contamination of surface waters. *Environ. Sci. Technol.*, 37(4): 691-700.

Buser, H. R., M. D. Muller, and N. Theobald. 1998. Occurrence of the pharmaceutical drug clofibric acid and the herbicide mecoprop in various Swiss lakes and in the North Sea. *Environ. Sci. Technol.*, 32(1): 188-192.

Buser, H. R., T. Poiger, and M. D. Muller. 1998. Occurrence and fate of the pharmaceutical drug diclofenac in surface waters: Rapid photodegradation in a lake. *Environ. Sci. Technol.*, 32(22): 3449-3456.

Buser, H. R., T. Poiger, and M. D. Muller. 1999. Occurrence and environmental behavior of the chiral pharmaceutical drug ibuprofen in surface waters and in wastewater. *Environ. Sci. Technol.*, 33(15): 2529-2535.

Cahill, J. D., E. T. Furlong, M. R. Burkhardt, D. Kolpin, and L. G. Anderson. 2004. Determination of pharmaceutical compounds in surface- and ground-water samples by solid-phase extraction and high-performance liquid chromatography-electrospray ionization mass spectrometry. *J. Chromatogr. A*, 1041(1-2): 171-180.

Calamari, D., E. Zuccato, S. Castiglioni, R. Bagnati, and R. Fanelli. 2003. Strategic survey of therapeutic drugs in the rivers Po and Lambro in northern Italy. *Environ. Sci. Technol.*, 37(7): 1241-1248.

Cargouët, M., D. Perdiz, A. Mouatassim-Souali, S. Tamisier-Karolak, and Y. Levi. 2004. Assessment of river contamination by estrogenic compounds in Paris area (France). *Sci. Total Environ.*, 324(1-3): 55-66.

Crescenzi, C., A. Di Corcia, and R. Samperi. 1995. Determination of nonionic polyethoxylate surfactants in environmental waters by liquid chromatography/electrospray mass spectrometry. *Anal. Chem.*, 67: 1797-1804.

Deng, A., M. Himmelsbach, Q. Z. Zhu, S. Frey, M. Sengl, W. Buchberger, R. Niessner, and D. Knopp. 2003. Residue analysis of the pharmaceutical dichlofenac in different water types using ELISA and GC-MS. *Environ. Sci. Technol.*, 37: 3422-3429.

Farré, M., I. Ferrer, A. Ginebreda, M. Figueras, L. Olivella, L. Tirapu, M. Vilanova, and D. Barceló. 2001. Determination of drugs in surface water and wastewater samples by liquid chromatography-mass spectrometry: methods and preliminary resutls including toxicity studies with *Vibrio fischeri*. *J. Chromatogr. A*, 2001: 187-197.

Fendinger, N. J., W. M. Begley, D. C. McAvoy, and W. S. Eckhoff. 1995. Measurement of Alkyl Ethoxylate Surfactants in Natural-Waters. *Environ. Sci. Technol.*, 29(4): 856-863.

Fono, L. J., and D. L. Sedlak. 2005. Use of the Chiral Pharmaceutical Propranolol to Identify Sewage Discharges into Surface Waters. 39: 9244-9252.

Fries, E., and W. Püttmann. 2003. Occurrence and behaviour of 4-nonylphenol in river water of Germany. *J. Environ. Monit.*, 5: 598-603.

Fromme, H., T. Kuchler, T. Otto, K. Pilz, J. Muller, and A. Wenzel. 2002. Occurrence of phthalates and bisphenol A and F in the environment. *Water Res.*, 36(6): 1429-1438.

Glassmeyer, S. T., E. T. Furlong, D. W. Kolpin, J. D. Cahill, S. D. Zaugg, S. L. Werner, M. T. Meyer, and D. D. Kryak. 2005. Transport of chemical and microbial compounds from known wastewater discharges: Potential for use as indicators of human fecal contamination. *Environ. Sci. Technol.*, 39(14): 5157-5169.

Göbel, A., C. S. McArdell, M. J. F. Suter, and W. Giger. 2004. Trace determination of macrolide and sulfonamide antimicrobials, a human sulfonamide metabolite, and trimethoprim in wastewater using liquid chromatography coupled to electrospray tandem mass spectrometry. *Anal. Chem.*, 76: 4756-4764.

Golet, E. M., A. C. Alder, and W. Giger. 2002. Environmental exposure and risk assessment of fluoroquinolone antibacterial agents in wastewater and river water of the Glatt Valley Watershed, Switzerland. *Environ. Sci. Technol.*, 36(17): 3645-3651.

González-Barreiro, C., M. Lores, M. C. Casais, and R. Cela. 2003. Simultaneous determination of neutral and acidic pharmaceuticals in wastewater by high-performance liquid chromatography-post-column photochemically induced fluorimetry. *J. Chromatogr. A*, 993(1-2): 29-37.

Halden, R. U., and D. H. Paull. 2005. Co-occurrence of triclocarban and triclosan in US water resources. *Environ. Sci. Technol.*, 39(6): 1420-1426.

Hartig, C., T. Storm, and M. Jekel. 1999. Detection and identification of sulphonamide drugs in municipal waste water by liquid chromatography coupled with electrospray ionisation tandem mass spectrometry. *J. Chromatogr. A*, 854(1-2): 163-173.

Heberer, T. 2002a. Occurrence, fate, and removal of pharmaceutical residues in the aquatic environment: a review of recent research data. *Toxicol. Lett.*, 131(1-2): 5-17.

Heberer, T. 2002b. Tracking persistent pharmaceutical residues from municipal sewage to drinking water. *Journal of Hydrology*, 266(3-4): 175-189.

Heberer, T., U. Dunnbier, C. Reilich, and H. J. Stan. 1997. Detection of drugs and drug metabolites in ground water samples of a drinking water treatment plant. *Fresenius Environmental Bulletin*, 6(7-8): 438-443.

Hernando, M. D., M. Mezcua, M. J. Gómez, O. Malato, A. Agüeera, and A. R. Fernández-Alba. 2004a. Comparative study of analytical methods involving gas chromatography-mass spectrometry after derivatization and gas chromatography-tandem mass spectrometry for the determination of selected endocrine disrupting compounds in wastewaters. *J. Chromatogr. A*, 1047(1): 129-135.

Hernando, M. D., M. Petrovic, A. R. Fernández-Alba, and D. Barceló. 2004b. Analysis by liquid chromatography-electrospray ionization tandem mass spectrometry and acute toxicity evaluation for β-blockers and lipid-regulating agents in wastewater samples. *J. Chromatogr. A*, 1046(1-2): 133-140.

Hilton, M. J., and K. V. Thomas. 2003. Determination of selected human pharmaceutical compounds in effluent and surface water samples by high-performance liquid chromatography-electrospray tandem mass spectrometry. *J. Chromatogr. A*, 1015: 129-141.

Hirsch, R., T. Ternes, K. Haberer, and K.-L. Kratz. 1999. Occurrence of antibiotics in the aquatic environment. *Sci. Total Environ.*, 225: 109-118.

Holm, J. V., K. Rügge, P. J. Bjerg, and T. H. Christensen. 1995. Occurence and distribution of pharmaceutical organic compounds in the groundwater downgradient of a landfill (Grindsted, Denmark). *Environ. Sci. Technol.*, 29(5): 1415-1420.

Huang, C.-H., and D. L. Sedlak. 2001. Analysis of estrogenic hormones in municipal wastewater effluent and surface water using enzyme-linked immunosorbent assay and gas chromatography/tandem mass spectrometry. *Environ. Toxicol. Chem.*, 20(1): 133-139.

Kawaguchi, M., K. Inoue, M. Yoshimura, R. Ito, N. Sakui, and H. Nakazawa. 2004a. Determination of 4-nonylphenol and 4-tert.-octylphenol in water samples by stir bar sorptive extraction and thermal desorption-gas chromatography-mass spectrometry. *Anal. Chim. Acta*, 505(2): 217-222.

Kawaguchi, M., K. Inoue, M. Yoshimura, N. Sakui, N. Okanouchi, R. Ito, Y. Yoshimura, and H. Nakazawa. 2004b. Trace analysis of phenolic xenoestrogens in water samples by stir bar sorptive extraction with in situ derivatization and thermal desorption-gas chromatography-mass spectrometry. *J. Chromatogr. A*, 1041: 19-26.

Kolpin, D. W., E. T. Furlong, M. T. Meyer, E. M. Thurman, S. D. Zaugg, L. B. Barber, and H. T. Buxton. 2002. Pharmaceuticals, Hormones, and Other Organic Waste Contaminants in U.S. Streams, 1999-2000: A National Reconnaissance. *Environ. Sci. Technol.*, 36(6): 1202-1211.

Kolpin, D. W., M. Skopec, M. T. Meyer, E. T. Furlong, and S. D. Zaugg. 2004. Urban contribution of pharmaceuticals and other organic wastewater contaminants to streams during differing flow conditions. *Sci. Total Environ.* , 328(1-3): 119-130.

Kuch, H. M., and K. Ballschmiter. 2001. Determination of endocrine-disrupting phenolic compounds and estrogens in surface and drinking water by HRGC-(NCI)-MS in the picogram per liter range. *Environ. Sci. Technol.*, 35(15): 3201-3206.

Laganà, A., A. Bacaloni, I. De Leva, A. Faberi, G. Fago, and A. Marino. 2004. Analytical methodologies for determining the occurrence of endocrine disrupting chemicals in sewage treatment plants and natural waters. *Anal. Chim. Acta*, 501(1): 79-88.

Lamas, J. P., C. Salgado-Petinal, C. García-Jares, M. Llompart, R. Cela, and M. Gómez. 2004. Solid-phase microexraction-gas chromatography-mass spectrometry for the analysis of selective serotonin reuptake inhibitors in environmental water. *J. Chromatogr. A*, 1046: 241-247.

Lin, W. C., H. C. Chen, and W. H. Ding. 2005. Determination of pharmaceutical residues in waters by solid-phase extraction and large-volume on-line derivatization with gas chromatography-mass spectrometry. *J. Chromatogr. A*, 1065: 279-285.

Lindqvist, N., T. Tuhkanen, and L. Kronberg. 2005. Occurrence of acidic pharmaceuticals in raw and treated sewages and in receiving waters. *Water Res.*, 39: 2219-2228.

Lindsey, M. E., M. Meyer, and E. M. Thurman. 2001. Analysis of trace levels of sulfonamide and tetracycline antimicrobials in groundwater and surface water using solid-phase extraction and liquid chromatography/mass spectrometry. *Anal. Chem.*, 73(19): 4640-4646.

Lindstrom, A., I. J. Buerge, T. Poiger, P. A. Bergqvist, M. D. Muller, and H. R. Buser. 2002. Occurrence and environmental behavior of the bactericide triclosan and its methyl derivative in surface waters and in wastewater. *Environ. Sci. Technol.*, 36(11): 2322-2329.

Liu, R., J. L. Zhou, and A. Wilding. 2004. Simultaneous determination of endocrine disrupting phenolic compounds and steroids in water by solid-phase extraction-gas chromatography-mass spectrometry. *J. Chromatogr. A*, 1022(1-2): 179-189.

López-Roldán, P., M. J. L. de Alda, and D. Barceló. 2004. Simultaneous determination of selected endocrine disrupters (pesticides, phenols and phthalates) in water by in-field solid-phase extraction (SPE) using the prototype PROFEXS followed by on-line SPE (PROSPEKT) and analysis by liquid chromatography-atmospheric pressure chemical ionisation-mass spectrometry. *Anal. Bioanal. Chem.*, 378(3): 599-609.

Macià, A., F. Borrull, M. Calull, and C. Aguilar. 2004. Determination of some acidic drugs in surface and sewage treatment plant waters by capillary electrophoresis-electrospray ionization-mass spectrometry. *Electrophoresis*, 25: 3441-3449.

Meyer, M. T., J. E. Bumgarner, J. L. Varns, J. V. Daughtridge, E. M. Thurman, and K. A. Hostetler. 2000. Use of radioimmunoassay as a screen for antibiotics in confined animal feeding operations and confirmation by liquid chromatography/mass spectrometry. *Sci. Total Environ.*, 248(2-3): 181-187.

Miao, X.-S., and C. D. Metcalfe. 2003. Determination of pharmaceuticals in aqueous samples using positive and negative voltage switching microbore liquid chromatography/electrospray ionization tandem mass spectrometry. *J. Mass Spectrom.*, 38(1): 27-34.

Mitjans, D., and F. Ventura. 2004. Determination of musks and other fragrance compounds at ng/L levels using CLSA (closed loop stripping analysis) and GC/MS detection. *Water Sci. Technol.*, 50(5): 119-123.

Müller, S., M. Möder, S. Schrader, and P. Popp. 2003. Semi-automated hollow-fibre membrane extraction, a novel enrichment technique for the determination of biologically active compounds in water samples. *J. Chromatogr. A*, 985(1-2): 99-106.

Nakata, H., K. Kurunthachalam, P. D. Jones, and J. P. Giesy. 2005. Determination of fluoroquinolone antibiotics in wastewater effluents by liquid chromatography-mass spectrometry and fluorescence detection. *Chemosphere*, 58: 759-766.

Naylor, C. G., J. P. Mieure, W. J. Adams, J. A. Weeks, F. J. Castaldi, L. D. Ogle, and R. R. Romano. 1992. Alkylphenol ethoxylates in the environment. *JAOCS*, 69(7): 695-703.

Öllers, S., H. P. Singer, P. Fassler, and S. R. Muller. 2001. Simultaneous quantification of neutral and acidic pharmaceuticals and pesticides at the low-ng/1 level in surface and waste water. *J. Chromatogr. A*, 911(2): 225-234.

Peñalver, A., E. Pocurull, F. Borrull, and R. M. Marcé. 2002. Method based on solid-phase microextraction-high-performance liquid chromatography with UV and electrochemical detection to determine estrogenic compounds in water samples. *J. Chromatogr. A*, 964: 153-160.

Pitarch, E., F. Hernandez, J. ten Hove, H. Meiring, W. Niesing, E. Dijkman, L. Stolker, and E. Hogendoorn. 2004. Potential of capillary-column-switching liquid chromatography-tandem mass spectrometry for the quantitative trace analysis of small molecules. Application to the on-line screening of drugs in water. *J. Chromatogr. A*, 1031: 1-9.

Putschew, A., S. Schittko, and M. Jekel. 2001. Quantification of triiodinated benzene derivatives and X-ray contrast media in water samples by liquid chromatography-electrospray tandem mass spectrometry. *J. Chromatogr. A*, 930(1-2): 127-134.

Quintana, J. B., J. Carpinteiro, I. Rodríguez, R. A. Lorenzo, A. M. Carro, and R. Cela. 2004. Determination of natural and synthetic estrogens in water by gas chromatography with mass spectrometric detection. *J. Chromatogr. A*, 1024(1-2): 177-185.

Quintana, J. B., and T. Reemtsma. 2004. Sensitive determination of acidic drugs and triclosan in surface and wastewater by ion-pair reverse-phase liquid chromatography/tandem mass spectrometry. *Rapid Commun. Mass Spectrom.*, 18: 765-774.

Quintana, J. B., R. Rodil, and T. Reemtsma. 2004. Suitability of hollow fibre liquid-phase microextraction for the determination of acidic pharmaceuticals in wastewater by liquid chromatography-electrospary tandem mass spectrometry without matrix effects. *J. Chromatogr. A*, 1061: 19-26.

Reddersen, K., and T. Heberer. 2003. Multi-compound methods for the detection of pharmaceutical residues in various waters applying solid phase extraction (SPE) and gas chromatography with mass spectrometric (GC-MS) detection. *J. Sep. Sci.*, 26: 1443-1450.

Reverté, S., F. Borrull, E. Pocurull, and R. M. Marcé. 2003. Determination of antibiotic compounds in water by solid-phase extraction-high-performance liquid chromatography-(electrospray) mass spectrometry. *J. Chromatogr. A*, 1010(2): 225-232.

Rodriguez-Mozaz, S., M. J. L. de Alda, and D. Barcéló. 2004. Monitoring of estrogens, pesticides and bisphenol A in natural waters and drinking water treatment plants by solid-phase extraction-liquid chromatography-mass spectrometry. *J. Chromatogr. A*, 1045(1-2): 85-92.

Rodríguez, I., J. Carpinteiro, J. B. Quintana, A. M. Carro, R. A. Lorenzo, and R. Cela. 2004. Solid-phase microextraction with on-fiber derivatization for the analysis of anti-inflammatory drugs in water samples. *J. Chromatogr. A*, 1024(1-2): 1-8.

Sabaliunas, D., S. F. Webb, A. Hauk, M. Jacob, and W. S. Eckhoff. 2003. Environmental fate of Triclosan in the River Aire Basin, UK. *Water Res.*, 37(13): 3145-3154.

Sacher, F., F. T. Lange, H.-J. Brauch, and I. Blankenhorn. 2001. Pharmaceuticals in groundwaters: Analytical methods and results of a monitoring program in Baden-Wurttemberg, Germany. *J. Chromatogr. A*, 938(1-2): 199-210.

Sedlak, D. L., and K. E. Pinkston. 2001. Factors affecting the concentrations of pharmaceuticals released to the aquatic environment. *Water Resouces Update*: 56-64.

Singer, H., S. Mueller, C. Tixier, and L. Pillonel. 2002. Triclosan: Occurrence and Fate of a Widely Used Biocide in the Aquatic Environment: Field Measurements in Wastewater Treatment Plants, Surface Waters, and Lake Sediments. *Environ. Sci. Technol.*, 36(23): 4998-5004.

Soliman, M. A., J. A. Pedersen, and I. H. Suffet. 2004. Rapid gas chromatography-mass spectrometry screening method for human pharmaceuticals, hormones, antioxidants and plasticizers in water. *J. Chromatogr. A*, 1029(1-2): 223-237.

Sørensen, L. K., and T. H. Elbæk. 2004. Simultaneous determination of trimethoprim, sulfadiazine, florfenicol and oxolinic acid in surface water by liquid chromatography tandem mass spectrometry. *Chromatographia*, 60: 287-291.

Stackelberg, P. E., E. T. Furlong, M. T. Meyer, S. D. Zaugg, A. K. Henderson, and D. B. Reissman. 2004. Persistence of pharmaceutical compounds and other organic wastewater contaminants in a conventional drinking-water-treatment plant. *Sci. Total Environ.*, 329(1-3): 99-113.

Stolker, A. A. M., W. Niesing, E. A. Hogendoorn, J. F. M. Versteegh, R. Fuchs, and U. A. T. Brinkman. 2004. Liquid chromatography with triple-quadrupole or quadrupole-time of flight mass spectrometry for screening and confirmation of residues of pharmaceuticals in water. *Anal. Bioanal. Chem.*, 378(4): 955-963.

Stumpf, M., T. A. Ternes, W. Rolf-Dieter, S. V. Rodrigues, and W. Baumann. 1999. Polar drug residues in sewage and natural waters in the state of Rio de Janeiro, Brazil. *Sci. Total Environ.*, 225: 135-141.

Suzuki, T., Y. Nakagawa, I. Takano, K. Yaguchi, and K. Yasuda. 2004. Environmental fate of bisphenol A and its biological metabolites in river water and their xeno-estrogenic activity. *Environ. Sci. Technol.*, 38(8): 2389-2396.

Ternes, T., M. Bonerz, and T. Schmidt. 2001. Determination of neutral pharmaceuticals in wastewater and rivers by liquid chromatography-electrospray tandem mass spectrometry. *J. Chromatogr. A.* , 938(1-2): 175-185.

Ternes, T. A. 1998. Occurrence of drugs in German sewage treatment plants and rivers. *Water Res.*, 32(11): 3245-3260.

Ternes, T. A. 2001. Analytical methods for the determination of pharmaceuticals in aqueous environmental samples. *Trends Anal. Chem.*, 20(8): 419-434.

Tixier, C., H. P. Singer, S. Canonica, and S. R. Mueller. 2002. Phototransformation of triclosan in surface waters: a relevant elimination process for this widely used biocide-laboratory studies, field measurements, and modeling. *Environ. Sci. Technol.*, 36(16): 3482-3489.

Tixier, C., H. P. Singer, S. Oellers, and S. R. Mueller. 2003. Occurrence and fate of carbamazepine, clofibric acid, diclofenac, ibuprofen, ketoprofen, and naproxen in surface waters. *Environ. Sci. Technol.*, 37(6): 1061-1068.

van der Ven, K., W. V. Dongen, B. U. W. Maes, E. L. Esmans, R. Blust, and W. M. D. Coen. 2004. Determination of diazepam in aquatic samples by capillary liquid chormatography-electrospray tandem mass spectrometry. *Chemosphere*, 57: 967-973.

Vanderford, B. J., R. A. Pearson, D. J. Rexing, and S. A. Snyder. 2003. Analysis of Endocrine Disruptors, Pharmaceuticals, and Personal Care Products in Water Using Liquid Chromatography/Tandem Mass Spectrometry. *Anal. Chem.*, 75(22): 6265-6274.

Vieno, N. M., T. Tuhkanen, and L. Kronberg. 2005. Seasonal variation in the occurrence of pharmaceuticals in effluents from a sewage treatment plant and in the recipient water. *Environ. Sci. Technol.*, 39(21): 8220-8226.

Watabe, Y., K. Hosoya, N. Tanaka, T. Kondo, M. Morita, and T. Kubo. 2005. LC/MS determination of bisphenol A in river water using a surface-modified molecularly-imprinted polymer as an on-line pretreatment device. *Anal. Bioanal. Chem.*, 381: 1193-1198.

Watabe, Y., T. Kondo, M. Morita, N. Tanaka, J. Haginaka, and K. Hosoya. 2004. Determination of bisphenol A in environmental water at ultra-low level by high-performance liquid chromatography with an effective on-line pretreatment device. *J. Chromatogr. A*, 1032: 45-49.

Weigel, S., R. Kallenborn, and H. Hühnerfuss. 2004. Simultaneous solid-phase extraction of acidic, neutral and basic pharmaceuticals from aqueous samples at ambient (neutral) pH and their determination by gas chromatography-mass spectrometry. *J. Chromatogr. A*, 1023(2): 183-195.

Yang, S., and K. H. Carlson. 2004. Solid-phase extraction-high-performance liquid chromatography-ion trap mass spectrometry for analysis of trace concentrations of macrolide antibiotics in natural and waste water matrices. *J. Chromatogr. A*, 1038(1-2): 141-155.

Yang, S., J. Cha, and K. Carlson. 2004. Quantitative determination of trace concentrations of tetracycline and sulfonamide antibiotics in surface water using solid-phase extraction and liquid chromatography/ion trap tandem mass spectrometry. *Rapid Commun. Mass Spectrom.*, 18(18): 2131-2145.

Yuan, T., J. Y. Hu, S. L. Ong, and W. J. Ng. 2004. Simultaneous analysis of five selected phenolic xenoestrogens in water sample using solid phase extraction-gas chromatography-mass spectrometry. *Chromatographia*, 58(9-10): 643-648.

Zoller, U., I. Plaut, and M. Hushan. 2004. The case of the nonionic alkylphenol ethoxylates in the Mediterranean Sea region: Is there a problem? *Water Sci. Technol.*, 50(5): 79-84.

Zuccato, E., D. Calamari, M. Natangelo, and R. Fanelli. 2000. Presence of therapeutic drugs in the environment. *Lancet*, 355: 1789-1790.

Zuehlke, S., U. Duennbier, and T. Heberer. 2004. Determination of polar drug residues in sewage and surface water applying liquid chromatography-tandem mass spectrometry. *Anal. Chem.*, 76: 6548-6554.

Zuehlke, S., U. Duennbier, T. Heberer, and B. Fritz. 2004. Analysis of endocrine disrupting steroids: investigation of their release into the environment and their behavior during back filtration. *Ground Water Monit. R.*, 24: 78-85.

Zühlke, S., U. Dünnbier, and T. Heberer. 2004. Detection and identification of phenazone-type drugs and their microbial metabolites in ground and drinking water applying solid-phase extraction and gas chromatography with mass spectrometric detection. *J. Chromatogr. A*, 1050: 201-209.

APPENDIX B
PILOT PLANT SCHEMATICS

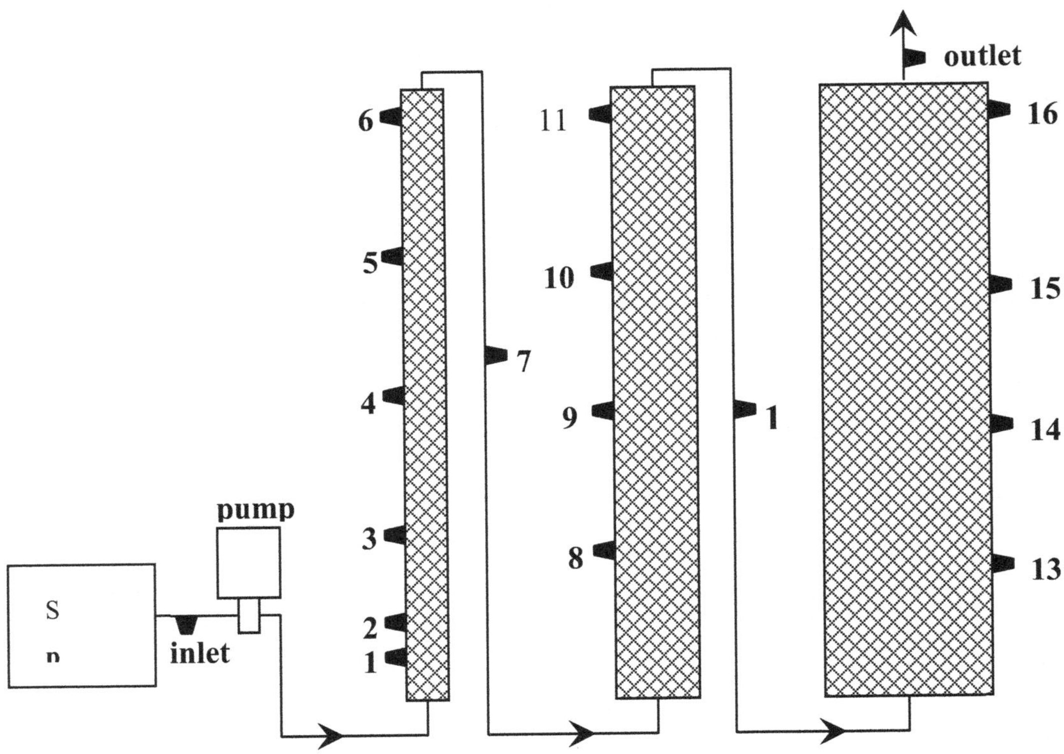

Figure B.1 Configuration of the aquifer storage and recovery columns. Sample ports marked accordingly

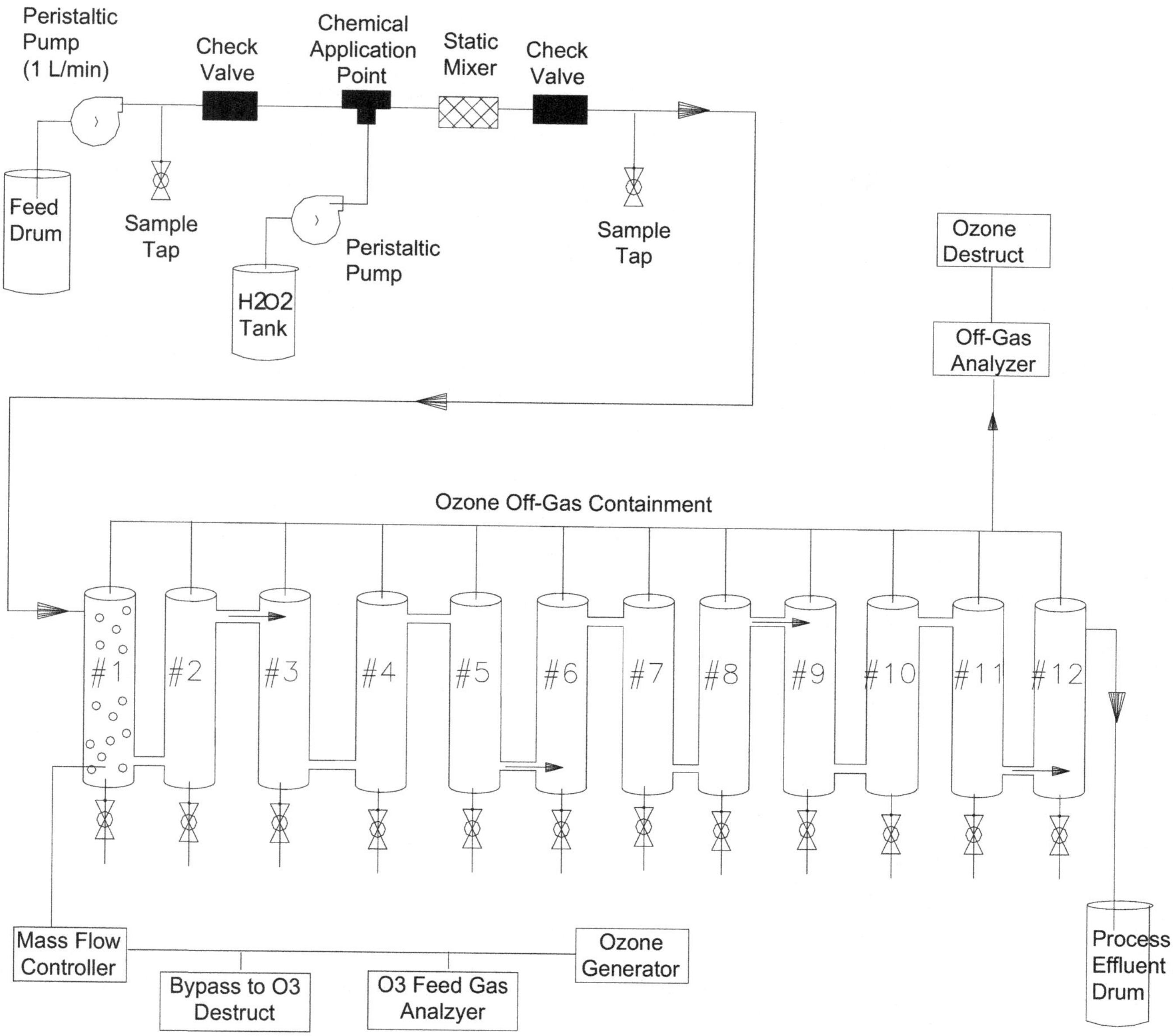

Figure B.2 Process flow schematic for SNWA bench-top ozone pilot plant

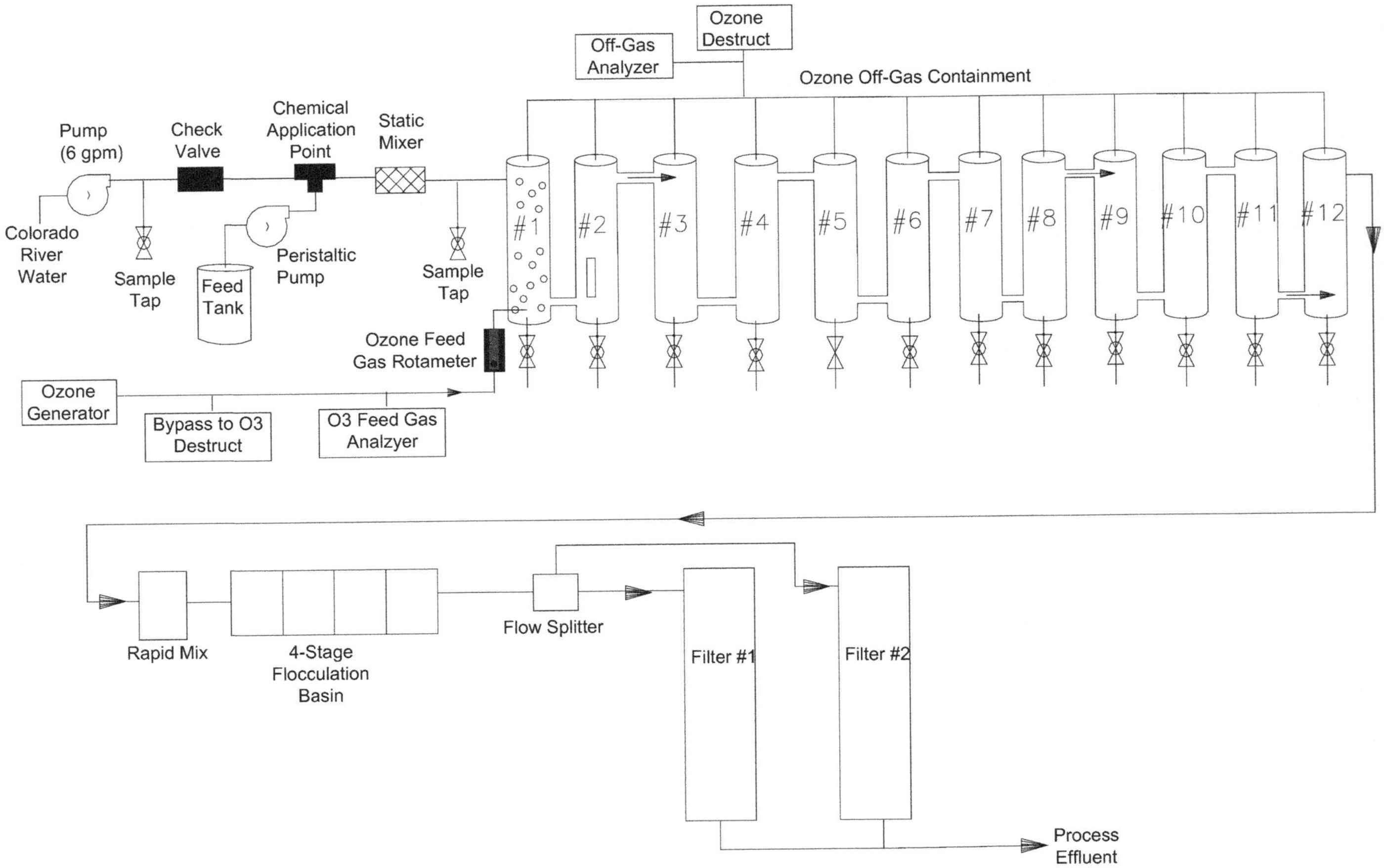

Figure B.3 SNWA pilot-plant

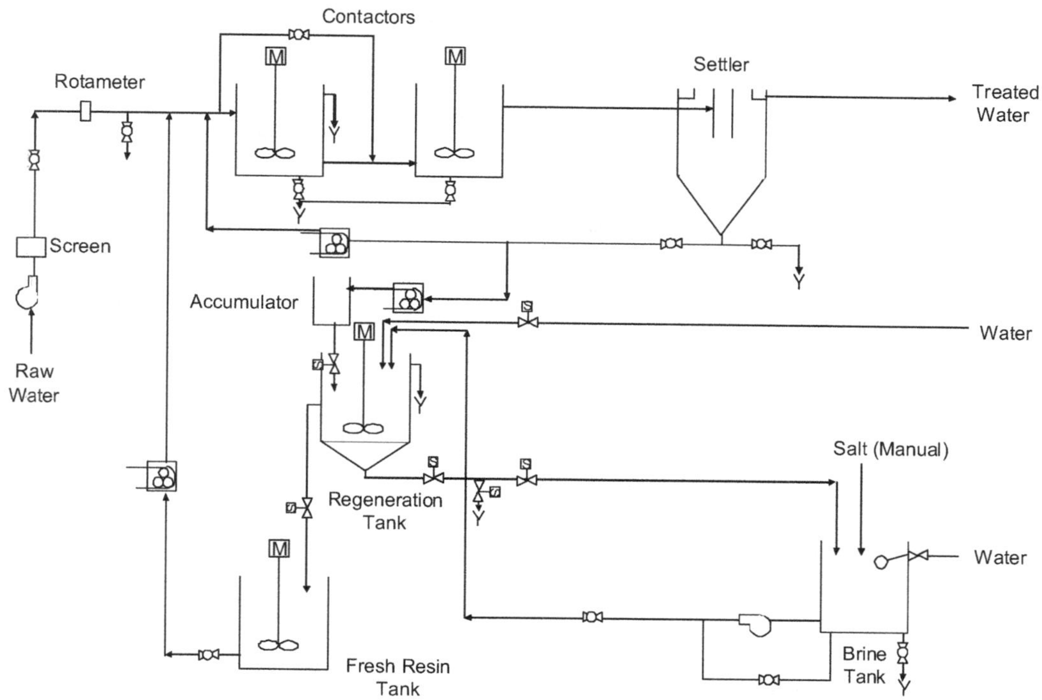

Figure B.4 Schematic for MIEX pilot plant set-up

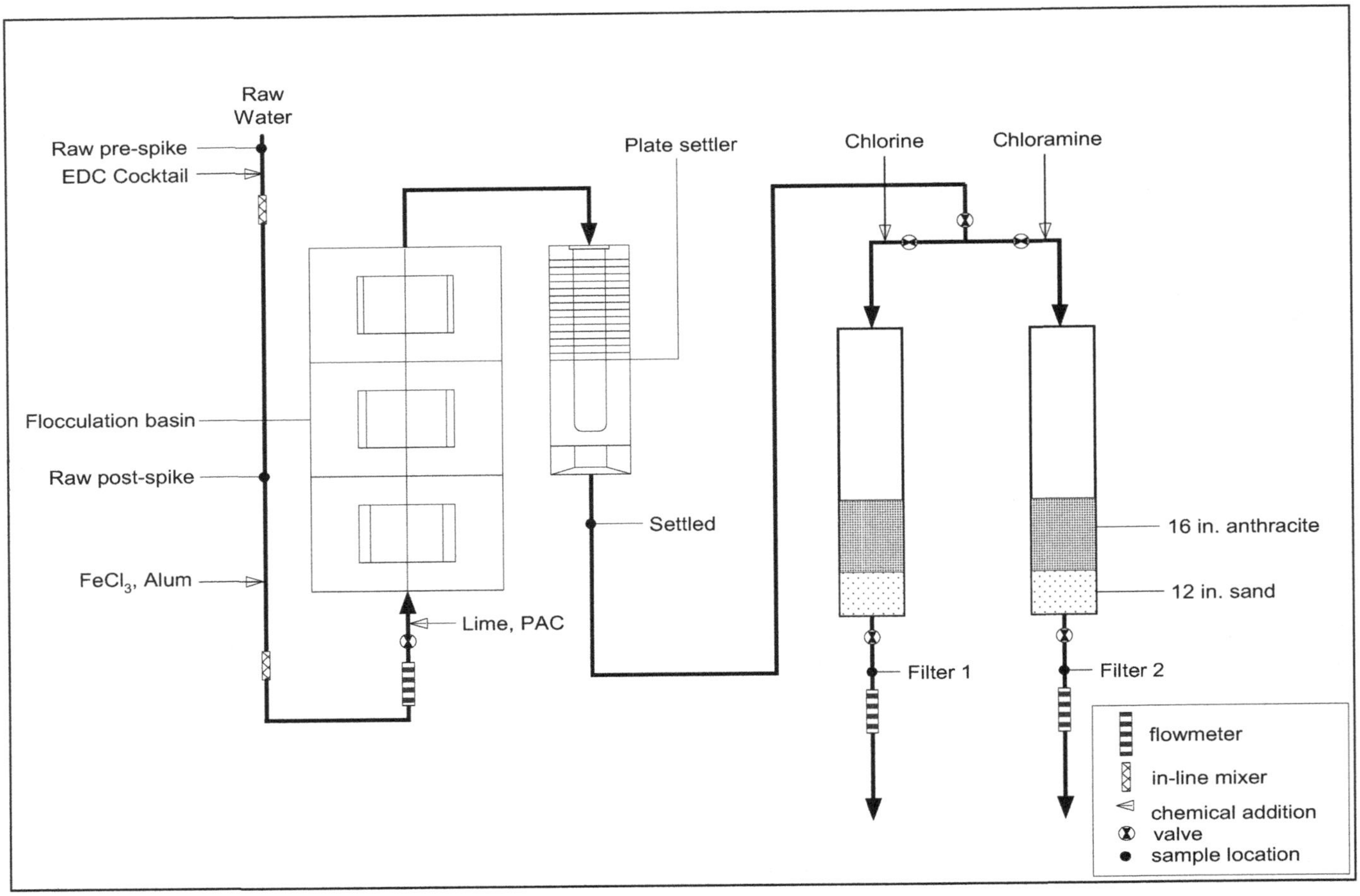

Figure B.5 Louisville conventional pilot plant schematic

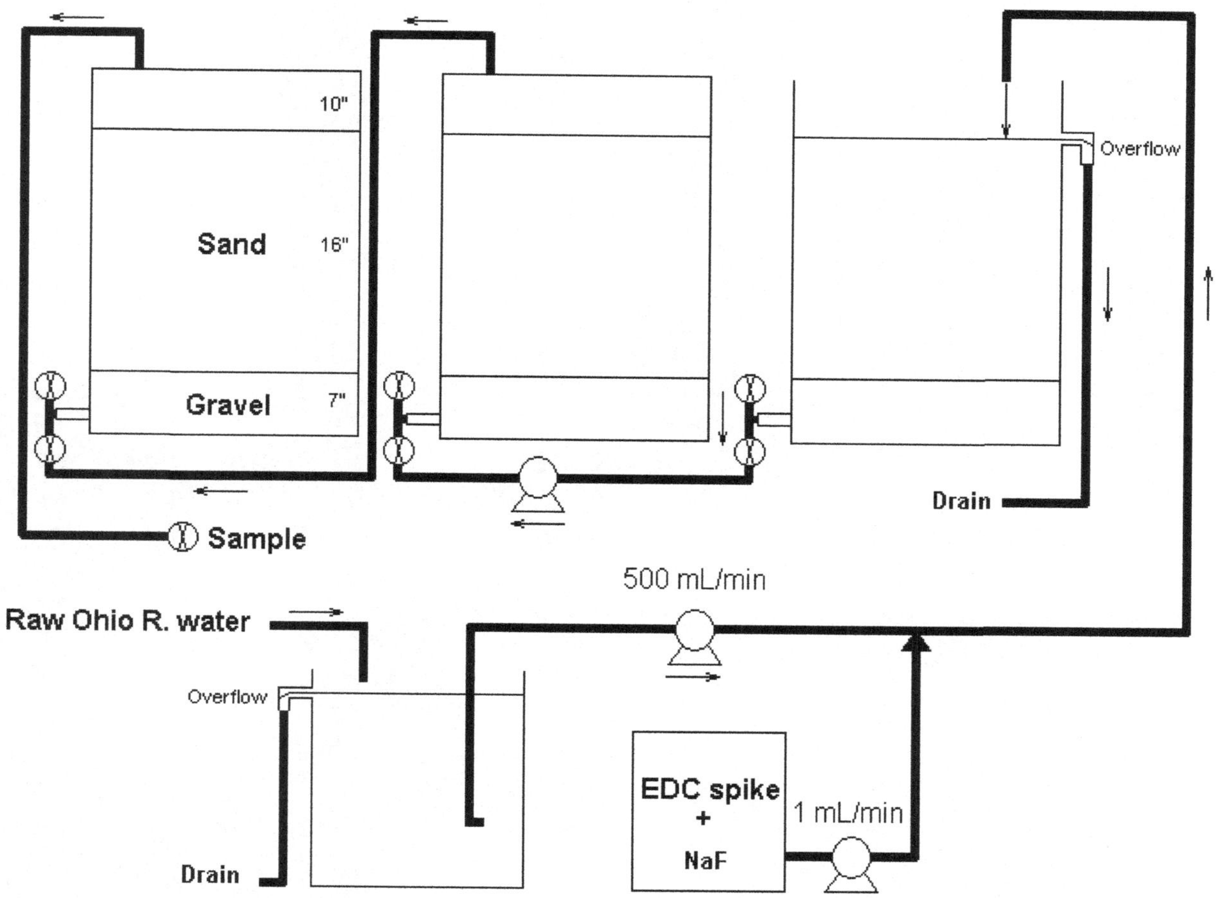

Figure B.6 Louisville river bank filtration pilot plant schematic

Figure B.7 Photograph of Trojan Technologies medium pressure UV pilot plant

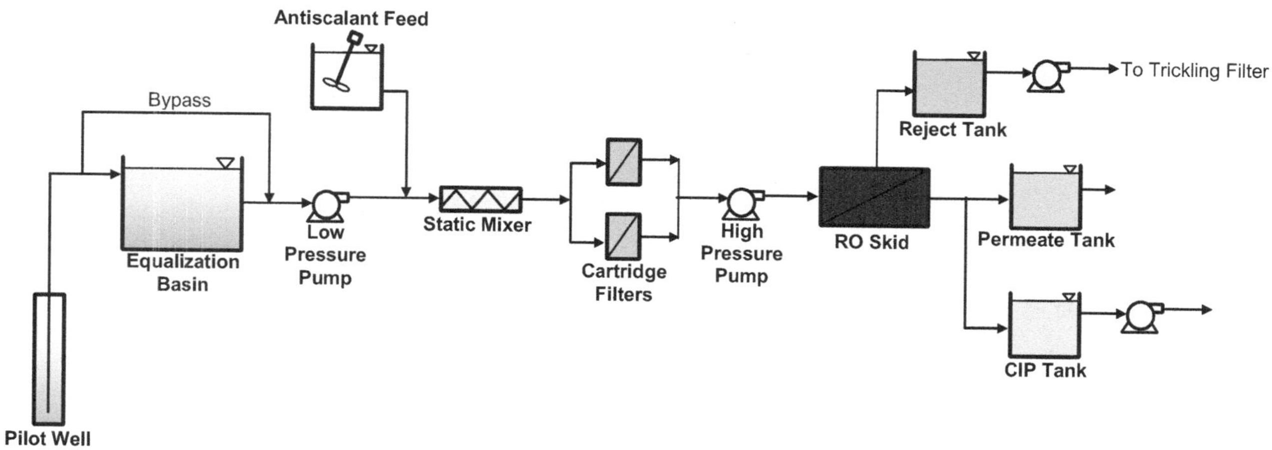

Figure B.8 Black & Veatch saline groundwater reverse osmosis pilot plant schematic

Figure B.9 Black & Veatch saline groundwater reverse osmosis pressure vessel array

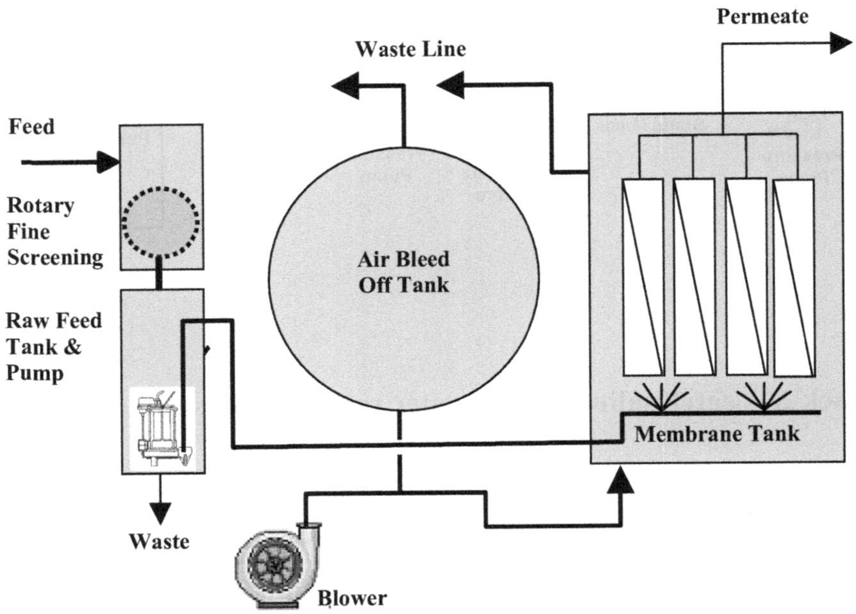

Figure B.10 Schematic of US Filter Jet Tech MBR MWH pilot system

Table B.1
Specifications for US Filter MemJet B10R microfiltration membrane

	Units	Value
Manufacturer	---	US Filter
Approximate Size of Element (L x Dia)	mm	1850 x 100
Active Membrane Area (outside) per module	m2	9.2
Number of Fibers		~2000
Inside Diameter of Fiber	mm	0.65
Outside Diameter of Fiber	mm	1.0
Approximate Length of Fiber	m	1.5
Flow Direction	---	outside-in
Nominal Membrane Pore size	micron	0.08
Absolute Membrane Pore Size	micron	0.2
Membrane Material/Construction	---	PV dF / Hollow Fiber
Membrane Surface Characteristics	---	hydrophilic
Membrane Charge	---	Neutral
Design Flux	gfd	14.4
Acceptable Range of Operating pH Values	---	2-11
Vacuum Pressure for System	kPa	50
Chlorine/Oxidant Tolerance	ppm hr/yr	100,000

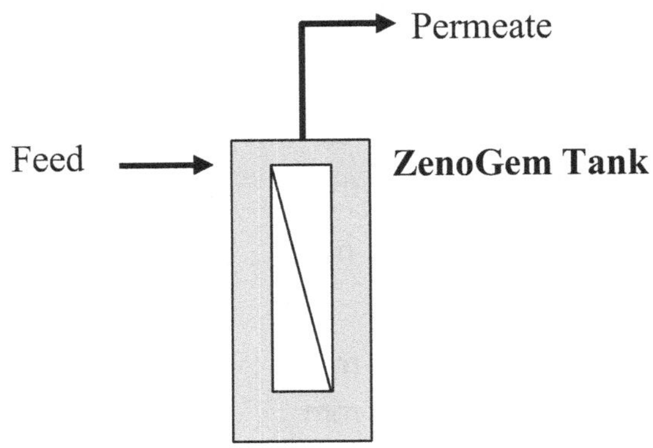

Figure B.11 Schematic of Zenon UF MWH pilot system

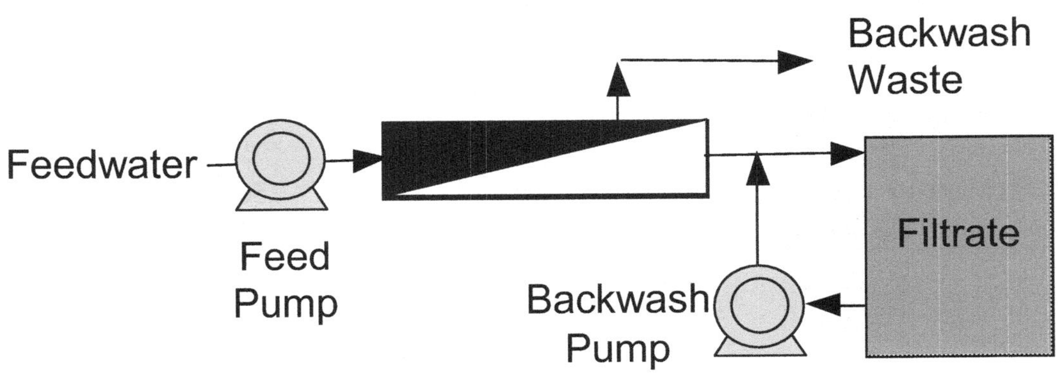

Figure B.12 Schematic of the Ionics UF MWH pilot system

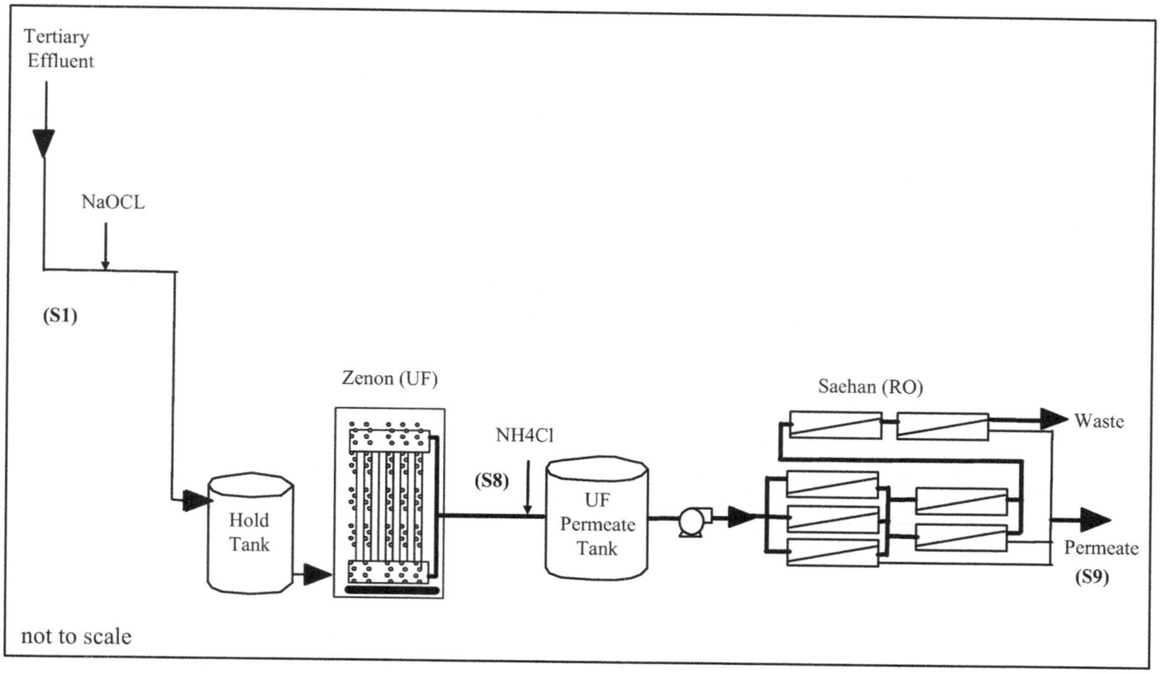

Figure B.13 Process flow diagram of MWH UF/RO pilot

Table B.2
Specifications of the ZW1000 UF membrane pilot

Parameter	Unit	Value
Manufacturer		ZENON Environmental
Membrane Model and ID Number		E1000-0061, 0066 & 0068
Membrane Commercial Designation		ZeeWeed® 1000
Approximate Size of Membrane Element	ft (m)	2.2 (0.68) x 2.0 (0.62) x 0.34 (0.104)
Active Membrane Area per Membrane Element	ft^2 (m^2)	350 ft^2 (32.5 m^2)
Number of Fibers per Element		30,000
Number of Elements (Operational)		3
Inside Diameter of Fiber	mm	0.35
Outside Diameter of Fiber	mm	0.6
Approximate Length of Fiber	ft (m)	2 (0.6)
Flow Direction		Outside-in
Nominal Molecular Weight Cutoff	Daltons	100,000
Absolute Molecular Weight Cutoff	Daltons	NA
Nominal Membrane Pore Size	micron	0.02
Absolute Membrane Pore Size	micron	0.10
Membrane Material/Construction		PVDF
Membrane Surface Characteristics		Hydrophilic
Membrane Charge		Non-ionic
Design Operating Pressure	psi	1.0 - 10.0
Design Flux at Design Pressure	gfd (l/hr-sq m)	5 - 50 (8.5 - 85)
Maximum Transmembrane Pressure	psi (bar)	12 (0.83)
Standard Testing pH		7
Acceptable Range of Operating pH Values		2 - 9.5
Standard Testing Temperature	degF (degC)	77 (25)
Acceptable Range of Operating Temperatures	degF (degC)	33.8 - 95 (1 - 35)
Maximum Permissible Turbidity	NTU	> 2,000
Chlorine/Oxidant Tolerance	ppm	> 2,000

Figure B.14 Photograph of the Zenon UF pilot system (MWH)

Figure B.15 Photograph of MWH RO pilot system

Table B.3
Operating parameters for MWH UF/RO pilot systems

Process	**Operating Parameters**
Ultrafiltration	Flux = 35 gfd @ 20°C
	Transmembrane pressure = 1 -10 psi
	Backwash frequency = 30 min
	Backwash pressure = 90 psi (Air)
	Flow mode = direct flow (no recirculation)
	Free chlorine dose = 1 - 2 mg/L
	Free chlorine dose during backwash = 0 mg/L
	Chemical cleaning: when P_{tm} = 7 psi
Reverse Osmosis	Flux = 12 gfd @ 25 °C
	Recovery 85%
	Feed pH = 7 - 8
	Antiscalant dose = 2 mg/L
	[1] Combined chlorine dose = 1-2 mg/L
	Chemical cleaning (per mfg recommendation)

[1] Formed by dosing ammonium chloride and sodium hypochlorite; no chlorine used during EDC/PCPP sampling.

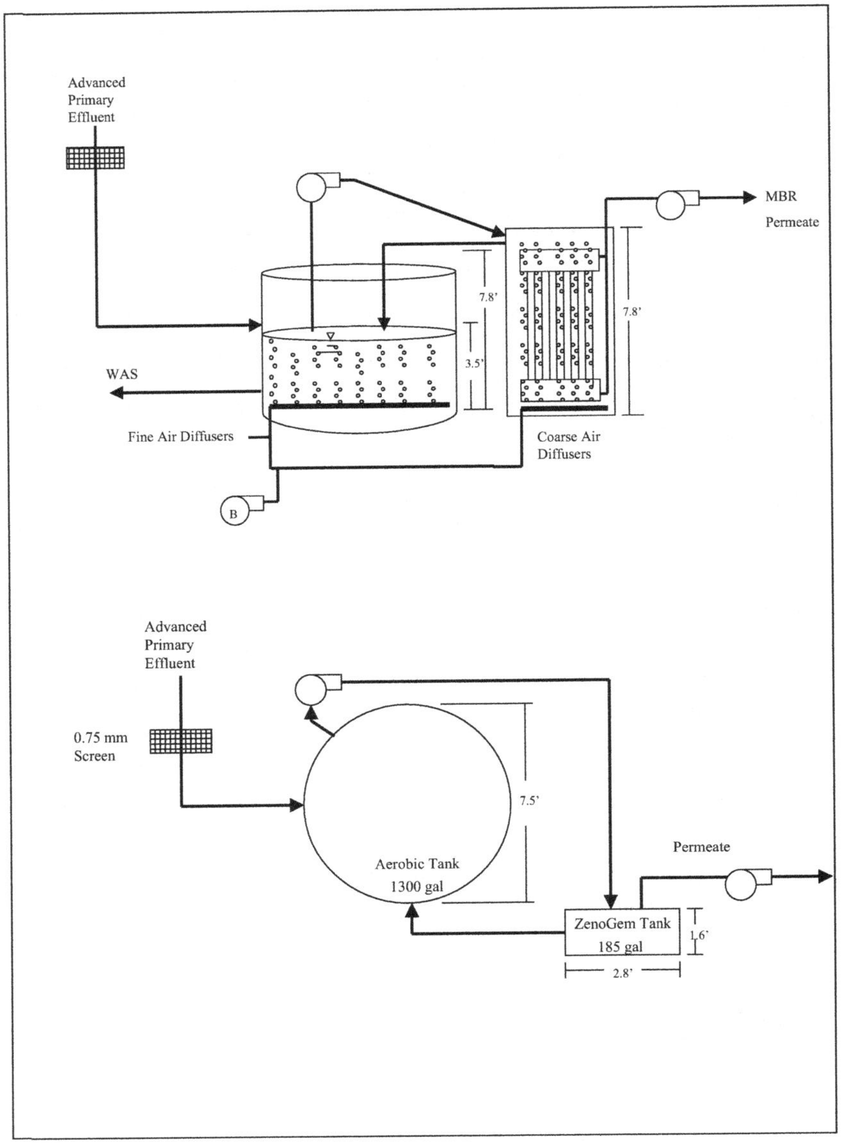

Figure B.16 MWH Zenon MBR: Side view (top); Plan view (bottom)

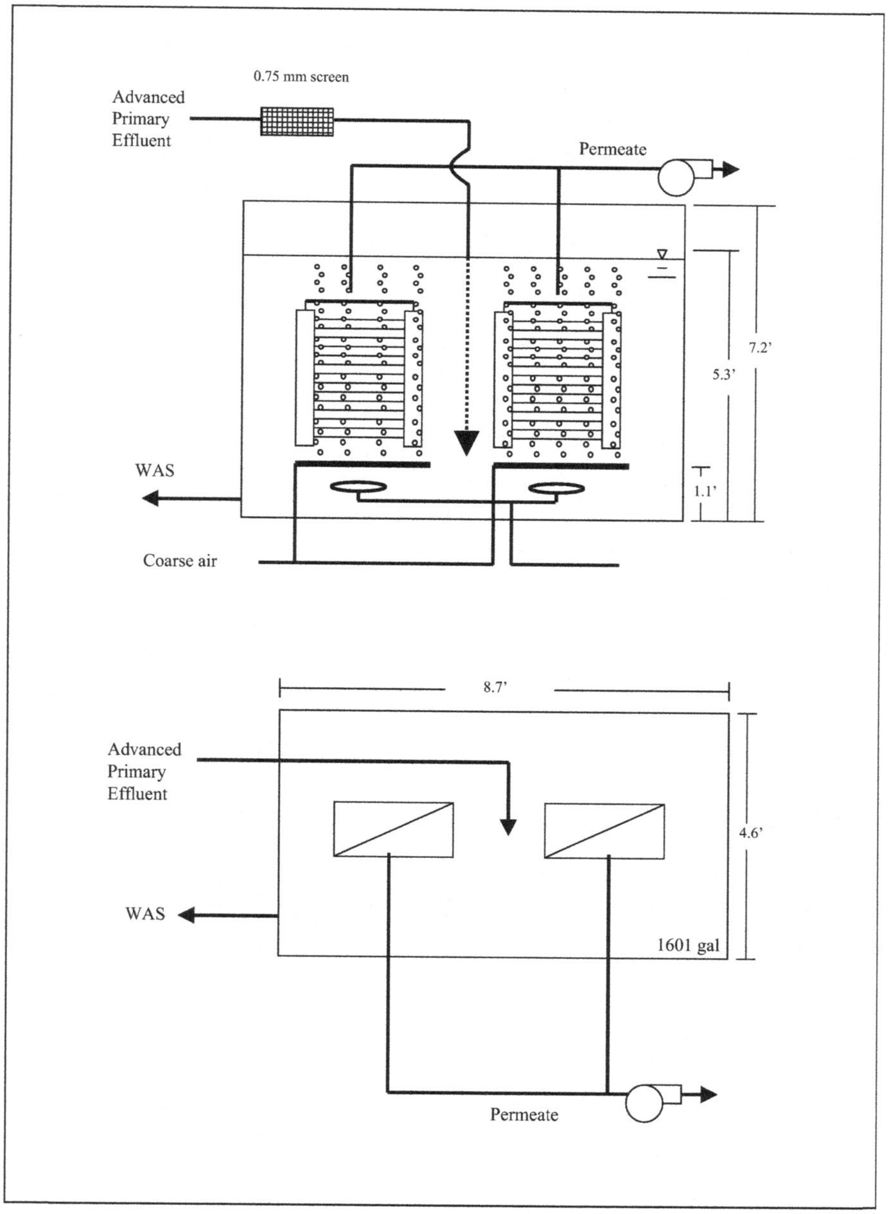

Figure B.17 MWH Mitsubishi MBR: Side view (top); Plan view (bottom)

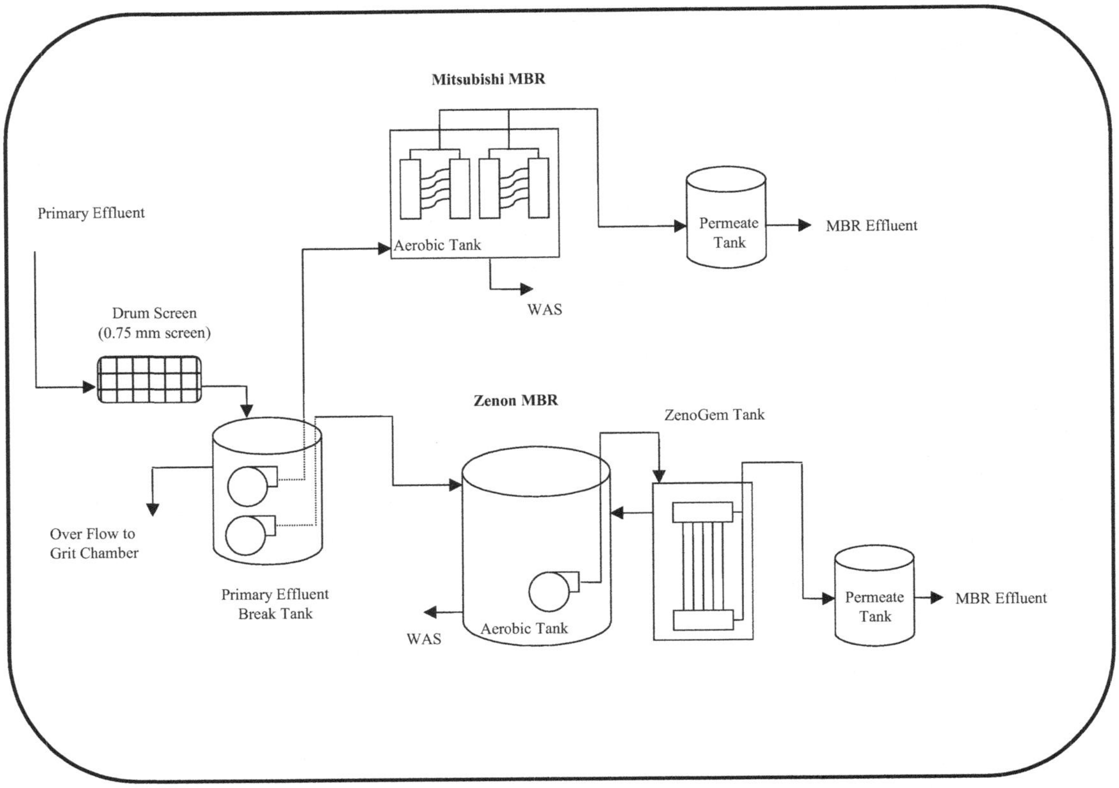

Figure B.18 Schematic diagram of MWH MBR pilot treatment train in southern california

Table B.4
MWH pilot RO membrane specifications

	OSMONICS	**SAEHAN**
Commerical designation	AK4040	RE 4040-FRM
Membrane material	Polyamide	Polyamide
	(thin-film composite)	(thin-film composite)
Operating pH range	4-11	3-10
Maximum feedwater turbidity	1 NTU	< 1 NTU
Maximum feedwater chlorine concentration	<0.1 ppm	<0.1 ppm
Maximum operating pressure	600 psig	600 psig
Nominal membrane surface area	85 ft^2	85 ft^2
Configuration	Spiral wound	Spiral wound
Element length	40.0 inches	40.0 inches
Element diameter	4.0 inches	4.0 inches

Figure B.19 MWH reverse osmosis pilot system

Figure B.20 Photographs of the MWH EDR pilot plant

Table B.5
Specifications of the Trojan UV system

Parameter	Unit	Value
Manufacturer		Trojan Technologies Inc.
Model and ID Number		UVLogic 8AL30
Inside Diameter of Reactor	cm	30
Lamp Type		LPHO
Lamp Configuration		Parallel to flow
Lamp Diameter (ceramic base)	cm	2.31
Quartz Sleeve Diameter (O.D.)	cm	2.8±0.3
Quartz Sleeve Diameter (I.D.)	cm	2.5
Number of Lamps		8
Total Lamp Power	Watts	2000
Output Power per Lamp	Watts	250
Length of Lamp	cm	156.21
Arc Length	cm	147
Lamp spacing (lamp radius)	cm	6.35

APPENDIX C
SUMMARY OF BENCH-SCALE DATA

Table C.1
Removal of EDCs/PPCPs by coagulation in bench-scale tests

Source Water	Colorado River Water			Ohio River Water				Passaic River Water			Suwannee River NOM			
Alum Dose (mg/L $Al_2(SO_4)_3$)		30			35 (pH=5.5)	35 (pH=8.5)			40			40 (pH=5.5)	40 (pH=8.5)	
Ferric Chloride Dose (mg/L $FeCl_3$)			28				8.3			38				9.5
Target Compound	Raw Conc. (ng/L)	Percent Removal		Raw Conc. (ng/L)	Percent Removal			Raw Conc. (ng/L)	Percent Removal		Raw Conc. (ng/L)	Percent Removal		
Acetaminophen (Tylenol)	30	<1	<1	11	<1	<1	<1	42	8	<1	2.4	1	<1	5
Androstenedione	99	17	<1	19	<1	<1	<1	91	<1	<1	10	32	15	12
Atrazine	56	<1	<1	60	<1	<1	<1	47	<1	<1	61	14	2	4
Benzo(a)pyrene	105	68	67	39	>74	>74	>74	60	72	80	115	>91	43	57
Caffeine	64	2	<1	38	<1	<1	<1	123	<1	<1	5.4	<1	<1	<1
Carbamazepine	64	7	<1	51	<1	<1	<1	127	<1	<1	47	19	4	9
DDT	106	46	45	58	9	<1	14	72	36	26	67	58	41	38
DEET	78	6	<1	57	<1	<1	<1	79	<1	<1	60	17	8	12
Diazepam (Valium)	56	4	<1	40	<1	<1	<1	57	<1	<1	44	15	6	6
Diclofenac	33	<1	<1	26	<1	<1	<1	11	<1	<1	3.8	<1	<1	<1
Dilantin	40	<1	<1	35	<1	<1	<1	44	<1	<1	3.7	11	8	14
Erythromycin	38	2	<1	28	<1	3	8	30	33	47	*			
Estradiol	98	13	<1	59	<1	<1	<1	94	2	0	9.5	24	3	11
Estriol	92	3	<1	*				89	<1	<1	*			
Estrone	98	9	<1	*				98	5	7	*			
Ethinyl Estradiol	102	17	<1	139	<1	<1	<1	99	0	<1	10	24	14	12
Fluorene	207	8	4	102	<1	<1	<1	159	<1	<1	*			
Fluoxetine (Prozac)	49	20	0	8.6	<1	<1	<1	46	15	24	3.8	<1	<1	<1
Galaxolide	143	19	16	137	<1	<1	<1	154	6	<1	91	34	20	18
Gemfibrozil	64	20	1	67	<1	<1	<1	55	<1	<1	6.8	6	0	3
Hydrocodone	66	6	0	*				77	24	18	*			
Ibuprofen (Advil)	38	<1	<1	38	<1	<1	<1	29	<1	<1	45	<1	<1	<1
Iopromide	63	12	<1	43	<1	<1	<1	100	<1	<1	3.3	<1	<1	<1
Lindane (γ-BHC)	171	22	5	138	<1	<1	<1	154	<1	<1	72	19	14	7
Meprobamate	68	0	<1	57	<1	<1	<1	59	<1	<1	48	10	2	6
Metolachlor	215	27	9	320	<1	<1	<1	211	<1	<1	95	37	20	13
Musk Ketone	182	18	<1	150	<1	<1	<1	176	10	<1	119	32	14	21
Naproxen	17	<1	<1	19	<1	<1	<1	13	<1	<1	1.7	<1	<1	<1
Oxybenzone	33	<1	<1	29	<1	<1	<1	31	<1	<1	4.9	<1	0	<1
Pentoxifylline	54	1	<1	29	<1	<1	<1	58	<1	<1	3.6	15	1	4
Progesterone	51	6	<1	9.8	<1	<1	24	45	<1	<1	4.3	23	16	18
Sulfamethoxazole	32	<1	<1	12	<1	<1	<1	7.7	<1	<1	1.1	<1	<1	<1
TCEP	33	<1	<1	30	<1	<1	<1	32	<1	<1	2.6	8	4	4
Testosterone	95	17	<1	11	<1	<1	<1	95	9	9	98	30	12	14
Triclosan	46	14	<1	33	<1	<1	<1	38	<1	<1	6.0	<1	<1	<1
Trimethoprim	76	4	<1	51	<1	<1	<1	83	13	2	5.9	<1	5	8

* Target Compound Not Analyzed

Table C.2
Removal of EDCs/PPCPs by softening in bench-scale tests

Source Water	Colorado River Water		Ohio River Water		Passaic River Water		Suwannee River NOM	
Lime (mg/L $Ca(OH)_2$)		238		135		133		53
Soda Ash (mg/L Na_2CO_3)		177		41		32		0
Target Compound	Raw Conc. (ng/L)	Percent Removal	Raw Conc. (ng/L)	Percent Removal	Raw Conc. (ng/L)	Percent Removal	Raw Conc. (ng/L)	Percent Removal
Acetaminophen (Tylenol)	30	<1	11	<1	42	29	2.4	<1
Androstenedione	99	4	19	<1	91	7	10	13
Atrazine	56	<1	60	<1	47	2	61	12
Benzo(a)pyrene	105	>90	39	>74	60	>83	115	47
Caffeine	64	<1	38	<1	123	14	5.4	<1
Carbamazepine	64	<1	51	<1	127	19	47	11
DDT	106	41	58	9	72	39	67	49
DEET	78	<1	57	<1	79	16	60	0
Diazepam (Valium)	56	<1	40	<1	57	15	44	8
Diclofenac	33	<1	26	<1	11	<1	3.8	<1
Dilantin	40	<1	35	<1	44	14	3.7	11
Erythromycin	38	<1	28	<1	30	33	*	
Estradiol	98	2	59	<1	94	4	9.5	7
Estriol	92	9	*		89	8	*	
Estrone	98	0	*		98	13	*	
Ethinyl Estradiol	102	4	139	<1	99	7	10	11
Fluorene	207	7	102	<1	159	<1	*	
Fluoxetine (Prozac)	49	8	8.6	<1	46	48	3.8	<1
Galaxolide	143	7	137	<1	154	<1	91	22
Gemfibrozil	64	3	67	1	55	<1	6.8	7
Hydrocodone	66	<1	*		77	23	*	
Ibuprofen (Advil)	38	<1	38	<1	29	<1	45	3
Iopromide	63	1	43	<1	100	14	3.3	<1
Lindane (γ-BHC)	171	8	138	11	154	<1	72	83
Meprobamate	68	<1	57	<1	59	12	48	6
Metolachlor	215	26	320	<1	211	<1	95	15
Musk Ketone	182	<1	150	<1	176	<1	119	9
Naproxen	17	<1	19	<1	13	<1	1.7	<1
Oxybenzone	33	<1	29	<1	31	12	4.9	<1
Pentoxifylline	54	<1	29	<1	58	18	3.6	6
Progesterone	51	<1	9.8	1	45	23	4.3	6
Sulfamethoxazole	32	<1	12	<1	7.7	<1	1.1	17
TCEP	33	<1	30	<1	32	7.2	2.6	15
Testosterone	95	<1	11	<1	95	15	98	14
Triclosan	46	1	33	<1	38	<1	6.0	<1
Trimethoprim	76	<1	51	<1	83	13	5.9	<1

* Target Compound Not Analyzed

Table C.3
Removal of EDCs/PPCPs by chlorination in bench-scale tests

Source Water	Colorado River Water			Ohio River Water			Passaic River Water			Suwannee River NOM		
Chlorine Type		Free	Free		Free	Free		Free	Free		Free	Free
Chlorine Dose [mg/L]		3.5	3.5		2.8	2.8		3.8	3.8		6.75	6.75
pH		5.5	8.2		5.5	7.9		5.5	6.8		5.5	8.5
Reaction Time [hrs]		24	24		24	24		24	24		24	24
Target Compound	Raw Conc. (n=1)	Percent Removal		Raw Conc. (n=1)	Percent Removal		Raw Conc. (n=1)	Percent Removal		Raw Conc. (n=1)	Percent Removal	
Acetaminophen (Tylenol)	28	>96	>96	29	>97	>97	41	97	>98	2	>90	>90
Androstenedione	94	40	23	80	86	25	100	57	44	10	22	13
Atrazine	59	15	2	56	<1	<1	45	24	11	61	25	17
Benzo(a)pyrene	34	>70	>70	82	>87	>87	30	>66	>66	115	91	91
Caffeine	78	68	17	84	>99	11	140	60	28	5	10	<1
Carbamazepine	65	>98	22	51	>98	20	122	92	33	47	40	15
DDT	68	25	8	83	11	<1	44	27	22	67	37	11
DEET	85	16	14	54	2	<1	83	33	31	60	15	2
Diazepam (Valium)	67	81	13	48	83	25	62	79	27	44	>99	13
Diclofenac	28	>96	>96	35	>97	>97	7.3	>86	>86	4	>93	>93
Dilantin	46	33	18	32	22	<1	43	53	35	4	54	22
Erythromycin	25	>96	>96	35	82	>97	25	>96	>96	*		
Estradiol	95	>99	>99	*			107	>99	>99	10	>97	>97
Estriol	104	>99	>99	*			99	>99	>99	*		
Estrone	95	>99	>99	160	>99	>99	112	>99	>99	*		
Ethinyl Estradiol	100	>99	>99	191	>99	>99	113	>99	>99	10	>98	>98
Fluorene	187	30	24	205	41	20	213	21	14	*		
Fluoxetine (Prozac)	50	20	19	32	13	<1	46	46	42	4	<1	<1
Galaxolide	106	40	15	134	25	4	147	56	39	91	54	40
Gemfibrozil	67	>99	70	69	>99	>99	68	>99	58	7	>96	56
Hydrocodone	63	>98	>98	*			70	>99	>99	*		
Ibuprofen (Advil)	47	45	37	43	37	23	35	78	59	45	47	47
Iopromide	58	7	28	58	5	21	94	27	27	3	<1	<1
Lindane (γ-BHC)	159	21	11	154	8	1	163	13	7	72	19	17
Meprobamate	71	15	18	50	18	<1	60	22	22	48	4	10
Metolachlor	183	32	35	480	36	21	248	50	41	95	12	6
Musk Ketone	198	26	>95	188	32	73	197	30	88	119	55	92
Naproxen	17	>94	>94	24	>96	>96	12	>92	>92	2	>86	>86
Oxybenzone	38	>97	>97	31	>97	>97	32	>97	>97	5	>95	92
Pentoxifylline	64	86	20	56	>98	30	62	74	27	4	12	<1
Progesterone	70	50	29	70	93	31	89	62	51	9	34	18
Sulfamethoxazole	38	>97	>97	23	>96	>96	4.7	>79	>79	1	>78	>78
TCEP	45	4	<1	26	<1	<1	41	22	12	3	<1	<1
Testosterone	90	52	26	64	95	28	97	61	48	98	39	19
Triclosan	44	>98	>98	34	>97	>97	46	>98	>98	6	>96	>96
Trimethoprim	88	>99	>99	64	>98	>98	90	>99	>99	6	>96	>96

* Target compound not analyzed

Table C.4
Removal of EDCs/PPCPs by chloramines and comparison with free chlorine in bench-scale tests

Source Water			Colorado River Water					
Chlorine Type			Free		Combined		Combined	
Chlorine Dose [mg/L]			3	3	2	2	3	3
pH			7.9	7.9	7.9	7.9	7.9	7.9
Reaction Time [hrs]			24	168	24	168	24	168
Target Compound	Raw Conc. (n=1)	Raw Conc. (n=1) (t=24 hrs)	Percent Removal		Percent Removal		Percent Removal	
Acetaminophen (Tylenol)	104	85	>99	>99	>99	>99	>99	>99
Androstenedione	328	301	4	<1	<1	<1	<1	<1
Atrazine	401	408	14	7	0	<1	<1	0
Benzo(a)pyrene	83	95	89	85	65	>89	79	>89
Caffeine	420	427	19	56	1	7	1	10
Carbamazepine	415	416	17	47	1	<1	<1	5
DDT	108	115	<1	<1	10	23	9	8
DEET	423	421	9	2	0	2	<1	6
Diazepam (Valium)	325	340	20	16	4	6	4	8
Diclofenac	329	331	>99	>99	47	96	62	98
Dilantin	391	399	15	4	1	3	<1	4
Erythromycin	14	15	>93	>93	7	42	20	63
Estradiol	352	342	>99	>99	99	>99	>99	>99
Estriol	378	389	>99	>99	>99	>99	>99	>99
Estrone	388	394	93	93	90	92	91	93
Ethinyl Estradiol	367	387	>99	>99	99	>99	>99	>99
Fluorene	211	185	<1	<1	<1	<1	<1	<1
Fluoxetine (Prozac)	223	224	14	14	8	14	<1	8
Galaxolide	168	199	33	47	22	49	26	38
Gemfibrozil	374	377	55	100	<1	<1	<1	9
Hydrocodone	330	329	>99	>99	35	87	50	96
Ibuprofen (Advil)	369	379	9	9	4	<1	<1	7
Iopromide	388	379	13	1	8	<1	<1	11
Lindane (γ-BHC)	336	358	<1	1	<1	11	<1	6
Meprobamate	429	422	28	50	<1	4	<1	9
Metolachlor	485	517	8	<1	6	14	<1	<1
Musk Ketone	529	507	>98	>98	17	44	4	8
Naproxen	317	321	98	>99	12	24	12	45
Oxybenzone	31	54	>98	>98	86	>98	84	>98
Pentoxifylline	383	385	17	58	0	<1	<1	4
Progesterone	285	242	0	<1	<1	<1	<1	<1
Sulfamethoxazole	380	373	>99	>99	11	35	14	55
TCEP	358	370	8	4	1	<1	2	5
Testosterone	305	276	2	<1	<1	<1	<1	<1
Triclosan	261	266	99	98	67	99	89	98
Trimethoprim	340	339	>99	>99	1	7	1	14

Table C.5
Removal of EDCs/PPCPs by MIEX in bench-scale tests

Source Water		Colorado River Water						
MIEX Dose (mL/L)		5	15	20	5	10	15	20
Contact Time (min)		10	10	10	20	20	20	20
Target Compound	Raw Conc. (n=2)	Percent Removal				Percent Removal		
Acetaminophen (Tylenol)	51	<1	<1	<1	<1	<1	<1	6.1
Androstenedione	183	<1	<1	<1	<1	<1	<1	<1
Atrazine	141	<1	<1	<1	<1	<1	1.0	<1
Benzo(a)pyrene	97	5.7	10	13	8.8	21	18	30
Caffeine	141	<1	<1	<1	<1	<1	<1	<1
Carbamazepine	137	<1	<1	<1	<1	4.0	4.8	<1
DDT	95	7.9	14	14	16	25	20	27
DEET	160	<1	<1	<1	<1	<1	<1	<1
Diazepam (Valium)	138	<1	5.7	4.3	<1	6.5	8.6	6.5
Diclofenac	125	61	86	90	78	86	91	93
Dilantin	98	22	36	41	23	34	44	45
Erythromycin*								
Estradiol	197	4.4	21	30	4.9	22	30	30
Estriol	183	<1	4.8	3.1	<1	3.1	1.5	<1
Estrone	183	<1	13	16	<1	12	20	20
Ethinyl Estradiol	187	12	35	43	19	34	45	48
Fluorene	169	31	55	63	43	54	63	69
Fluoxetine (Prozac)	115	<1	4.2	5.9	1.6	6.8	11	10
Galaxolide	129	11	35	40	22	41	46	53
Gemfibrozil	133	6.8	33	41	19	24	38	43
Hydrocodone	132	<1	<1	<1	<1	8.1	3.5	<1
Ibuprofen (Advil)	130	16	37	45	26	29	40	43
Iopromide	117	<1	<1	<1	<1	<1	<1	<1
Lindane (γ-BHC)	103	<1	5.4	16	<1	6.3	19	28
Meprobamate	146	<1	1.1	<1	1.8	1.1	<1	<1
Metolachlor	119	<1	<1	<1	<1	<1	<1	<1
Musk Ketone	120	4.2	20	37	<1	34	37	68
Naproxen	63	25	56	62	41	43	60	67
Oxybenzone	90	22	42	50	31	45	55	57
Pentoxifylline	131	<1	<1	<1	<1	<1	<1	<1
Progesterone	178	1.6	15	20	1.0	14	21	23
Sulfamethoxazole	99	<1	20	25	<1	20	28	32
TCEP	103	<1	<1	<1	3.8	<1	<1	<1
Testosterone	184	<1	<1	<1	<1	<1	<1	<1
Triclosan	106	73	91	95	89	93	96	96
Trimethoprim	162	<1	<1	<1	<1	11	<1	<1

* Target compound not analyzed

Table C.6
Removal of EDCs/PPCPs by ozone/AOP in bench-scale tests

Source Water	Colorado River Water			Ohio River Water				Passaic River Water			Suwannee River NOM			
Ozone Dose [mg/L]		2.5	2.5		3.5	7	3.5		3	3		4	8	8
Peroxide Dose [mg/L]			0.0625				0.0875			0.075				0.2
Reaction Time [min]		5	5		5	5	5		5	5		5	5	5
Target Compound	Raw Conc. (n=1)	Percent Removal		Raw Conc. (n=1)	Percent Removal			Raw Conc. (n=1)	Percent Removal		Raw Conc. (n=1)	Percent Removal		
Acetaminophen (Tylenol)	28	>96	>96	34	>97	>97	>97	41	96	96	2.4	>90	>90	>90
Androstenedione	94	>99	>99	49	>98	>98	>98	100	78	95	10	85	>98	>98
Atrazine	59	45	52	74	71	68	80	45	17	24	61	60	96	96
Benzo(a)pyrene	34	>71	>71	112	>91	>91	>91	30	>67	>67	115	>91	>91	>91
Caffeine	78	>99	>99	94	>99	>99	>99	140	83	87	5.4	87	>95	>95
Carbamazepine	65	>98	>98	61	>98	>98	>98	122	>99	>99	47	>99	>99	>99
DDT	68	57	61	84	58	65	71	44	35	27	67	77	>85	>85
DEET	85	76	84	65	95	94	96	83	63	68	60	90	99	>99
Diazepam (Valium)	67	79	86	51	96	94	96	62	66	69	44	90	96	>99
Diclofenac	28	>96	>96	34	>97	>97	>97	7.3	>86	>86	3.8	>93	>93	>93
Dilantin	46	82	89	44	>98	>98	>98	43	72	73	3.7	92	>93	>93
Erythromycin	25	>96	>96	42	>98	>98	>98	25	>96	>96	*			
Estradiol	95	>99	>99	136	99	99	>99	107	>99	>99	10	>97	>97	>97
Estriol	104	>99	>99	*				99	>99	>99	*			
Estrone	95	>99	>99	*				112	>99	>99	*			
Ethinyl Estradiol	100	>99	>99	169	>99	>99	>99	113	>99	>99	10	>98	>98	>98
Fluorene	187	94	>95	163	>94	>94	>94	213	83	83	*			
Fluoxetine (Prozac)	50	>98	>98	36	92	>97	>97	46	82	81	3.8	89	>93	>93
Galaxolide	106	87	89	150	91	90	91	147	76	77	91	79	>89	>89
Gemfibrozil	67	>99	>99	65	>98	>98	>98	68	>99	>99	6.8	>96	>96	>96
Hydrocodone	63	>98	>98	*				70	>99	>99	*			
Ibuprofen (Advil)	47	82	88	38	94	94	96	35	56	60	45	91	99	>99
Iopromide	58	46	60	47	75	77	86	94	52	57	3.3	69	>92	>92
Lindane (γ-BHC)	159	4	14	162	24	19	26	163	1	3	72	<1	27	32
Meprobamate	71	50	61	65	84	80	85	60	40	43	48	72	99	99
Metolachlor	183	80	86	588	92	88	94	248	73	82	95	81	86	86
Musk Ketone	198	30	34	232	50	41	68	197	20	3	119	47	76	79
Naproxen	17	>94	>94	20	>95	>95	>95	12	>92	>92	1.7	>86	>86	>86
Oxybenzone	38	>97	>97	33	>97	>97	>97	32	>97	>97	4.9	>95	>95	>95
Pentoxifylline	64	>98	>98	59	>98	>98	>98	62	86	90	3.6	89	>93	>93
Progesterone	70	>98	>99	38	>97	>97	>97	89	83	96	9.3	90	>97	>97
Sulfamethoxazole	38	>97	>97	20	>95	>95	>95	4.7	>79	>79	1.1	>78	>78	>78
TCEP	45	13	16	38	5	<1	16	41	<1	<1	2.6	<1	<1	<1
Testosterone	90	>99	>99	40	>98	>98	>98	97	83	96	98	95	>99	>99
Triclosan	44	79	82	35	>97	>97	>97	46	>98	>98	6.0	>96	91	>96
Trimethoprim	88	>99	>99	79	>99	>99	>99	90	>99	>99	5.9	>96	>96	>96

* Target compound not analyzed

Table C.7
Removal of EDCs/PPCPs by UV in bench-scale tests conducted at MWD

Source Water	Colorado River Water				Ohio River Water				Passaic River Water				Suwannee River NOM			
UV Dose [mJ/cm^2]		40	1000	1000		40	1000	1000		40	1000	1000		40	1000	1000
H_2O_2 Dose (mg/L)				5				5				5				5
Bovine Catalase (mg/L)				0.2				0.2				0.2				0.2
Target Compound	Raw Spiked Concentration (ng/L) (n=1)	Percent Removal			Raw Spiked Concentration (ng/L) (n=1)	Percent Removal			Raw Spiked Concentration (ng/L) (n=2)	Percent Removal			Raw Spiked Concentration (ng/L) (n=1)	Percent Removal		
Acetaminophen (Tylenol)	210	37	>99	68	213	13	99	92	193 ± 6	16	>98	71	198	26	>99	76
Androstenedione	391	<1	51	53	339	<1	41	51	404 ± 9	<1	41	48	452	15	51	59
Atrazine	329	5	70	65	338	<1	62	67	277 ± 18	2	56	58	298	17	75	75
Benzo(a)pyrene	104	29	63	77	176	<1	74	82	147 ± 64	<1	71	64	332	60	64	75
Caffeine	309	<1	9	4	316	<1	<1	8	720 ± 10	<1	<1	8	316	14	3	19
Carbamazepine	348	<1	9	8	322	<1	<1	22	355 ± 33	<1	<1	13	317	14	12	25
DDT	171	19	77	79	164	<1	83	85	132 ± 41	<1	77	73	176	63	80	81
DEET	339	1	6	3	318	<1	<1	14	323 ± 33	<1	<1	11	330	10	<1	14
Diazepam (Valium)	312	<1	5	4	275	<1	<1	15	274 ± 4	<1	<1	7	299	14	8	23
Diclofenac	300	55	>99	>99	247	36	>99	>99	259 ± 20	34	>99	>99	298	43	>99	>99
Dilantin	288	7	72	67	261	<1	61	62	308 ± 17	<1	63	64	333	17	61	67
Erythromycin	386	<1	15	6	263	<1	12	19	204 ± 8	<1	<1	<1	228	13	11	16
Estradiol	399	31	>99	88	328	8	99	89	413 ± 18	12	97	83	445	30	97	97
Estriol	431	26	>99	86	360	8	>99	88	446 ± 54	20	98	82	446	30	>99	97
Estrone	405	30	>99	92	325	8	>99	96	415 ± 16	14	>99	96	443	31	>99	>99
Ethinyl Estradiol	422	30	>99	87	354	9	99	88	406 ± 9	7	97	82	432	27	95	96
Fluorene	426	4	8	31	442	13	17	54	325 ± 124	<1	<1	12	358	27	<1	31
Fluoxetine (Prozac)	294	15	97	95	247	<1	92	93	281 ± 4	5	92	91	338	41	87	87
Galaxolide	142	<1	<1	23	148	7	19	42	141 ± 46	<1	11	17	114	38	34	35
Gemfibrozil	379	<1	<1	2	336	<1	<1	12	354 ± 14	<1	<1	<1	382	10	21	38
Hydrocodone	257	6	79	74	170	<1	62	42	238 ± 15	<1	44	36	245	7	48	44
Ibuprofen (Advil)	288	5	35	41	248	<1	27	34	356 ± 135	14	32	43	291	8	5	46
Iopromide	232	21	77	56	411	<1	76	82	400 ± 24	<1	70	72	318	34	79	82
Lindane (γ-BHC)	269	33	<1	<1	316	24	15	3	299 ± 17	<1	<1	<1	311	9	<1	13
Meprobamate	338	<1	4	3	310	<1	<1	5	298 ± 11	<1	<1	<1	346	16	13	16
Metolachlor	296	10	77	74	319	9	76	68	355 ± 2	7	65	64	364	11	85	85
Musk Ketone	194	<1	<1	<1	270	<1	33	30	370 ± 36	<1	32	33	305	3	37	39
Naproxen	190	25	99	94	269	15	97	96	205 ± 75	<1	96	92	248	30	98	98
Oxybenzone	206	5	63	23	175	<1	25	27	164 ± 11	<1	6	7	186	12	13	21
Pentoxifylline	300	<1	9	20	256	<1	<1	22	239 ± 1	<1	<1	9	273	12	11	29
Progesterone	245	3	55	57	237	<1	50	60	198 ± 6	<1	41	49	186	10	48	57
Sulfamethoxazole	185	54	>98	>98	52	54	>96	>96	14 ± 0	44	>85	79	149	32	96	93
TCEP	302	<1	5	<1	292	<1	<1	15	234 ± 2	<1	<1	5	226	2	<1	5
Testosterone	386	<1	50	53	338	<1	41	52	389 ± 12	<1	42	47	441	14	52	60
Triclosan	259	67	>99	>99	220	40	>99	>99	293 ± 29	21	>99	>99	302	52	>99	>99
Trimethoprim	357	<1	<1	56	264	<1	<1	4	265 ± 6	<1	<1	<1	545	15	25	33

Table C.8
Removal of EDCs/PPCPs by UV in bench-scale tests conducted at Trojan UV Technologies

Source Water						Lake Huron Water							
UV Dose [mJ/cm2]		39.5	439	787	1318	0	372	671	969	0	371	668	966
Peroxide Dose [mg/L]						5	5	5	5	7.5	7.5	7.5	7.5
Target Compound	Raw Conc. (n=1)	Percent Removal				Percent Removal				Percent Removal			
Acetaminophen (Tylenol)	57	81	>98	>98	97	>98	97	>98	>98	>98	>98	>98	>98
Androstenedione	125	<1	21	11	88	<1	77	89	89	<1	88	95	95
Atrazine	155	3	57	88	95	<1	67	90	96	<1	77	92	97
Benzo(a)pyrene*													
Caffeine	156	<1	16	51	61	<1	64	88	91	<1	79	93	95
Carbamazepine	163	<1	28	57	77	53	90	97	99	72	97	99	>99
DDT*													
DEET	148	<1	12	51	55	<1	68	89	95	<1	82	94	97
Diazepam (Valium)	96	<1	15	32	63	<1	73	90	94	<1	85	95	97
Diclofenac	107	88	>99	97	>99	92	>99	>99	>99	92	>99	>99	>99
Dilantin	126	3	77	90	>99	<1	93	99	>99	<1	97	>99	>99
Erythromycin	219	16	52	66	68	10	69	87	94	62	93	95	96
Estradiol	121	74	99	96	>99	>99	>99	>99	>99	>99	>99	>99	>99
Estriol	144	71	>97	>97	>97	>97	>97	>97	>97	>97	>97	>97	>97
Estrone	100	70	92	87	>99	>99	>99	>99	>99	>99	>99	>99	>99
Ethinyl Estradiol	113	73	98	94	>99	>99	>99	>99	>99	>99	>99	>99	>99
Fluorene*													
Fluoxetine (Prozac)	211	17	93	92	>99	<1	99	>99	99	13	>99	>99	>99
Galaxolide*													
Gemfibrozil	87	<1	21	43	72	<1	82	94	96	<1	90	96	98
Hydrocodone	136	25	95	>99	>99	89	98	>99	>99	>99	>99	>99	>99
Ibuprofen (Advil)	130	5	49	80	89	<1	88	97	>99	<1	93	99	>99
Iopromide	185	15	69	94	98	<1	77	94	98	<1	86	95	99
Lindane (γ-BHC)*													
Meprobamate	156	<1	<1	22	26	<1	42	69	81	<1	56	78	86
Metolachlor*													
Musk Ketone*													
Naproxen	112	29	>99	>99	>99	9	>99	>99	>99	54	>99	>99	>99
Oxybenzone	12	55	>92	>92	>92	<1	91	>92	>92	17	>92	>92	>92
Pentoxifylline	157	<1	20	51	64	<1	70	89	95	<1	81	94	98
Progesterone	443	<1	34	37	93	<1	84	90	87	<1	91	97	94
Sulfamethoxazole	130	78	>99	>99	>99	<1	>99	>99	>99	<1	>99	>99	>99
TCEP	144	<1	<1	<1	<1	<1	<1	8	5	<1	<1	3	<1
Testosterone	100	<1	44	58	92	<1	85	95	96	<1	93	97	98
Triclosan	168	83	97	78	>99	95	>99	>99	>99	>99	>99	>99	>99
Trimethoprim	153	<1	35	66	82	<1	83	95	98	<1	90	98	>99

* Target compound not analyzed

Table C.9
Removal of EDCs/PPCPs by PAC in bench-scale tests

Source Water	Colorado River Water		Ohio River Water			Passaic River Water		Suwannee River NOM				
PAC Dose (mg/L)		5		5	5		5		5	1	5	20
PAC Name		WPM		AC 800	WPM		WPM		AC 800	WPM	WPM	WPM
Contact Time (hrs)		4hrs @100 rpm		4hrs @100 rpm			4hrs @100 rpm		4hrs @100 rpm			
Target Compound	Raw Conc. (n=1)	Percent Removal	Raw Conc. (n=1)	Percent Removal		Raw Conc. (n=1)	Percent Removal	Raw Conc. (n=1)	Percent Removal			
Acetaminophen (Tylenol)	89	88	21	61	70	52	68	2.4	85	30	86	>90
Androstenedione	60	58	62	89	84	102	68	10	>98	63	>98	>98
Atrazine	66	55	70	69	60	52	53	61	90	25	85	99
Benzo(a)pyrene	155	>94	105	>90	>90	66	>85	115	69	52	72	87
Caffeine	31	16	70	74	73	128	64	5.4	89	<1	83	95
Carbamazepine	52	55	52	80	79	132	62	47	94	40	92	99
DDT	149	81	84	78	74	55	56	67	>85	77	>85	>85
DEET	36	<1	56	49	46	84	40	60	83	24	74	99
Diazepam (Valium)	60	56	39	69	69	60	52	44	96	43	92	99
Diclofenac	58	64	38	49	38	17	45	3.8	36	-2	51	>93
Dilantin	13	<1	36	44	40	47	46	3.7	85	26	81	>93
Erythromycin	31	44	32	48	50	35	64	*				
Estradiol	23	2	131	93	91	106	71	9.5	>97	64	>97	97
Estriol	85	55	*			96	61	*				
Estrone	90	78	*			110	75	*				
Ethinyl Estradiol	91	67	154	87	83	108	61	10	>98	57	>98	>98
Fluorene	186	>95	143	>93	>93	188	94	*				
Fluoxetine (Prozac)	82	93	25	94	93	49	89	3.8	>93	85	>93	>93
Galaxolide	162	63	125	60	44	141	46	91	88	56	81	89
Gemfibrozil	22	<1	74	51	38	58	18	6.8	60	17	57	>96
Hydrocodone	81	70	*			78	74	*				
Ibuprofen (Advil)	65	48	43	20	2	33	17	45	12	3	37	80
Iopromide	75	33	45	6	9	113	30	3.3	39	14	53	>92
Lindane (γ-BHC)	154	66	160	86	81	148	63	72	>86	50	>86	>86
Meprobamate	36	<1	54	26	31	63	32	48	58	9	45	94
Metolachlor	193	56	396	34	32	200	31	95	85	37	79	>89
Musk Ketone	195	71	143	76	58	186	54	119	>92	52	91	>92
Naproxen	64	86	18	41	44	17	49	1.7	43	<1	62	>86
Oxybenzone	45	93	30	>97	>97	33	92	4.9	91	76	92	>95
Pentoxifylline	53	65	41	82	82	62	66	3.6	>93	45	>93	>93
Progesterone	100	89	25	92	87	54	77	4.3	>94	66	>94	>94
Sulfamethoxazole	49	42	16	6	40	9.3	43	1.1	41	<1	46	>78
TCEP	66	70	29	47	50	35	50	2.6	81	15	70	>90
Testosterone	42	36	49	87	83	105	67	98	99	60	98	99
Triclosan	97	93	44	>98	95	41	83	6.0	>96	75	>96	95
Trimethoprim	41	38	59	92	91	87	79	5.9	>96	81	>96	>96

* Target compound not analyzed

Table C.10
Removal of EDCs/PPCPs by UF in bench-scale tests

Source Water	Colorado River Water		Ohio River Water		Passaic River Water		Suwannee River NOM	
Membrane	Ultrafiltration		Ultrafiltration		Ultrafiltration		Ultrafiltration	
Target Compound	Raw Conc. (n=1)	Percent Removal	Raw Conc. (n=1)	Percent Removal	Raw Conc. (n=1)	Percent Removal	Raw Conc. (n=1)	Percent Removal
Acetaminophen (Tylenol)	89	64	25	12	31	29	2.4	<1
Androstenedione	60	<1	51	4	84	93	10.2	6
Atrazine	66	6	78	6	54	70	61.3	<1
Benzo(a)pyrene	155	>94	89	>88	40	>75	115.0	>91
Caffeine	31	<1	80	1	127	66	5.4	<1
Carbamazepine	52	<1	53	6	112	66	47.0	<1
DDT	149	>93	91	>89	47	>78	66.5	>85
DEET	36	<1	63	6	77	77	60.0	<1
Diazepam (Valium)	60	7	43	14	59	78	43.7	<1
Diclofenac	58	50	42	2	27	59	3.8	<1
Dilantin	13	<1	39	5	42	67	3.7	<1
Erythromycin	31	61	42	48	58	93	*	
Estradiol	23	<1	119	<1	87	75	9.5	<1
Estriol	85	-26	*		85	74	*	
Estrone	90	14	*		84	71	*	
Ethinyl Estradiol	91	12	152	18	92	80	10.2	16
Fluorene	186	>95	114	89	171	>94	*	
Fluoxetine (Prozac)	82	98	39	>97	39	93	3.8	>93
Galaxolide	162	78	172	78	92	60	90.5	55
Gemfibrozil	22	<1	70	<1	67	79	6.8	<1
Hydrocodone	81	19	*		57	77	*	
Ibuprofen (Advil)	65	28	46	<1	50	68	45.3	<1
Iopromide	75	36	50	<1	112	86	3.3	<1
Lindane (γ-BHC)	154	24	159	18	137	75	72.0	8
Meprobamate	36	<1	58	3	59	71	48.0	<1
Metolachlor	193	19	535	41	161	80	95.0	12
Musk Ketone	195	63	179	63	170	78	118.5	54
Naproxen	64	70	20	<1	16	25	1.7	<1
Oxybenzone	45	84	32	78	31	95	4.9	43
Pentoxifylline	53	<1	45	7	56	70	3.6	<1
Progesterone	100	76	19	48	40	86	4.3	48
Sulfamethoxazole	49	20	21	19	11	20	1.1	<1
TCEP	66	32	32	22	36	69	2.6	4
Testosterone	42	<1	38	0	84	93	97.7	<1
Triclosan	97	93	46	88	43	92	6.0	84
Trimethoprim	41	<1	71	20	74	77	5.9	71

* Target compound not analyzed

Table C.11
Removal of EDCs/PPCPs by NF in bench-scale tests

Source Water	Colorado River Water		Ohio River Water		Passaic River Water		Suwannee River NOM	
Membrane	Nanofiltration		Nanofiltration		Nanofiltration		Nanofiltration	
Target Compound	Raw Conc. (n=1)	Percent Removal	Raw Conc. (n=1)	Percent Removal	Raw Conc. (n=1)	Percent Removal	Raw Conc. (n=1)	Percent Removal
Acetaminophen (Tylenol)	90	83	21	14	31	-16	2.4	1
Androstenedione	67	64	62	53	84	46	10	62
Atrazine	63	65	70	79	54	11	61	25
Benzo(a)pyrene	154	>94	105	>90	40	>75	115	>91
Caffeine	45	31	70	56	127	9	5.4	<1
Carbamazepine	58	60	52	77	112	6	47	28
DDT	123	>92	84	>88	47	>78	67	<1
DEET	47	57	56	75	77	10	60	30
Diazepam (Valium)	67	76	39	79	59	14	44	38
Diclofenac	59	75	38	74	27	<1	3.8	42
Dilantin	17	18	36	69	42	14	3.7	8
Erythromycin	38	78	32	96	58	36	*	
Estradiol	25	<1	131	71	87	0	9.5	<1
Estriol	95	63	*		85	<1	*	
Estrone	95	66	*		84	<1	*	
Ethinyl Estradiol	104	77	154	77	92	9	10	65
Fluorene	198	>95	143	>93	171	>94	*	
Fluoxetine (Prozac)	70	92	25	<1	39	>97	3.8	>93
Galaxolide	159	82	125	86	92	54	91	85
Gemfibrozil	28	14	74	82	67	<1	6.8	50
Hydrocodone	88	81	*		57	2	*	
Ibuprofen (Advil)	78	77	43	80	50	8	45	47
Iopromide	85	90	45	86	112	38	3.3	12
Lindane (γ-BHC)	160	71	160	86	137	18	72	61
Meprobamate	38	32	54	70	59	5	48	8
Metolachlor	187	78	396	74	161	<1	95	64
Musk Ketone	188	76	143	>93	170	45	119	85
Naproxen	65	87	18	17	16	6	1.7	8
Oxybenzone	50	96	30	88	31	83	4.9	86
Pentoxifylline	64	66	41	66	56	2	3.6	4
Progesterone	100	92	25	52	40	60	4.3	74
Sulfamethoxazole	48	71	16	56	11	29	1.1	<1
TCEP	71	82	29	84	36	11	2.6	46
Testosterone	46	50	49	76	84	45	98	54
Triclosan	94	96	44	93	43	92	6.0	87
Trimethoprim	44	41	59	76	74	8	5.9	54

* Target compound not analyzed

APPENDIX D
SUMMARY OF PILOT-SCALE DATA

Table D.1
Removal of EDCs/PPCPs by coagulation for ORW in pilot-scale tests

Source Water	Ohio River Water							
Coagulant Dose	8 mg/L $FeCl_3$		26.5 mg/L $FeCl_3$		35 mg/L $Al_2(SO_4)_3$		35 mg/L $Al_2(SO_4)_3$	
Target Compound	Raw Conc (n=2) (ng/L)	Percent Removal	Raw Conc (n=3) (ng/L)	Percent Removal	Raw Conc (n=2) (ng/L)	Percent Removal	Raw Conc (n=2) (ng/L)	Percent Removal
Acetaminophen (Tylenol)	57	<1	35	<1	58	<1	64	<1
Androstenedione	111	27	51	<1	112	<1	135	<1
Atrazine	687	<1	1057	4.4	473	<1	433	<1
Benzo(a)pyrene*								
Caffeine	116	<1	188	2.3	168	<1	162	<1
Carbamazepine	57	<1	396	54	120	<1	116	<1
DDT*								
DEET	74	<1	411	48	140	<1	133	<1
Diazepam (Valium)*								
Diclofenac	52	<1	83	<1	66	<1	66	<1
Dilantin	58	<1	188	46	78	<1	74	<1
Erythromycin	60	<1	108	24	153	1.3	125	<1
Estradiol	127	<1	336	34	197	<1	190	<1
Estriol	135	<1	124	<1	177	<1	178	<1
Estrone	116	<1	204	<1	171	<1	170	<1
Ethinyl Estradiol	129	<1	270	9.3	189	<1	180	<1
Fluorene	74	<1	42	<1	70	2.9	106	12
Fluoxetine (Prozac)	40	<1	108	42	66	<1	66	<1
Galaxolide	41	<1	26	<1	43	<1	55	4.6
Gemfibrozil	75	<1	84	<1	93	<1	94	0.5
Hydrocodone*								
Ibuprofen (Advil)	49	<1	142	18	86	<1	82	3.7
Iopromide	108	3.3	70	<1	109	5.1	106	16
Lindane (γ-BHC)*								
Meprobamate*								
Metolachlor	253	<1	365	<1	220	<1	247	24
Musk Ketone	67	<1	38	6.1	64	<1	79	4
Naproxen	21	<1	57	<1	28	<1	32	1.6
Oxybenzone	37	<1	24	<1	39	<1	45	<1
Pentoxifylline	67	<1	372	58	126	<1	123	<1
Progesterone	62	40	18	<1	44	<1	66	<1
Sulfamethoxazole	31	<1	40	<1	35	<1	38	<1
TCEP	37	<1	311	58	84	<1	82	<1
Testosterone*								
Triclosan	20	<1	8	<1	15	<1	24	23
Trimethoprim	120	<1	163	14	144	<1	144	<1

* Target Compound Not Analyzed

Table D.2
Removal of EDCs/PPCPs by coagulation for CRW in pilot-scale tests

Source Water	Colorado River Water (CRW)	
Coagulant Dose	0.25 mg/L $FeCl_3$	
Target Compound	Raw Conc (n=3) (ng/L)	Percent Removal
Acetaminophen (Tylenol)	39	1.1
Androstenedione	175	10
Atrazine	130	<1
Benzo(a)pyrene	107	12
Caffeine	104	1.5
Carbamazepine	122	8.0
DDT*		
DEET	135	<1
Diazepam (Valium)	107	<1
Diclofenac	111	3.4
Dilantin	89	<1
Erythromycin	111	<1
Estradiol	173	15
Estriol	164	13
Estrone	182	18
Ethinyl Estradiol	169	14
Fluorene	175	5.0
Fluoxetine (Prozac)	82	3.0
Galaxolide	57	16
Gemfibrozil	139	3.1
Hydrocodone	81	7.7
Ibuprofen (Advil)	117	<1
Iopromide	112	<1
Lindane (γ-BHC)*		
Meprobamate	127	<1
Metolachlor	143	9.0
Musk Ketone	161	<1
Naproxen	58	14
Oxybenzone	77	13
Pentoxifylline	101	3.3
Progesterone	106	9.6
Sulfamethoxazole	57	<1
TCEP	88	<1
Testosterone	164	12
Triclosan	102	17
Trimethoprim	110	6.8

* Target Compound Not Analyzed

Table D.3
Removal of EDCs/PPCPs by MIEX in pilot-scale tests

Source Water	Colorado River Water		
MIEX(R) Dose [mL/L]		20	20
Contact Time [min]		10	20
Target Compound	Raw Conc. (n=2) (ng/L)	Percent Removal	
Acetaminophen (Tylenol)	17	<1	29
Androstenedione	83	27	25
Atrazine	67	1	3
Benzo(a)pyrene*			
Caffeine	69	6	<1
Carbamazepine	38	<1	3
DDT	17	24	18
DEET	71	<1	<1
Diazepam (Valium)	44	<1	9
Diclofenac	53	90	90
Dilantin	54	22	26
Erythromycin*			
Estradiol	114	46	43
Estriol	116	20	22
Estrone	109	39	39
Ethinyl Estradiol	111	53	50
Fluorene	52	69	67
Fluoxetine (Prozac)	59	15	24
Galaxolide	39	51	46
Gemfibrozil	61	46	41
Hydrocodone	51	<1	14
Ibuprofen (Advil)	52	50	33
Iopromide	56	23	<1
Lindane (γ-BHC)	68	71	53
Meprobamate	60	<1	<1
Metolachlor	86	52	17
Musk Ketone	52	52	46
Naproxen	27	59	26
Oxybenzone	17	41	56
Pentoxifylline	34	<1	12
Progesterone	34	35	41
Sulfamethoxazole	37	49	30
TCEP	35	<1	<1
Testosterone	103	21	21
Triclosan	19	91	>90
Trimethoprim	74	<1	41

* Target Compound Not Analyzed

Table D.4
Removal of EDCs/PPCPs by ozonation in pilot-scale tests

Source Water		Colorado River Water							
Ozone Dose*		1.25 mg/L				2.50 mg/L			
Reaction Time [min]		2	6	14	24	2	6	14	24
Target Compound	Raw Conc. (n=9)	Percent Removal				Percent Removal			
Acetaminophen (Tylenol)	117	95	99	99	>99	>99	>99	>99	>99
Androstenedione	343	48	77	81	89	73	98	98	>99
Atrazine	367	9	7	8	26	27	40	55	57
Benzo(a)pyrene	113	76	68	68	62	†	†	†	41
Caffeine	363	64	94	96	98	84	>97	>97	>97
Carbamazepine	373	>99	>99	>99	>99	>99	>99	>99	>99
DDT	83	13	<1	<1	24	†	†	†	14
DEET	375	25	36	40	45	42	62	77	76
Diazepam (Valium)	296	28	41	44	56	43	63	77	82
Diclofenac	313	>99	>99	>99	>99	>99	>99	>99	>99
Dilantin	359	26	43	48	52	48	71	81	86
Erythromycin	12	>89	>89	>89	>89	>92	>92	>92	>92
Estradiol	347	99	100	>99	>99	>99	>99	>99	>99
Estriol	361	>98	>98	>98	>98	>99	>99	>99	>99
Estrone	356	95	99	99	>99	97	>99	>99	>99
Ethinyl Estradiol	348	>99	>99	>99	>99	>99	>99	>99	>99
Fluorene	349	48	74	76	74	†	†	†	72
Fluoxetine (Prozac)	208	99	>99	>99	>99	>99	>99	>99	>99
Galaxolide	121	29	34	43	60	†	†	†	57
Gemfibrozil	357	>99	>99	>99	>99	>99	>99	>99	>99
Hydrocodone	303	>99	>99	>99	>99	>99	>99	>99	>99
Ibuprofen (Advil)	363	31	42	47	51	51	71	82	87
Iopromide	354	4	11	18	21	38	45	54	61
Lindane (γ-BHC)	280	<1	<1	<1	<1	†	†	†	9
Meprobamate	375	14	19	22	25	32	45	57	59
Metolachlor	454	43	37	41	51	†	†	†	68
Musk Ketone	260	<1	<1	<1	<1	†	†	†	17
Naproxen	298	>99	>99	>99	>99	>99	>99	>99	>99
Oxybenzone	72	>99	>99	>99	>99	>99	>99	>99	>99
Pentoxifylline	347	67	97	98	99	88	>99	100	>99
Progesterone	293	54	84	86	93	78	99	98	>99
Sulfamethoxazole	327	99	>99	>99	>99	>99	>99	>99	>99
TCEP	341	<1	<1	<1	2	7	6	15	0
Testosterone	337	58	87	89	94	80	99	99	>99
Triclosan	259	98	99	99	98	98	99	99	99
Trimethoprim	312	>99	>99	>99	>99	>99	>99	>99	>99

* Testing conducted using a 1-Lpm pilot plant

† Target compound not analyzed

Table D.5
Removal of EDCs/PPCPs by ozonation in bench-top pilot-scale tests

Source Water		Colorado River Water							
Ozone Dose*		1.25 mg/L				2.50 mg/L			
Reaction Time [min]		2	6	14	24	2	6	14	24
Target Compound	Raw Conc. (n=9)	Percent Removal				Percent Removal			
Acetaminophen (Tylenol)	117	95	99	99	>99	>99	>99	>99	>99
Androstenedione	343	48	77	81	89	73	98	98	>99
Atrazine	367	9	7	8	26	27	40	55	57
Benzo(a)pyrene	113	76	68	68	62	†	†	†	41
Caffeine	363	64	94	96	98	84	>97	>97	>97
Carbamazepine	373	>99	>99	>99	>99	>99	>99	>99	>99
DDT	83	13	<1	<1	24	†	†	†	14
DEET	375	25	36	40	45	42	62	77	76
Diazepam (Valium)	296	28	41	44	56	43	63	77	82
Diclofenac	313	>99	>99	>99	>99	>99	>99	>99	>99
Dilantin	359	26	43	48	52	48	71	81	86
Erythromycin	12	>89	>89	>89	>89	>92	>92	>92	>92
Estradiol	347	99	100	>99	>99	>99	>99	>99	>99
Estriol	361	>98	>98	>98	>98	>99	>99	>99	>99
Estrone	356	95	99	99	>99	97	>99	>99	>99
Ethinyl Estradiol	348	>99	>99	>99	>99	>99	>99	>99	>99
Fluorene	349	48	74	76	74	†	†	†	72
Fluoxetine (Prozac)	208	99	>99	>99	>99	>99	>99	>99	>99
Galaxolide	121	29	34	43	60	†	†	†	57
Gemfibrozil	357	>99	>99	>99	>99	>99	>99	>99	>99
Hydrocodone	303	>99	>99	>99	>99	>99	>99	>99	>99
Ibuprofen (Advil)	363	31	42	47	51	51	71	82	87
Iopromide	354	4	11	18	21	38	45	54	61
Lindane (γ-BHC)	280	<1	<1	<1	<1	†	†	†	9
Meprobamate	375	14	19	22	25	32	45	57	59
Metolachlor	454	43	37	41	51	†	†	†	68
Musk Ketone	260	<1	<1	<1	<1	†	†	†	17
Naproxen	298	>99	>99	>99	>99	>99	>99	>99	>99
Oxybenzone	72	>99	>99	>99	>99	>99	>99	>99	>99
Pentoxifylline	347	67	97	98	99	88	>99	100	>99
Progesterone	293	54	84	86	93	78	99	98	>99
Sulfamethoxazole	327	99	>99	>99	>99	>99	>99	>99	>99
TCEP	341	<1	<1	<1	2	7	6	15	0
Testosterone	337	58	87	89	94	80	99	99	>99
Triclosan	259	98	99	99	98	98	99	99	99
Trimethoprim	312	>99	>99	>99	>99	>99	>99	>99	>99

* Testing conducted using a 1-Lpm pilot plant

† Target compound not analyzed

Table D.6
Removal of EDCs/PPCPs by ozonation in traditional pilot-scale tests

Source Water		Colorado River Water		
Ozone Dose [mg/L]*		2.4 mg/L		
Reaction Time [min]		2	6	24
Target Compound	Raw Conc. (n=9) (ng/L)	Percent Removal		
Acetaminophen (Tylenol)	39	92	>97	>97
Androstenedione	175	87	99	>99
Atrazine	130	16	43	54
Benzo(a)pyrene	107	73	83	87
Caffeine	104	94	>99	>99
Carbamazepine	122	>99	>99	>99
DDT	77	25	12	35
DEET	135	41	62	76
Diazepam (Valium)	107	46	70	82
Diclofenac	111	>99	>99	>99
Dilantin	89	42	72	83
Erythromycin	111	>99	>99	>99
Estradiol	173	>99	>99	>99
Estriol	164	>99	99	>99
Estrone	182	>99	>99	>99
Ethinyl Estradiol	169	>99	>99	>99
Fluorene	175	65	82	89
Fluoxetine (Prozac)	82	>99	>99	>99
Galaxolide	57	69	77	81
Gemfibrozil	139	>99	>99	>99
Hydrocodone	81	>99	>99	>99
Ibuprofen (Advil)	117	40	69	83
Iopromide	112	30	44	64
Lindane (γ-BHC)	135	5.3	8.2	19
Meprobamate	127	23	41	54
Metolachlor	143	56	73	83
Musk Ketone	161	19	24	34
Naproxen	58	>98	>98	>98
Oxybenzone	77	>99	>99	>99
Pentoxifylline	101	97	>99	>99
Progesterone	163	91	99	>99
Sulfamethoxazole	57	>98	>98	>98
TCEP	88	<1	4.9	<1
Testosterone	164	91	99	>99
Triclosan	102	>99	>99	>99
Trimethoprim	110	>99	>99	>99

* Testing conducted using a 6-gpm pilot plant

Table D.7
Removal of EDCs/PPCPs by ozone AOP in pilot-scale tests

Source Water		Colorado River Water															
Ozone Dose [mg/L]*		1.25				2.5				1.25				2.5			
Peroxide Dose [mg/L]		0.0625				0.0625				0.25				0.5			
Reaction Time [min]		2	6	14	24	2	6	14	24	2	6	14	24	2	6	14	24
Target Compound	Raw Conc. (n=9)	Percent Removal				Percent Removal				Percent Removal				Percent Removal			
Acetaminophen (Tylenol)	117	>97	98	96	>99	>99	>99	>99	>99	>98	97	97	>98	>99	>99	99	>99
Androstenedione	343	50	87	78	80	70	95	98	99	57	66	72	73	80	93	94	94
Atrazine	367	21	46	42	37	23	39	52	57	19	25	28	32	37	57	62	56
Benzo(a)pyrene	113	†	†	†	64	68	†	†	35	<1	<1	20	30	†	†	†	24
Caffeine	363	63	95	91	93	85	>97	>97	>97	68	80	81	82	86	96	97	97
Carbamazepine	373	>99	>99	>99	>99	>99	>99	>99	>99	>99	>99	>99	>99	>99	>99	>99	>99
DDT	83	†	†	†	32	8	†	†	<1	<1	<1	<1	14	†	†	†	<1
DEET	375	35	66	59	55	48	67	76	82	44	55	57	57	70	84	87	85
Diazepam (Valium)	296	37	67	59	63	51	71	81	85	50	61	61	63	73	89	91	90
Diclofenac	313	>99	>99	>99	>99	>99	>99	>99	>99	>99	>99	>99	>99	>99	>99	>99	>99
Dilantin	359	44	73	67	64	55	77	87	88	56	65	64	69	80	93	94	93
Erythromycin	12	>90	>90	>90	>90	>92	>92	>92	>92	>92	>92	>92	>92	>92	>92	>92	>92
Estradiol	347	>99	>99	>99	100	>99	>99	>99	>99	>99	99	>99	>99	>99	>99	>99	>99
Estriol	361	>99	>99	>99	>99	>99	>99	>99	>99	>99	>99	>99	>99	>98	>98	>98	>98
Estrone	356	95	>99	99	98	97	100	>99	>99	96	97	98	>99	98	>99	>99	>99
Ethinyl Estradiol	348	>99	>99	>99	>99	>99	>99	>99	>99	>99	>99	>99	>99	>99	>99	>99	>99
Fluorene	349	†	†	†	79	70	†	†	19	<1	<1	14	27	†	†	†	82
Fluoxetine (Prozac)	208	99	>99	>99	>99	>99	>99	>99	>99	99	>99	99	>99	99	>99	>99	>99
Galaxolide	121	†	†	†	68	73	†	†	38	30	47	61	70	†	†	†	52
Gemfibrozil	357	>99	>99	>99	>99	>99	>99	>99	>99	>99	>99	>99	>99	>99	>99	>99	>99
Hydrocodone	303	>99	>99	>99	>99	>99	>99	>99	>99	>99	>99	>99	>99	>99	>99	>99	>99
Ibuprofen (Advil)	363	39	72	65	61	54	78	86	88	51	60	65	65	76	89	90	89
Iopromide	354	30	52	49	39	34	50	54	58	28	34	36	38	42	58	63	58
Lindane (γ-BHC)	280	†	†	†	18	20	†	†	<1	<1	-69	<1	<1	†	†	†	<1
Meprobamate	375	26	46	42	38	34	48	61	60	25	37	38	40	51	68	71	65
Metolachlor	454	†	†	†	66	65	†	†	76	3.2	<1	1.4	14	†	†	†	83
Musk Ketone	260	†	†	†	24	26	†	†	15	1.6	<1	4.9	11	†	†	†	14
Naproxen	298	>99	>99	>99	>99	>99	>99	>99	>99	>99	>99	>99	>99	>99	>99	>99	>99
Oxybenzone	72	>99	>99	>99	>99	>99	>99	>99	>99	>98	>98	>98	>98	>98	>98	>98	>98
Pentoxifylline	347	65	97	93	96	86	>99	>99	>99	71	83	84	85	88	98	98	98
Progesterone	293	54	90	83	86	76	96	99	99	65	74	79	80	85	96	96	96
Sulfamethoxazole	327	>99	>99	>99	>99	>99	>99	>99	>99	>99	>99	>99	>99	>99	>99	>99	>99
TCEP	341	6.0	11	12	16	<1	<1	<1	8.9	<1	<1	<1	9.3	4.5	9.2	17	<1
Testosterone	337	59	91	86	88	78	97	99	>99	65	74	79	80	86	96	97	97
Triclosan	259	98	99	99	98	98	98	99	98	98	98	98	99	98	99	99	99
Trimethoprim	312	>99	>99	>99	>99	>99	>99	>99	>99	>99	>99	>99	>99	>99	>99	>99	>99

* Testing conducted using a 1-Lpm pilot plant

† Target compound not analyzed

Table D.8
Removal of EDCs/PPCPs by PAC in pilot-scale tests

Source Water		Ohio River Water							
PAC Dose [mg/L]*		5 mg/L				35 mg/L			
Ferric Chloride Dose [mg/L]		8 mg/L				8 mg/L			
Reaction Time [min]		5 Hours		20 Hours		5 Hours		20 Hours	
Target Compound	log Kow	Avg (n=2)	Percent Removal	Avg (n=2)	Percent Removal	Avg (n=2)	Percent Removal	Avg (n=2)	Percent Removal
Acetaminophen (Tylenol)	0.46	76	58	56	42	66	97	73	98
Androstenedione	2.75	86	36	11	<1	43	92	50	96
Atrazine	2.61	406	47	403	38	413	94	475	96
Benzo(a)pyrene*	6.13								
Caffeine	-0.07	150	45	142	41	147	95	167	96
Carbamazepine	2.45	118	51	188	52	120	96	142	97
DDT	6.91	12	<1	12	>17	12	>13	12	<1
DEET	2.18	140	38	215	35	142	93	183	95
Diazepam (Valium)*	2.82								
Diclofenac	0.70	66	18	63	<1	66	83	86	89
Dilantin	2.47	88	45	111	21	81	90	84	93
Erythromycin	3.06	130	23	128	13	128	86	147	89
Estradiol	4.01	174	58	193	44	165	97	207	98
Estriol	2.45	175	43	143	33	155	97	166	>99
Estrone	3.13	216	71	257	63	240	>99	279	>99
Ethinyl Estradiol	3.67	197	55	213	36	195	96	245	98
Fluorene	4.18	117	83	49	67	92	>89	70	>86
Fluoxetine (Prozac)	4.05	55	56	63	35	49	94	73	97
Galaxolide	4.86	78	46	38	<1	47	45	38	39
Gemfibrozil	4.77	84	21	71	<1	85	85	104	91
Hydrocodone*	2.16								
Ibuprofen (Advil)	3.97	61	<1	76	<1	62	74	78	81
Iopromide	-2.05	78	<1	58	<1	71	65	109	80
Lindane (γ-BHC)*	3.72								
Meprobamate	0.7	2	37	2	17	2	>44	2	>49
Metolachlor	3.13	229	32	220	20	225	>95	214	>95
Musk Ketone	3.82	100	55	46	<1	75	80	58	>83
Naproxen	3.18	19	<1	25	10	20	89	27	91
Oxybenzone	3.79	56	77	36	72	54	>98	50	>98
Pentoxifylline	0.29	116	52	188	53	124	97	129	98
Progesterone	3.87	24	45	1	<1	8	>87	10	>90
Sulfamethoxazole	0.89	31	<1	30	17	35	90	37	87
TCEP	1.44	81	39	138	40	79	91	95	93
Testosterone*	3.32								
Triclosan	4.76	27	82	9	41	21	84	18	93
Trimethoprim	0.91	192	73	159	69	172	>99	190	>99

* Target compound not analyzed

APPENDIX E
SUMMARY OF FULL-SCALE DATA

Table E.1
Removal of EDCs/PPCPs by coagulation in full-scale tests

Utility Number	1		2		3		4		7		8 - Coag.		8 - Soft.		9		10		12		17		18		19		20	
Pre or Post Coagulation	*Pre*	*Post*	*Pre*	*Post*	*Pre*	*Post*	*Pre*	*Post*	*Pre*	*Post*	*Pre*	*Post*	*Pre*	*Post*	*Pre*	*Post*	*Pre*	*Post*	*Pre*	*Post*	*Pre*	*Post*	*Pre*	*Post*	*Pre*	*Post*	*Pre*	*Post*
Coagulant Type																												
Coagulant Dose																												
Reaction pH																												
Target Compound																												
Acetaminophen (Tylenol)	†	†	†	†	†	†	†	†	†	†	1.9	1.4	1.4	†	9.2	†	†	†	1.5	†	†	†	†	†	†	†	1.1	1.1
Androstenedione	†	†	†	†	†	†	†	†	†	†	†	†	†	†	†	†	†	†	†	†	†	†	†	†	†	†	1.9	†
Atrazine	571	571	427	428	377	427	54	51	7.6	8.0	44	42	42	36	†	†	41	41	25	26	34	34	40	38	†	†	†	†
Benzo(a)pyrene*																												
Caffeine	7.1	27	31	43	23	87	9.9	25	26	25	81	63	63	23	22	24	11	9.1	74	110	26	30	14	12	36	36	30	27
Carbamazepine	2.5	2.4	1.2	1.2	2.2	2.3	†	†	2.4	2.2	3.9	4.1	4.1	3.5	†	†	9.6	4.1	20	3.9	2.8	3.1	2.9	2.5	24	16	5.8	4.8
DDT*																												
DEET	4.0	3.6	17	17	12	18	5.6	7.1	6.2	3.9	4.4	3.7	3.7	2.6	5.7	5.6	8.8	7.0	28	30	12	14	8.3	7.0	12	13	15	14
Diazepam (Valium)*																												
Diclofenac*																												
Dilantin	1.4	2.2	3.7	3.0	2.5	3.2	1.1	†	2.9	1.5	4.1	3.8	3.8	3.8	†	†	5.1	3.8	13	8.6	3.2	3.1	4.3	3.9	4.8	4.8	2.2	2.0
Erythromycin	1.4	1.9	†	†	†	†	†	†	2.1	2.5	1.3	1.5	1.5	1.4	†	†	3.0	†	3.2	1.4	†	†	†	†	1.8	2.5	†	†
Estradiol*																												
Estriol*																												
Estrone	†	†	†	†	†	†	†	†	†	†	†	†	†	†	†	†	†	†	1.7	3.3	†	†	†	†	†	†	†	†
Ethinyl Estradiol*																												
Fluorene	†	†	†	†	†	†	†	†	†	†	14	11	11	11	†	†	12	‡	†	†	†	†	†	†	†	†	15	‡
Fluoxetine (Prozac)*																												
Galaxolide	†	†	†	†	†	†	†	†	†	†	†	†	†	†	†	†	25	§	28	§	25	§	25	§	30	29	25	§
Gemfibrozil	4.8	4.5	1.7	2.3	1.3	2.0	†	†	13	9.5	4.8	3.9	3.9	4.1	†	†	11	3.0	5.8	1.2	1.3	†	†	†	4.0	4.8	†	†
Hydrocodone	†	†	†	†	†	†	†	†	†	†	†	†	†	†	†	†	†	†	1.9	†	†	†	†	†	†	†	†	†
Ibuprofen (Advil)	2.4	2.9	†	†	†	†	†	†	4.8	6.2	3.8	3.1	3.1	3.2	5.4	5.2	24	28	10	22	4.8	4.5	2.7	2.7	12	11	1.9	1.9
Iopromide	2.2	2.4	6.6	11	11	7.9	11	11	10	7.3	12	10	10	9.6	†	†	†	†	46	27	4.3	5.5	2.5	2.2	48	48	†	†
Lindane (γ-BHC)*																												
Meprobamate	2.0	2.0	2.6	2.6	2.3	2.8	†	†	13	10	1.5	1.6	1.6	1.6	†	†	4.0	3.6	10	8.1	2.4	2.9	2.6	2.3	3.9	4	†	†
Metolachlor	122	121	155	174	156	176	†	†	†	†	†	†	†	†	†	†	12	‡	15	16	11	14	15	13	†	†	†	†
Musk Ketone	†	†	†	†	†	†	†	†	†	†	14	14	14	14	†	†	†	†	†	†	16	16	16	‡	†	†	†	†
Naproxen	1.9	1.9	1.8	2.0	1.4	2.4	†	†	1.9	†	9.3	7.8	7.8	7.3	1.4	†	10	†	16	†	†	†	†	†	11	12	1.1	†
Oxybenzone	†	†	†	†	†	†	†	†	†	†	†	†	†	†	1.5	†	†	†	†	†	†	†	1.1	†	2.3	†	†	†
Pentoxifylline*																												
Progesterone*																												
Sulfamethoxazole	1.2	1.6	5.6	7.2	6.6	8.8	†	†	11	1.7	41	30	30	27	†	†	27	†	35	†	8.1	†	9.7	9.2	8.8	10	44	†
TCEP	2.0	1.7	42	43	53	15	7.9	4.8	22	21	7.3	18	18	6.4	37	3.1	17	8.6	19	20	7.4	5.2	13	2.3	13	5.4	11	10
Testosterone*																												
Triclosan	†	†	†	†	†	†	†	†	†	†	36	50	50	25	1.1	†	1.2	†	2.7	†	†	†	†	†	†	†	†	†
Trimethoprim	†	†	†	†	†	†	†	†	†	†	†	1.1	1.1	1.3	†	†	2.4	†	2.2	†	†	†	†	†	†	†	†	†

* Target compound not analyzed
† Below Reporting Limit (<1.0 ppt)
‡ Below Reporting Limit (<10 ppt)
§ Below Reporting Limit (<25 ppt)
** Impacted by Blank Contamination

Table E.2
Removal of EDCs/PPCPs by chlorination in full-scale tests

	Utility																											
	1		2		3		4		6		7		8		9		10		11		12		15		16		20	
Target Compound	Pre	Post	Pre	Post	Pre	Post	Pre	Post	Pre	Post	Pre	Post	Pre	Post	Pre	Post	Pre	Post	Pre	Post	Pre	Post	Pre	Post	Pre	Post	Pre	Post
Acetaminophen																												
(Tylenol)			1.6		2																							
Androstenedione*																												
Atrazine	6.1	4.4	428	431	427	432	48	43	11	14.0	8.5	8.2	48	38			42	40	5.3	5.7	24	26	1.4	1.5	1.3	1.4		
Benzo(a)pyrene*																												
Caffeine			43	20	87	29	16	14	28	29	27	25	19	35	24		11				81	83	3.6	2.4	2.8	2.9	29	25
Carbamazepine			1.2		2.3						2.8	3.7	4.3	3.7			9.6	5.7	1.4	1.5	4.1	3.5	3.7	1.7	2.4	1.8	2.3	1.8
DDT*																												
DEET			17	16	18	18	4.4	4	1.7	3.3	4.6	6.2	2.9	3	5.6	3.9	8.8	7.5	4.6	5.6	30	30	4.1	4.0	2.8	3.1	13	11
Diazepam (Valium)*																												
Diclofenac*																												
Dilantin			3	2.2	3.2	2.2					2.1	2.7	3.8	3.7			5.1	4.4			10	6.7	2.6	2.4	1.6	1.7	2	1.6
Erythromycin											2.6		2	1.3							2.4							
Estradiol*																												
Estriol*																												
Estrone				1.1																							2.5	2.3
Ethinyl Estradiol*																												
Fluorene			<10										11		18	14	10		13									
Fluoxetine (Prozac)*																												
Galaxolide			10		10						11	15	14				25	21			20	19						
Gemfibrozil			2.3		2						10	6.5	4.5	5.1			11	4.2			1.3		1.2					
Hydrocodone*																												
Ibuprofen (Advil)			3.9	3	4.4	4.7			8.3	17	8.1	9.1	3.4	3.8	5.2	2.9	24	32			22	18					1.2	
Iopromide			11	7.8	7.9	9	13	10			7.8	5.9	14	11	5.5				2.4	3.6	27	16						
Lindane (γ-BHC)*																												
Meprobamate			2.6	2.3	2.8	2.9			2	3.1	12	13	1.7	1.6			4.0	3.8			7.7	7.6	12	12	7.7	9		
Metolachlor			174	160	176	156											12	10			16	14						
Musk Ketone*																												
Naproxen			2		2.4								7.1	8			10											
Oxybenzone			7.4		4.2																							
Pentoxifylline*																												
Progesterone*																												
Sulfamethoxazole			7.2		8.8						2.1		58	20							37		7.3		5.7			
TCEP			43		15	15	4.2	5.8	7	6.4	22	18					17				18	19						
Testosterone*																												
Triclosan			1.9		2								<50	<50			1.2				<1.0							
Trimethoprim													1.7	1.3			2.4				2.2							

* Target compound not analyzed

Table E.3
Occurrence of EDCs/PPCPs in raw water of full-scale plants

Utility Number	1	2	3	4	5	6	7	8	9	10	11	12	13	14	15	16	17	18	19	20
Acetaminophen	<1.0	1.6	1.6	<1.0	1.8	<1.0	<1.0	1.7	9.5	<1.0	<1.0	1.5	<1.0	<1.0	<1.0	<1.0	<1.0	<1.0	<1.0	1.1
Androstenedione	<1.0	<1.0	<1.0	<1.0	<1.0	<1.0	<1.0	<1.0	<1.0	<1.0	<1.0	<1.0	<1.0	<1.0	<1.0	<1.0	<1.0	<1.0	<1.0	1.9
Atrazine	571	428	427	54	28	9.7	8.0	42	<1.0	42	5.3	25	1.9	1.4	1.4	1.3	34	40	<1.0	<1.0
Benzo[a]pyrene	<10	<10	<10	<10	<10	<10	<10	<10	<10	<10	<10	<10	<10	<10	<10	<10	<10	<10	<10	<10
Caffeine	<10	43	87	<10	24	28	25	48	21	11	9.1	74	<10	<10	<10	<10	26	14	30	30
Carbamazepine	2.5	1.2	2.3	<1.0	1.2	3.2	2.2	3.8	<1.0	9.6	1.4	20	4.8	3.5	3.7	2.4	2.8	2.9	39	5.8
DDT	<10	<10	<10	<10	<10	<10	<10	<10	<10	<10	<10	<10	<10	<10	<10	<10	<10	<10	<10	<10
DEET	4.0	17	18	5.6	5.0	11	3.9	3.9	5.9	8.8	4.6	28	7.9	4.0	4.1	2.8	12	8.3	11	15
Diazepam	<1.0	<1.0	<1.0	<1.0	<1.0	<1.0	<1.0	<1.0	<1.0	<1.0	<1.0	<1.0	<1.0	<1.0	<1.0	<1.0	<1.0	<1.0	<1.0	<1.0
Diclofenac	<1.0	<1.0	<1.0	<1.0	<1.0	<1.0	<1.0	<1.0	<1.0	<1.0	<1.0	<1.0	<1.0	<1.0	<1.0	<1.0	<1.0	<1.0	<1.0	<1.0
Dilantin	1.4	3.0	3.2	1.1	1.3	3.3	1.5	3.7	<1.0	5.3	<1.0	13	4.2	3.1	2.6	1.6	3.2	4.3	5.1	2.2
Erythromycin	1.4	<1.0	<1.0	<1.0	1.0	3.5	2.5	1.2	<1.0	3.0	<1.0	3.2	<1.0	<1.0	<1.0	<1.0	<1.0	<1.0	1.9	<1.0
Estradiol	<1.0	<1.0	<1.0	<1.0	<1.0	<1.0	<1.0	<1.0	<1.0	<1.0	<1.0	<1.0	<1.0	<1.0	<1.0	<1.0	<1.0	<1.0	<1.0	<1.0
Estriol	<1.0	<5.0	<5.0	<5.0	<5.0	<5.0	<5.0	<5.0	<5.0	<5.0	<5.0	<5.0	<5.0	<5.0	<5.0	<5.0	<5.0	<5.0	<5.0	<5.0
Estrone	<1.0	<1.0	<1.0	<1.0	<1.0	<1.0	<1.0	<1.0	<1.0	<1.0	<1.0	<1.0	<1.0	1.4	<1.0	<1.0	<1.0	1.0	<1.0	<1.0
Ethynylestradiol	<1.0	<1.0	<1.0	<1.0	<1.0	<1.0	<1.0	<1.0	<1.0	<1.0	<1.0	<1.0	<1.0	<1.0	<1.0	<1.0	<1.0	<1.0	<1.0	<1.0
Fluorene	<10	<10	<10	<10	<10	<10	<10	<10	<10	<10	13	<10	<10	<10	<10	<10	<10	<10	<10	<10
Fluoxetine	<1.0	<1.0	<1.0	<1.0	<1.0	<1.0	<1.0	<1.0	<1.0	<1.0	<1.0	<1.0	<1.0	<1.0	<1.0	<1.0	<1.0	<1.0	<1.0	<1.0
Galaxolide	<25	<25	<25	<25	<25	<25	<25	<25	<25	25	<25	28	<25	<25	<25	<25	<25	<25	30	<25
Gemfibrozil	4.8	2.3	2.0	<1.0	7.4	11	9.5	5.2	<1.0	11	<1.0	5.8	2.0	<1.0	1.2	<1.0	1.3	<1.0	3.8	<1.0
Hydrocodone	<1.0	<1.0	<1.0	<1.0	<1.0	<1.0	<1.0	<1.0	<1.0	<1.0	<1.0	1.9	<1.0	<1.0	<1.0	<1.0	<1.0	<1.0	<1.0	<1.0
Ibuprofen	2.4	3.9	4.4	1.1	2.4	12	6.2	4.0	4.8	24	<1.0	10	1.7	<1.0	<1.0	<1.0	4.8	2.7	11	1.9
Iopromide	2.2	11	7.9	11	4.1	8.0	7.3	12	<1.0	<1.0	2.4	46	2.2	<1.0	<1.0	<1.0	4.3	2.5	46	<1.0
Lindane	<10	<10	<10	<10	<10	<10	<10	<10	<10	<10	<10	<10	<10	<10	<10	<10	<10	<10	<10	<10
Meprobamate	2.0	2.6	2.8	<1.0	8.0	13	10	1.4	<1.0	4.0	<1.0	10	16	13	12	7.7	2.4	2.6	3.8	<1.0
Metalochlor	122	174	146	<10	<10	<10	<10	<10	<10	12	<10	14	<10	<10	<10	<10	11	15	<10	<10
Musk Ketone	<10	<10	<10	<10	<10	<10	<10	14	<10	<10	<10	<10	<10	<10	<10	<10	17	16	<10	<10
Naproxen	1.9	2.0	2.4	<1.0	<1.0	1.6	<1.0	10	1.4	10	<1.0	16	<1.0	<1.0	<1.0	<1.0	<1.0	<1.0	11	1.1
Oxybenzone	<1.0	7.4	4.2	<1.0	<1.0	<1.0	<1.0	<1.0	1.5	<1.0	<1.0	<1.0	<1.0	<1.0	<1.0	<1.0	<1.0	1.1	<1.0	<1.0
Pentoxifylline	<1.0	<1.0	<1.0	<1.0	<1.0	<1.0	<1.0	<1.0	<1.0	<1.0	<1.0	<1.0	<1.0	<1.0	<1.0	<1.0	<1.0	<1.0	<1.0	<1.0
Progesterone	<1.0	<1.0	<1.0	<1.0	<1.0	<1.0	<1.0	<1.0	<1.0	<1.0	<1.0	<1.0	<1.0	<1.0	<1.0	<1.0	1.1	<1.0	<1.0	<1.0
Sulfamethoxazole	1.2	7.2	8.8	<1.0	5.8	3.6	1.7	40	<1.0	27	<1.0	35	12	11	7.3	5.7	8.1	9.7	7.5	44
TCEP	2.0	43	15	7.9	19	17	21	5.8	66	11	<1.0	19	<1.0	<1.0	<1.0	<1.0	7.4	13	7.2	11
Testosterone	<1.0	<1.0	<1.0	<1.0	<1.0	<1.0	<1.0	<1.0	<1.0	<1.0	<1.0	<1.0	<1.0	<1.0	<1.0	<1.0	<1.0	<1.0	<1.0	<1.0
Triclosan	<1.0	1.9	1.9	<1.0	<1.0	<1.0	<1.0	30	1.0	1.1	<1.0	2.7	<1.0	<1.0	<1.0	<1.0	<1.0	<1.0	<1.0	<1.0
Trimethoprim	<1.0	<1.0	<1.0	<1.0	<1.0	<1.0	<1.0	1.0	<1.0	2.3	<1.0	2.2	<1.0	<1.0	<1.0	<1.0	<1.0	<1.0	<1.0	<1.0

Table E.4
Occurrence of EDCs/PPCPs in finished water of full-scale plants

Utility Number	1	2	3	4	5	6	7	8	9	10	11	12	13	14	15	16	17	18	19	20
Acetaminophen	<1.0	<1.0	<1.0	<1.0	<1.0	<1.0	<1.0	<1.0	<1.0	<1.0	<1.0	<1.0	<1.0	<1.0	<1.0	<1.0	<1.0	<1.0	<1.0	<1.0
Androstenedione	<1.0	<1.0	<1.0	<1.0	<1.0	<1.0	<1.0	<1.0	<1.0	<1.0	<1.0	<1.0	<1.0	<1.0	<1.0	<1.0	<1.0	<1.0	<1.0	<1.0
Atrazine	4.4	431	396	43	29	14	8.2	38	<1.0	40	5.7	26	<1.0	<1.0	1.5	1.4	36	40	<1.0	<1.0
Benzo[a]pyrene	<10	<10	<10	<10	<10	<10	<10	<10	<10	<10	<10	<10	<10	<10	<10	<10	<10	<10	<10	<10
Caffeine	2.6	20	29	14	12	<10	25	35	<10	<10	<10	83	<10	<10	<10	<10	31	14	4.8	25
Carbamazepine	<1.0	<1.0	<1.0	<1.0	<1.0	<1.0	3.8	3.7	<1.0	5.7	1.5	3.5	<1.0	<1.0	1.7	1.8	3.6	2.8	1.1	1.8
DDT	<10	<10	<10	<10	<10	<10	<10	<10	<10	<10	<10	<10	<10	<10	<10	<10	<10	<10	<10	<10
DEET	<1.0	16	17	4.0	4.1	3.3	6.2	3.0	3.9	7.5	5.6	30	<1.0	2.1	4	3.1	14	7.4	4.5	11
Diazepam	<1.0	<1.0	<1.0	<1.0	<1.0	<1.0	<1.0	<1.0	<1.0	<1.0	<1.0	<1.0	<1.0	<1.0	<1.0	<1.0	<1.0	<1.0	<1.0	<1.0
Diclofenac	<1.0	<1.0	<1.0	<1.0	<1.0	<1.0	<1.0	<1.0	<1.0	<1.0	<1.0	<1.0	<1.0	<1.0	<1.0	<1.0	<1.0	<1.0	<1.0	<1.0
Dilantin	<1.0	2.2	2.1	<1.0	1.1	<1.0	2.7	3.7	<1.0	4.4	<1.0	6.7	1.8	1.5	2.4	1.7	3.1	3.4	<1.0	1.6
Erythromycin	<1.0	<1.0	<1.0	<1.0	<1.0	<1.0	<1.0	1.3	<1.0	<1.0	<1.0	<1.0	<1.0	<1.0	<1.0	<1.0	<1.0	<1.0	<1.0	<1.0
Estradiol	<1.0	<1.0	<1.0	<1.0	<1.0	<1.0	<1.0	<1.0	<1.0	<1.0	<1.0	<1.0	<1.0	<1.0	<1.0	<1.0	<1.0	<1.0	<1.0	<1.0
Estriol	<1.0	<5.0	<5.0	<5.0	<5.0	<5.0	<5.0	<5.0	<5.0	<5.0	<5.0	<5.0	<5.0	<5.0	<5.0	<5.0	<5.0	<5.0	<5.0	<5.0
Estrone	<1.0	1.1	<1.0	<1.0	<1.0	<1.0	<1.0	<1.0	<1.0	<1.0	<1.0	<1.0	<1.0	<1.0	<1.0	<1.0	<1.0	<1.0	<1.0	2.3
Ethynylestradiol	<1.0	<1.0	<1.0	<1.0	<1.0	<1.0	<1.0	<1.0	<1.0	<1.0	<1.0	<1.0	<1.0	<1.0	<1.0	<1.0	<1.0	<1.0	<1.0	<1.0
Fluorene	<10	<10	<10	<10	<10	<10	<10	<10	<10	<10	<10	<10	<10	<10	<10	<10	<10	<10	<10	<10
Fluoxetine	<1.0	<1.0	<1.0	<1.0	<1.0	<1.0	<1.0	<1.0	<1.0	<1.0	<1.0	<1.0	<1.0	<1.0	<1.0	<1.0	<1.0	<1.0	<1.0	<1.0
Galaxolide	<25	<25	<25	<25	<25	<25	<25	<25	<25	<25	<25	<25	<25	<25	<25	<25	<25	<25	<25	<25
Gemfibrozil	<1.0	<1.0	<1.0	<1.0	2.4	<1.0	6.5	5.1	<1.0	4.2	<1.0	<1.0	<1.0	<1.0	<1.0	<1.0	1.3	<1.0	<1.0	<1.0
Hydrocodone	<1.0	<1.0	<1.0	<1.0	<1.0	<1.0	<1.0	<1.0	<1.0	<1.0	<1.0	<1.0	<1.0	<1.0	<1.0	<1.0	<1.0	<1.0	<1.0	<1.0
Ibuprofen	<1.0	3.0	4.2	<1.0	2.7	17	9.1	3.8	2.9	32	1.0	18	<1.0	<1.0	<1.0	<1.0	4.5	3.1	1.2	<1.0
Iopromide	1.2	7.8	9.4	10	4.6	<1.0	5.9	11	<1.0	<1.0	3.6	16	1.1	<1.0	<1.0	<1.0	6.5	2.1	31	<1.0
Lindane	<10	<10	<10	<10	<10	<10	<10	<10	<10	<10	<10	<10	<10	<10	<10	<10	<10	<10	<10	<10
Meprobamate	<1.0	2.3	2.6	<1.0	8.0	3.1	13	1.6	<1.0	3.8	<1.0	7.6	12	9.4	12	9	3.2	2.1	1.6	<1.0
Metalochlor	<10	160	156	<10	<10	<10	<10	<10	<10	<10	<10	15	<10	<10	<10	<10	<10	14	<10	<10
Musk Ketone	<10	<10	<10	<10	<10	<10	<10	<10	<10	<10	<10	<10	<10	<10	<10	<10	17	<10	<10	<10
Naproxen	<1.0	<1.0	<1.0	<1.0	<1.0	<1.0	<1.0	8.0	<1.0	<1.0	<1.0	<1.0	<1.0	<1.0	<1.0	<1.0	<1.0	<1.0	<1.0	<1.0
Oxybenzone	1.1	<1.0	<1.0	<1.0	<1.0	<1.0	<1.0	<1.0	<1.0	<1.0	<1.0	<1.0	<1.0	<1.0	<1.0	<1.0	<1.0	<1.0	<1.0	<1.0
Pentoxifylline	<1.0	<1.0	<1.0	<1.0	<1.0	<1.0	<1.0	<1.0	<1.0	<1.0	<1.0	<1.0	<1.0	<1.0	<1.0	<1.0	<1.0	<1.0	<1.0	<1.0
Progesterone	1.1	<1.0	<1.0	<1.0	<1.0	<1.0	<1.0	<1.0	<1.0	<1.0	<1.0	<1.0	<1.0	<1.0	<1.0	<1.0	1.1	<1.0	<1.0	<1.0
Sulfamethoxazole	<1.0	<1.0	<1.0	<1.0	<1.0	<1.0	<1.0	20	<1.0	<1.0	<1.0	<1.0	<1.0	<1.0	<1.0	<1.0	<1.0	<1.0	<1.0	<1.0
TCEP	<10	<10	<10	<10	15	<10	18	<10	<10	<10	<10	19	<10	<10	<10	<10	5.2	5.5	4.7	3.0
Testosterone	<1.0	<1.0	<1.0	<1.0	<1.0	<1.0	<1.0	<1.0	<1.0	<1.0	<1.0	<1.0	<1.0	<1.0	<1.0	<1.0	<1.0	<1.0	<1.0	<1.0
Triclosan	<1.0	<1.0	<1.0	<1.0	<1.0	<1.0	<1.0	43	<1.0	<1.0	<1.0	<1.0	<1.0	<1.0	<1.0	<1.0	<1.0	<1.0	<1.0	<1.0
Trimethoprim	<1.0	<1.0	<1.0	<1.0	<1.0	<1.0	<1.0	1.3	<1.0	<1.0	<1.0	<1.0	<1.0	<1.0	<1.0	<1.0	<1.0	<1.0	<1.0	<1.0

APPENDIX F
COMPUTER MODEL PREDICTIONS

Table F.1
Prediction of GAC adsorption by Polanyi equation

Name	Formula	Molecular Weight	Structure	Aqueous Solubility	Freundlich Parameters K (mg/g)*(L/mg)^(1/n)	Freundlich Parameters 1/n
Acetaminophen	$C_8H_9NO_2$	151.16	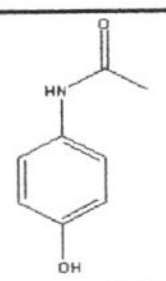	Value : 1.4E+004 mg/L Temp : 25 deg C Type : EXP Ref : YALKOWSKY,SH & DANNENFELSER, RM (1992)	1.84	0.746
Atrazine	$C_8H_{14}ClN_5$	215.1	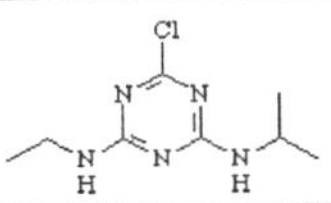	Value : 34.7 mg/L Temp : 26 deg C Type : EXP Ref : WARD,TM & WEBER,JB (1968)	421.3	0.443
Bisphenol A	$C_{15}H_{16}O_2$	228.12	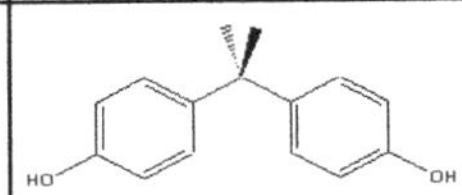	Value : 120 mg/L Temp : 25 deg C Type : EXP Ref : DORN,PB ET AL. (1987)	8.649	0.43
Caffeine	$C_8H_{10}N_4O_2$	194.1	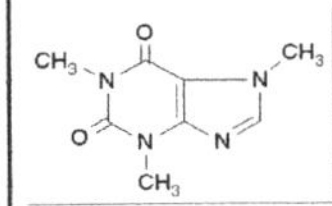	Value : 2.16E+004 mg/L Temp : 25 deg C Type : EXP Ref : YALKOWSKY,SH & DANNENFELSER, RM (1992)	10.8	0.551
Carbamazepine	$C_{15}H_{12}N_2O$	236.1	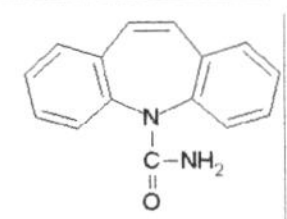	Value : 17.7 mg/L Temp : 25 deg C Type : EST Ref : MEYLAN,WM ET AL. (1996)	638	0.389
Clofibric Acid	$C_{10}H_{11}ClO_3$	214.04	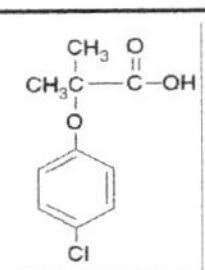	Value : 583 mg/L Temp : 25 deg C Type : EST Ref : MEYLAN,WM ET AL. (1996)	767.7	0.129
Cotinine	$C_{10}H_{12}N_2O$	176.1	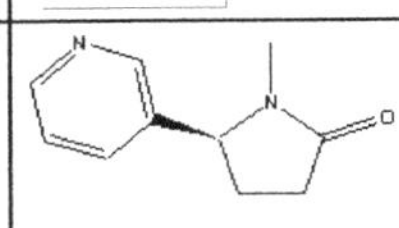	Value : 9.99E+005 mg/L Temp : 25 deg C Type : EST Ref : MEYLAN,WM ET AL. (1996)	0.4	0.691
DEET	C12H17NO	191.13	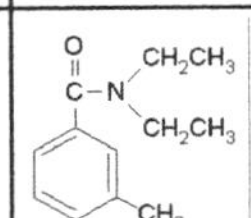	Value : 912 mg/L Temp : 25 deg C Type : EST Ref : MEYLAN,WM ET AL. (1996)	66.7	0.502
Diazepam	C16H13ClN2O	284.1	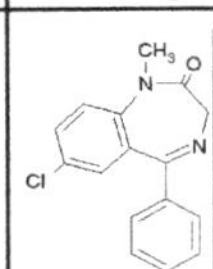	Value : 50 mg/L Temp : 25 deg C Type : EXP Ref : YALKOWSKY,SH & DANNENFELSER, RM (1992)	553.1	0.32
EE2	C20H24O2	296.18	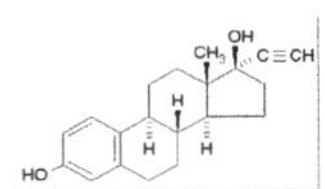	Value : 11.3 mg/L Temp : 27 deg C Type : EXP	9.301	0.23
Erythromycin (loss of H_2O)	C37H67NO13	733.46 (715)	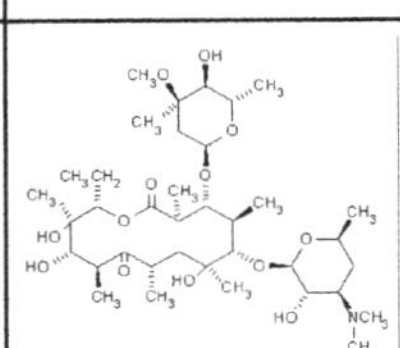	Value : 1.44 mg/L Temp : 25 deg C Type : EST Ref : MEYLAN,WM ET AL. (1996)	1075.8	0.094
Fluoxetine	C17H18F3NO	309.13 345.11 (salt)	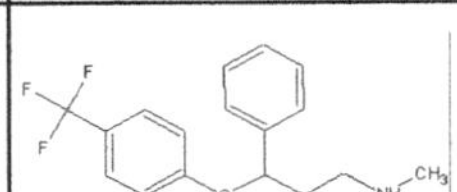	Value : 60.3 mg/L Temp : 25 deg C Type : EST Ref : MEYLAN,WM ET AL. (1996)	582	0.29

(continued)

Table F.1 (Continued)

Galaxolide	C18H26O	258.2	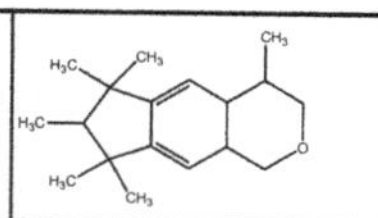		11.586	0.27
Gemfibrozil (Lopid)	C15H22O3	250.16	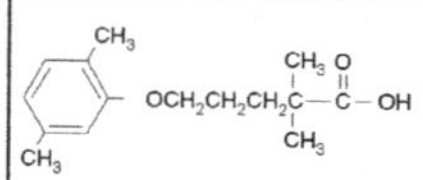		547	0.333
Ibuprofen	C13H18O2	206.13	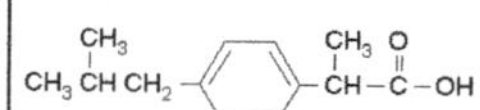	Value : 21 mg/L Temp : 25 deg C Type : EXP Ref : YALKOWSKY,SH & DANNENFELSER, RM (1992)	52.9	0.47
Iopromide	C18H24I3N3O8	790.87	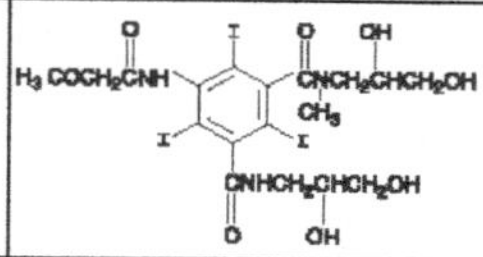	Value : 23.8 mg/L Temp : 25 deg C Type : EST Ref : MEYLAN,WM ET AL. (1996)		
Meprobamate	C9H18N2O4	218.25	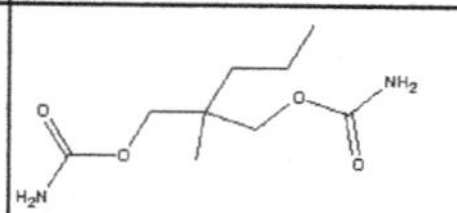	Value : 4700 mg/L Temp : 25 deg C Type : EXP Ref : YALKOWSKY,SH & DANNENFELSER, RM (1992)	48	0.446
Musk Ketone	C14H18N2O5	294.31	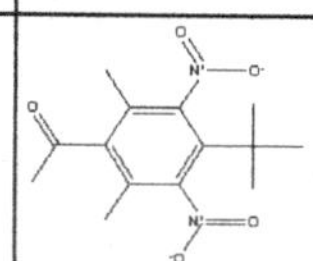		1147	0.306
Naproxen	C14H14O3	230.09	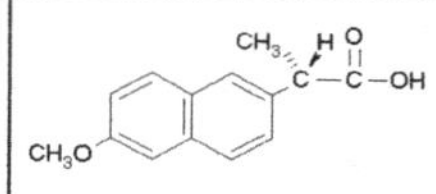	Value : 15.9 mg/L Temp : 25 deg C Type : EXP Ref : YALKOWSKY,SH & DANNENFELSER, RM (1992) Exper. Ref: YALKOWSKY,SH & DANNENFELSER,RM (1992)	189	0.524
NDMA	C2H6N2O	74.08	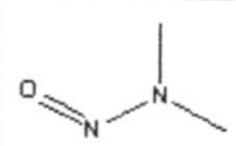	Value : 1E+006 mg/L Temp : 24 deg C Type : EXP Ref : YALKOWSKY,SH & DANNENFELSER, RM (1992)	0	1.963
Octylphenol	C14H22O	206.17	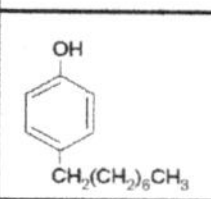	Value : 3.114 mg/L	293.6	0.346
Oxybenzone	C14H12O3	228.08	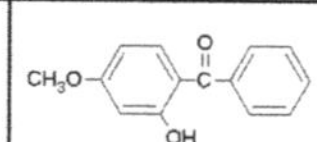	Value : 68.6 mg/L Temp : 25 deg C Type : EST Ref : MEYLAN,WM ET AL. (1996)	345.6	0.42
Pentoxifylline	C13H18N4O3	278.14	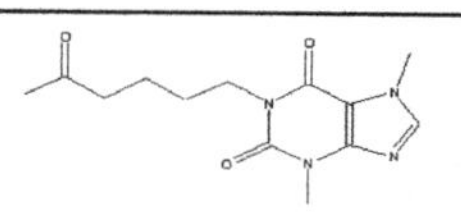	Value : 7.7E+004 mg/L Temp : 25 deg C Type : EXP Ref : MERCK INDEX (1996)	37.6	0.373
Primidone	C12H14N2O2	218.25	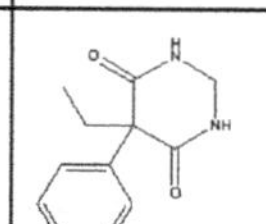	Value : 500 mg/L Temp : 22 deg C Type : EXP Ref : YALKOWSKY,SH & DANNENFELSER, RM (1992)	142.8	0.421
Sulfamethoxazole	C10H11N3O3S	253.05	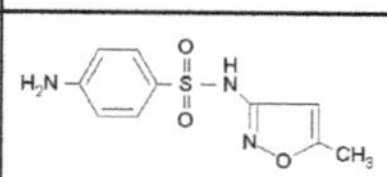	Value : 610 mg/L Temp : 37 deg C Type : EXP Ref : YALKOWSKY,SH & DANNENFELSER,	197.2	0.354
Triclosan	C12H7Cl3O2	289.55	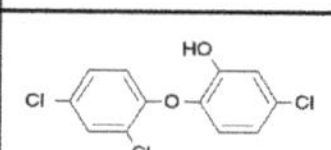	Value : 10 mg/L Temp : 20 deg C Type : EXP Ref : YALKOWSKY,SH & DANNENFELSER,RM (19 RM (1992)	932	0.301
Trimethoprim	C14H18N4O3	290.14	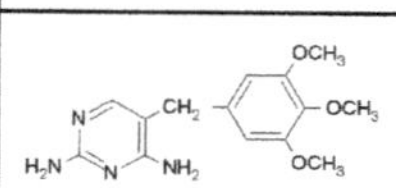	Value : 400 mg/L Temp : 25 deg C Type : EXP Ref : YALKOWSKY,SH & DANNENFELSER, RM (1992)	314	0.296

Table F.2
Prediction of biodegradation probabilities

Name	Structure	Biodeg	
		Biowin v4.0 EPA (EPI Suite)	Howard-Boethling
Acetaminophen	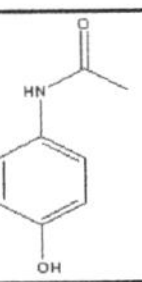	Linear Model Prediction: Fast (1.0015) Non-Linear Model Prediction: Fast (0.9886) Ultimate Biodegradation Timeframe: Weeks (2.8673) Primary Biodegradation Timeframe: Days (3.8748) MITI Linear Model Prediction : Not Readily Degradable (0.4866) MITI Non-Linear Model Prediction: Readily Degradable (0.5090)	Biodegradable 1.00 (Linear) Biodegradable 0.99 (Non-Linear)
Atrazine	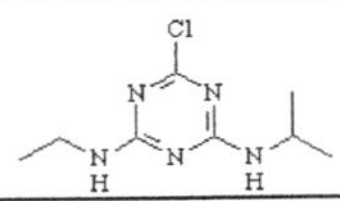	Linear Model Prediction : Does Not Biodegrade Fast (0.0045) Non-Linear Model Prediction: Does Not Biodegrade Fast (0.0000) Ultimate Biodegradation Timeframe: Months (2.0002) Primary Biodegradation Timeframe: Weeks (3.0969) MITI Linear Model Prediction : Not Readily Degradable (-0.1219) MITI Non-Linear Model Prediction: Not Readily Degradable (0.0000)	Indeterminate 0.52 (Linear) Persistent 0.18 (Non-Linear)
Bisphenol A	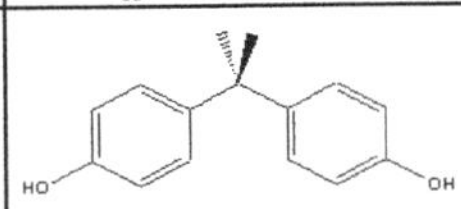	Linear Model Prediction : Biodegrades Fast (0.6866) Non-Linear Model Prediction: Does Not Biodegrade Fast (0.4653) Ultimate Biodegradation Timeframe: Weeks-Months (2.5953) Primary Biodegradation Timeframe: Days-Weeks (3.4443) MITI Linear Model Prediction : Not Readily Degradable (0.2956) MITI Non-Linear Model Prediction: Not Readily Degradable (0.1559)	Biodegradable 0.80 (Linear) Biodegradable 0.73 (Non-Linear)
Caffeine	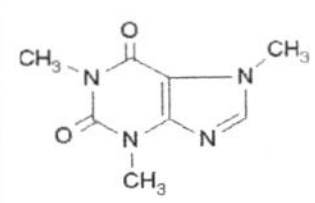	Linear Model Prediction : Biodegrades Fast (0.6551) Non-Linear Model Prediction: Biodegrades Fast (0.5625) Ultimate Biodegradation Timeframe: Weeks (2.7700) Primary Biodegradation Timeframe: Days-Weeks (3.5676) MITI Linear Model Prediction : Not Readily Degradable (0.1439) MITI Non-Linear Model Prediction: Not Readily Degradable (0.0521)	Biodegradable 0.92 (Linear) Biodegradable 0.95 (Non-Linear)
Carbamazepine	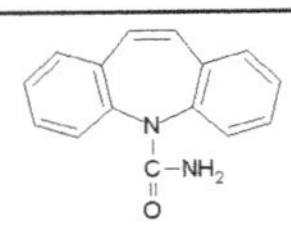	Linear Model Prediction : Biodegrades Fast (0.6351) Non-Linear Model Prediction: Does Not Biodegrade Fast (0.4143) Ultimate Biodegradation Timeframe: Weeks-Months (2.6770) Primary Biodegradation Timeframe: Days-Weeks (3.5068) MITI Linear Model Prediction : Not Readily Degradable (0.0873) MITI Non-Linear Model Prediction: Not Readily Degradable (0.0364)	Biodegradable 0.95 (Linear) Biodegradable 0.97 (Non-Linear)
Clofibric Acid	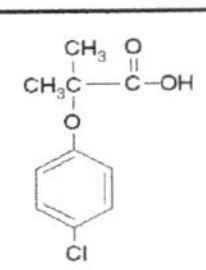	Linear Model Prediction : Does Not Biodegrade Fast (0.4836) Non-Linear Model Prediction: Does Not Biodegrade Fast (0.2918) Ultimate Biodegradation Timeframe: Weeks-Months (2.6126) Primary Biodegradation Timeframe: Days-Weeks (3.6820) MITI Linear Model Prediction : Readily Degradable (0.5574) MITI Non-Linear Model Prediction: Not Readily Degradable (0.3741)	Indeterminate 0.41 (Linear) Persistent 0.07 (Non-Linear)
Coprostanol	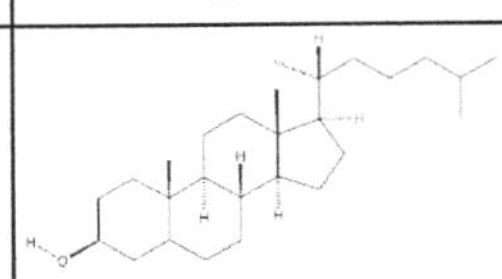	Linear Model Prediction : Does Not Biodegrade Fast (0.3534) Non-Linear Model Prediction: Does Not Biodegrade Fast (0.0079) Ultimate Biodegradation Timeframe: Months (2.0760) Primary Biodegradation Timeframe: Weeks (3.095) MITI Linear Model Prediction : Not Readily Degradable (0.1533) MITI Non-Linear Model Prediction: Not Readily Degradable (0.0127)	Biodegradable 0.72 (Linear) Persistent 0.18 (Non-Linear)
Cotinine	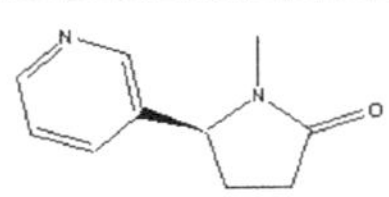	Linear Model Prediction : Biodegrades Fast (0.7192) Non-Linear Model Prediction: Biodegrades Fast (0.8263) Ultimate Biodegradation Timeframe: Weeks-Months (2.5414) Primary Biodegradation Timeframe: Days (3.7802) MITI Linear Model Prediction : Not Readily Degradable (0.3440) MITI Non-Linear Model Prediction: Not Readily Degradable (0.2890)	Biodegradable 0.77 (Linear) Biodegradable 0.89 (Non-Linear)
DEET	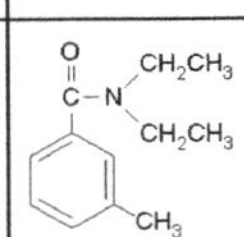	Linear Model Prediction : Biodegrades Fast (0.9213) Non-Linear Model Prediction: Biodegrades Fast (0.9724) Ultimate Biodegradation Timeframe: Weeks-Months (2.6474) Primary Biodegradation Timeframe: Days-Weeks (3.7087) MITI Linear Model Prediction : Not Readily Degradable (0.4437) MITI Non-Linear Model Prediction: Not Readily Degradable (0.3954)	Biodegradable 0.77 (Linear) Biodegradable 0.87 (Non-Linear)
Diazepam	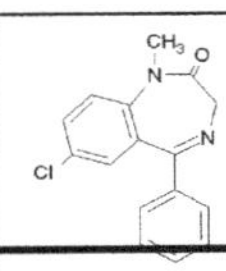	Linear Model Prediction : Biodegrades Fast (0.7678) Non-Linear Model Prediction: Biodegrades Fast (0.8085) Ultimate Biodegradation Timeframe: Weeks-Months (2.3311) Primary Biodegradation Timeframe: Days-Weeks (3.4819) MITI Linear Model Prediction : Not Readily Degradable (0.0837) MITI Non-Linear Model Prediction: Not Readily Degradable (0.0217)	Indeterminate 0.54 (Linear) Persistent 0.18 (Non-Linear)
Galaxolide	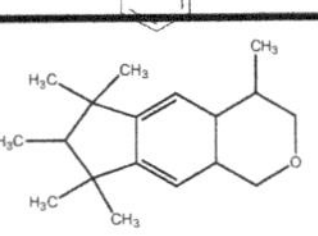	Linear Model Prediction : Does Not Biodegrade Fast (-0.0360) Non-Linear Model Prediction: Does Not Biodegrade Fast (0.0009) Ultimate Biodegradation Timeframe: Months (2.1204) Primary Biodegradation Timeframe: Weeks (3.0898) MITI Linear Model Prediction : Not Readily Degradable (0.0657) MITI Non-Linear Model Prediction: Not Readily Degradable (0.0255)	
Gemfibrozil (Lopid)	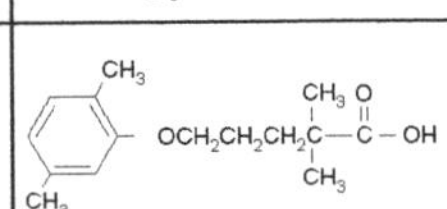	Linear Model Prediction : Biodegrades Fast (0.7584) Non-Linear Model Prediction: Biodegrades Fast (0.8552) Ultimate Biodegradation Timeframe: Weeks-Months (2.5906) Primary Biodegradation Timeframe: Days-Weeks (3.6587) MITI Linear Model Prediction : Readily Degradable (0.6680) MITI Non-Linear Model Prediction: Readily Degradable (0.6123)	
Glucophage	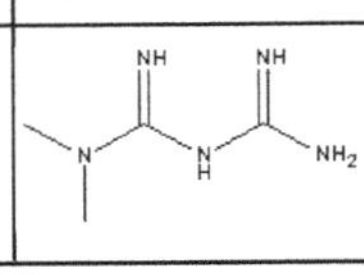	Linear Model Prediction : Biodegrades Fast (0.6861) Non-Linear Model Prediction: Biodegrades Fast (0.7640) Ultimate Biodegradation Timeframe: Weeks (2.9137) Primary Biodegradation Timeframe: Days-Weeks (3.6614) MITI Linear Model Prediction : Not Readily Degradable (0.3287) MITI Non-Linear Model Prediction: Not Readily Degradable (0.2379)	

(continued)

Table F.2 (Continued)

Hydrocodone (Loratab)	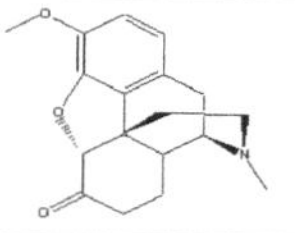	Linear Model Prediction : Biodegrades Fast (0.5412) Non-Linear Model Prediction: Does not Biodegrade Fast (0.3617) Ultimate Biodegradation Timeframe: Months (1.8571) Primary Biodegradation Timeframe: Weeks (3.0379) MITI Linear Model Prediction : Not Readily Degradable (0.3903) MITI Non-Linear Model Prediction: Not Readily Degradable (0.0858)	
Ibuprofen	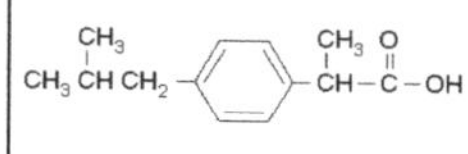	Linear Model Prediction : Biodegrades Fast (0.8314) Non-Linear Model Prediction: Biodegrades Fast (0.8672) Ultimate Biodegradation Timeframe: Weeks (2.9582) Primary Biodegradation Timeframe: Days (3.7986) MITI Linear Model Prediction : Not Readily Degradable (0.1976) MITI Non-Linear Model Prediction: Not Readily Degradable (0.1521)	Biodegradable 0.83 (Linear) Biodegradable 0.87 (Non-Linear)
Iopromide	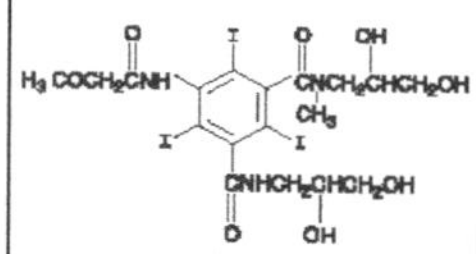	Linear Model Prediction : Does Not Biodegrade Fast (-0.9869) Non-Linear Model Prediction: Does Not Biodegrade Fast (0.0000) Ultimate Biodegradation Timeframe: Months (1.7847) Primary Biodegradation Timeframe: Days-Weeks (3.4495) MITI Linear Model Prediction : Not Readily Degradable (-1.6210) MITI Non-Linear Model Prediction: Not Readily Degradable (0.0000)	Persistent -1.34 (Linear) Persistent 0.00 (Non-Linear)
Meprobamate	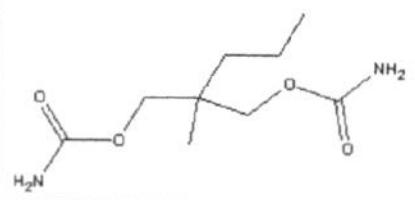	Linear Model Prediction : Biodegrades Fast (0.6188) Non-Linear Model Prediction: Biodegrades Fast (0.5511) Ultimate Biodegradation Timeframe: Weeks-Months (2.4113) Primary Biodegradation Timeframe: Days (3.7667) MITI Linear Model Prediction : Not Readily Degradable (0.2420) MITI Non-Linear Model Prediction: Not Readily Degradable (0.3140)	Biodegradable 0.91 (Linear) Biodegradable 0.98 (Non-Linear)
Musk Ketone	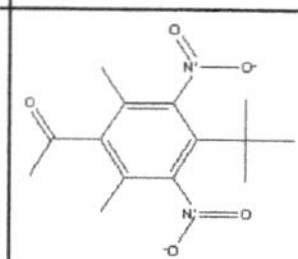	Linear Model Prediction : Does Not Biodegrade Fast (-0.0704) Non-Linear Model Prediction: Does Not Biodegrade Fast (0.0007) Ultimate Biodegradation Timeframe: Months (1.8253) Primary Biodegradation Timeframe: Weeks (2.8936) MITI Linear Model Prediction : Not Readily Degradable (-0.2687) MITI Non-Linear Model Prediction: Not Readily Degradable (0.0001)	Persistent 0.03 (Linear) Persistent 0.00 (Non-Linear)
Naproxen	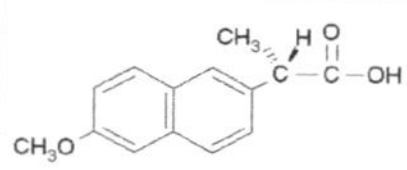	Linear Model Prediction : Biodegrades Fast (0.8972) Non-Linear Model Prediction: Biodegrades Fast (0.9611) Ultimate Biodegradation Timeframe: Weeks (2.9219) Primary Biodegradation Timeframe: Days (3.9097) MITI Linear Model Prediction : Not Readily Degradable (0.4439) MITI Non-Linear Model Prediction: Not Readily Degradable (0.3447)	Biodegradable 0.90 (Linear) Biodegradable 0.96 (Non-Linear)
NDMA	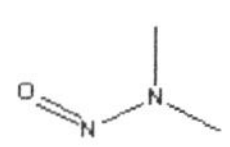	Linear Model Prediction : Does Not Biodegrade Fast (0.1878) Non-Linear Model Prediction: Does Not Biodegrade Fast (0.2138) Ultimate Biodegradation Timeframe: Weeks-Months (2.6503) Primary Biodegradation Timeframe: Days (3.7593) MITI Linear Model Prediction : Not Readily Degradable (0.2880) MITI Non-Linear Model Prediction: Not Readily Degradable (0.0000)	Persistent -0.02 (Linear) Persistent 0.03 (Non-Linear)
Octylphenol	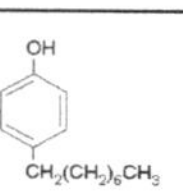	Linear Model Prediction : Biodegrades Fast (0.9282) Non-Linear Model Prediction: Biodegrades Fast (0.9680) Ultimate Biodegradation Timeframe: Weeks (3.0231) Primary Biodegradation Timeframe: Days (3.7903) MITI Linear Model Prediction : Not Readily Degradable (0.4366) MITI Non-Linear Model Prediction: Readily Degradable (0.5025)	Biodegradable 0.93 (Linear) Biodegradable 0.97 (Non-Linear)
Oxybenzone	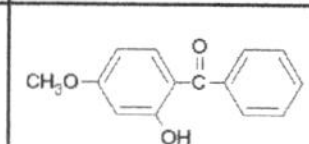	Linear Model Prediction : Biodegrades Fast (1.0215) Non-Linear Model Prediction: Biodegrades Fast (0.9862) Ultimate Biodegradation Timeframe: Weeks-Months (2.6926) Primary Biodegradation Timeframe: Days-Weeks (3.6179) MITI Linear Model Prediction : Not Readily Degradable (0.4764) MITI Non-Linear Model Prediction: Not Readily Degradable (0.3921)	Biodegradable 1.00 (Linear) Biodegradable 0.98 (Non-Linear)
Pentoxifylline	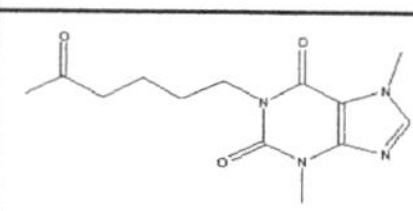	Linear Model Prediction : Biodegrades Fast (0.6219) Non-Linear Model Prediction: Does Not Biodegrade Fast (0.1984) Ultimate Biodegradation Timeframe: Weeks-Months (2.5617) Primary Biodegradation Timeframe: Days-Weeks (3.4240) MITI Linear Model Prediction : Not Readily Degradable (0.2090) MITI Non-Linear Model Prediction: Not Readily Degradable (0.0585)	Biodegradable 1.17 (Linear) Biodegradable 1.00 (Non-Linear)
Primidone	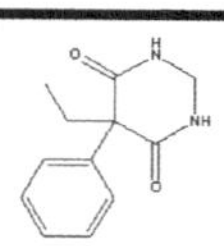	Linear Model Prediction : Biodegrades Fast (1.0081) Non-Linear Model Prediction: Biodegrades Fast (0.9954) Ultimate Biodegradation Timeframe: Weeks-Months (2.4183) Primary Biodegradation Timeframe: Days (3.7952) MITI Linear Model Prediction : Not Readily Degradable (0.4943) MITI Non-Linear Model Prediction: Not Readily Degradable (0.4211)	Biodegradable 1.04 (Linear) Biodegradable 0.99 (Non-Linear)
Sulfamethoxazole	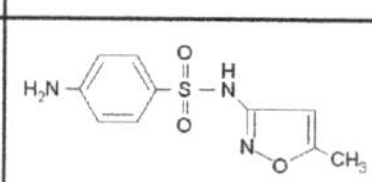	Linear Model Prediction : Does Not Biodegrade Fast (0.4479) Non-Linear Model Prediction: Does Not Biodegrade Fast (0.1281) Ultimate Biodegradation Timeframe: Weeks-Months (2.4297) Primary Biodegradation Timeframe: Days-Weeks (3.3054) MITI Linear Model Prediction : Not Readily Degradable (-0.1165) MITI Non-Linear Model Prediction: Not Readily Degradable (0.0060)	Indeterminate 0.45 (Linear) Persistent 0.13 (Non-Linear)
Triclosan	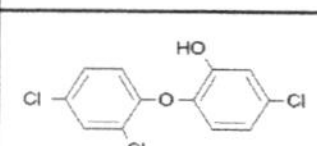	Linear Model Prediction : Does Not Biodegrade Fast (0.3102) Non-Linear Model Prediction: Does Not Biodegrade Fast (0.0181) Ultimate Biodegradation Timeframe: Months (1.9378) Primary Biodegradation Timeframe: Weeks (3.0508) MITI Linear Model Prediction : Not Readily Degradable (0.1780) MITI Non-Linear Model Prediction: Not Readily Degradable (0.0187)	Persistent 0.18 (Linear) Persistent 0.00 (Non-Linear)
Trimethoprim	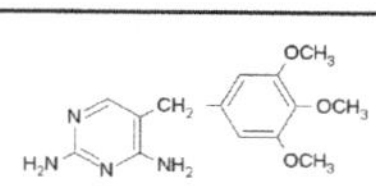	Linear Model Prediction : Biodegrades Fast (0.5922) Non-Linear Model Prediction: Biodegrades Fast (0.9164) Ultimate Biodegradation Timeframe: Months (2.0385) Primary Biodegradation Timeframe: Days-Weeks (3.3749) MITI Linear Model Prediction : Not Readily Degradable (0.0889) MITI Non-Linear Model Prediction: Not Readily Degradable (0.0172)	Biodegradable 0.88 (Linear) Biodegradable 0.99 (Non-Linear)
Tri(2-chloroethyl) phosphate (TCEP)	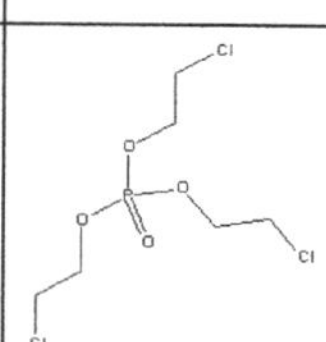	Linear Model Prediction : Biodegrades Fast (0.5914) Non-Linear Model Prediction: Biodegrades Fast (1.0000) Ultimate Biodegradation Timeframe: Months (2.2025) Primary Biodegradation Timeframe: Days-Weeks (3.5994) MITI Linear Model Prediction : Not Readily Degradable (0.3173) MITI Non-Linear Model Prediction: Not Readily Degradable (0.0193)	Indeterminate 0.59 (Linear) Biodegradable 1.00 (Non-Linear)

Table F.3

QikProp predicted and experimentally obtained physicochemical properties and molecular descriptors for target compounds

Compound name	mol MW	volume	SASA	FOSA	FISA	PISA	WPSA	#acid	#amide	#rtvFG	#metab	donorHB	hb accptHB	dipole	QPpolrz	IP(eV)	EA(eV)	logS (exp)	log Kow (exp)
Acenaphthene	154.2	589	370.2	106.2	0	264.0	0	0	0	0	2	0	0	0.78	19.9	8.37	0.56	-4.60	3.92
Acenaphthylene	152.2	566	357.6	0	0	357.6	0	0	0	0	0	0	0	0.59	19.9	8.85	1.23	-3.98	3.94
Acetaminophen	151.2	565	367.3	93.6	111.8	161.9	0	0	0	0	1	2	3.25	3.55	16.7	8.53	-0.09	-1.03	0.46
Aldrin	364.9	812	461.7	114.8	0	59.0	287.9	0	0	0	2	0	0	4.49	26.9	9.55	0.33	-7.33	6.50
Androstenedione	286.4	949	517.0	389.8	99.3	27.9	0	0	0	0	3	0	4	4.14	32.1	10.27	0.22	-3.70	2.75
Anthracene	178.2	656	404.3	0	0	404.3	0	0	0	0	1	0	0	0.00	23.9	7.95	1.25	-6.61	4.45
Atrazine	215.7	761	477.9	314.7	59.7	28.0	75.6	0	0	1	0	2	4	5.52	22.0	8.99	0.11	-3.79	2.61
Benzo[a]anthracene	228.3	790	471.0	0	0	471.0	0	0	0	0	0	0	0	0.06	30.0	8.05	1.19	-7.39	5.76
Benzo[a]pyrene	252.3	823	480.0	0	0	480.0	0	0	0	0	1	0	0	0.04	31.4	7.70	1.55	-8.19	6.13
Benzo[b]fluoranthene	252.3	831	485.6	0	0	485.6	0	0	0	0	0	0	0	0.98	31.7	8.38	1.48	-8.23	5.78
Benzo[k]fluoranthene	252.3	836	490.2	0	0	490.2	0	0	0	0	1	0	0	0.69	32.0	7.96	1.43	-8.50	6.11
Caffeine	194.2	646	398.1	239.0	102.8	56.3	0	0	0	0	0	0	5	4.83	20.2	8.90	0.43	-0.95	-0.07
Carbamazepine	236.3	775	459.9	43.2	91.9	324.7	0	0	1	0	2	2	2	5.35	27.9	8.54	0.59	**-2.58**	2.45
Chrysene	228.3	778	459.6	0	0	459.6	0	0	0	0	0	0	0	0.00	29.4	8.21	1.05	-8.06	5.81
DDD	320.0	887	522.1	16.7	0	246.5	258.9	0	0	0	1	0	0	0.57	30.4	9.48	0.38	-6.55	6.02
DDE	318.0	877	519.4	0	0	260.1	259.3	0	0	0	0	0	0	0.25	30.1	9.50	0.35	-6.90	6.51
DDT	354.5	910	526.6	5.3	0	230.3	290.9	0	0	0	1	0	0	1.38	31.2	9.50	0.51	-7.81	6.91
DEET	191.3	759	453.3	292.9	33.8	126.7	0	0	0	0	1	0	3	4.87	23.5	9.34	-0.07	**-2.82**	2.18
Diazepam	284.7	888	518.0	120.3	59.7	266.6	71.4	0	0	0	1	0	4	4.44	31.9	9.11	0.78	-3.76	2.82
Diclofenac	296.2	833	473.4	24.4	81.3	267.2	100.5	1	0	0	4	2	2.5	2.33	27.2	8.31	0.30	-5.10	4.51
Dieldrin	380.9	828	463.1	161.2	0	0	301.9	0	0	1	0	0	2	6.60	27.0	9.56	0.39	-6.29	5.40
Dilantin	252.3	782	450.9	0	111.1	339.8	0	0	0	0	2	2	3.5	3.62	28.4	9.87	0.21	-3.90	2.47
Endrin	380.9	812	447.7	147.0	0	0.0	300.7	0	0	1	0	0	2	1.21	26.4	9.68	0.52	-6.18	5.20
Erythromycin	733.9	1999	903.1	802.5	100.6	0	0	0	0	4	9	5	20.15	7.33	66.1	9.45	-0.81	**-2.65**	3.06
Estradiol	272.4	910	510.1	304.1	94.9	111.1	0	0	0	0	4	2	2.45	2.35	30.0	8.96	-0.24	-4.88	4.01
Estriol	288.4	930	519.0	266.9	141.1	111.1	0	0	0	0	5	3	4.15	2.55	30.2	8.92	-0.28	**-3.64**	2.45
Estrone	270.4	903	508.3	300.2	97.1	111.0	0	0	0	0	4	1	2.75	2.32	30.4	9.01	-0.19	-3.95	3.13
Ethynylestradiol	296.4	977	544.7	306.2	89.8	148.7	0	0	0	0	4	2.5	1.5	1.85	32.4	8.91	-0.29	-4.42	3.67
Fluoranthene	202.3	698	421.1	0	0	421.1	0	0	0	0	0	0	0	0.89	25.8	8.42	1.45	-5.89	5.16
Fluorene	166.2	625	388.7	50.4	0	338.4	0	0	0	0	1	0	0	0.52	22.1	8.71	0.47	-4.99	4.18
Fluoxetine	309.3	1016	597.9	156.9	24.0	298.3	118.7	0	0	0	4	1	2.25	4.55	33.5	9.39	0.36	**-4.66**	4.05
Galaxolide	258.4	918	502.8	475.4	0	27.4	0	0	0	0	3	0	1.7	2.03	30.8	9.13	-0.26	-5.17	5.90
Gemfibrozil	250.3	945	557.7	366.2	86.0	105.5	0	1	0	0	3	1	2.75	2.09	28.8	8.91	-0.29	-4.12	4.77
Heptachlor	373.3	784	449.2	51.3	0	62.4	335.4	0	0	1	2	0	0	4.85	25.8	9.67	0.43	-6.32	6.10
Heptachlor_epoxide	389.3	788	444.0	91.4	0	0.2	352.4	0	0	2	0	0	2	4.30	25.4	9.71	0.49	-6.29	4.98
Hydrocodone	299.4	906	501.3	383.5	49.8	68.1	0	0	0	1	5	0	5.5	5.80	30.1	8.81	-0.09	**-1.41**	**1.78**
Ibuprofen	206.3	798	477.7	278.5	92.5	106.8	0	1	0	0	2	1	2	2.60	24.2	9.51	-0.09	-3.99	3.97
Iopromide	791.1	1458	789.4	363.4	261.0	15.0	150.0	0	0	0	6	6	16.5	8.67	42.7	8.97	1.90	**-3.65**	-2.05
Meprobamate	218.3	781	471.4	234.4	237.0	0	0	0	0	0	0	4	5	7.82	21.2	10.02	-0.85	-1.67	0.70
Methoxychlor	345.7	974	549.6	189.8	0	211.8	148.0	0	0	0	3	0	1.5	3.94	32.3	9.11	0.25	-6.54	5.08
Metolachlor	283.8	948	529.9	337.8	26.4	103.5	62.2	0	0	0	5	0	4.7	6.05	29.6	9.45	-0.07	-2.73	3.13
Mirex	545.5	938	506.4	0	0	0.2	506.2	0	0	3	0	0	0	5.72	26.9	9.21	2.91	-6.75	6.89
Musk_ketone	294.3	907	508.1	339.6	156.2	12.3	0	0	0	0	2	0	4	3.87	27.7	10.94	1.69	-5.81	4.31
Naphthalene	128.2	517	334.8	0	0	334.8	0	0	0	0	0	0	0	0.00	17.7	8.60	0.63	-3.62	3.30
Naproxen	230.3	785	470.3	179.9	92.9	197.5	0	1	0	0	2	1	2.75	3.86	25.2	8.51	0.74	-4.16	3.18
Octylphenol-4t	206.3	806	466.5	282.3	54.5	129.6	0	0	0	0	1	1	0.75	1.90	24.8	9.02	-0.30	**-3.83**	**3.92**
Oxybenzone	228.2	772	465.5	92.2	79.4	293.9	0	0	0	0	2	0	2.5	5.21	25.0	9.25	0.35	**-3.15**	3.79
Pentoxfylline	278.3	949	559.8	366.2	136.4	57.2	0	0	0	1	1	0	7	6.99	29.2	8.93	0.46	-0.56	0.29
Phenanthrene	178.2	649	398.4	0	0	398.4	0	0	0	0	0	0	0	0.03	23.6	8.50	0.76	-5.19	4.46
Progesterone	314.5	1034	547.3	430.0	88.0	29.3	0	0	0	1	3	0	4	5.71	34.9	10.18	0.13	-4.55	3.87
Pyrene	202.3	690	414.5	0	0	414.5	0	0	0	0	1	0	0	0.00	25.4	7.95	1.29	-6.18	4.88
Sulfamethoxazole	253.3	788	486.8	96.4	183.8	206.0	0.6	0	0	0	2	2.5	7.5	12.01	24.8	8.87	0.49	-2.62	0.89
TCEP	285.5	819	503.5	241.0	38.2	0	224.3	0	0	3	0	0	5	2.33	22.8	10.62	0.08	-1.61	1.44
Testosterone	288.4	956	518.7	393.9	96.9	27.9	0	0	0	0	3	1	3.7	5.28	31.7	10.17	0.12	-4.09	3.32
Triclosan	289.5	793	478.8	0	48.2	223.9	206.6	0	0	0	1	1	1.25	2.63	25.8	9.12	0.43	-4.46	4.76
Trimethoprim	290.3	922	530.3	295.5	154.3	80.5	0	0	0	0	5	4	5.25	4.31	27.0	8.51	-0.08	-2.86	0.91
α_BHC	290.8	663	399.3	76.9	0	0	322.4	0	0	6	0	0	0	2.77	20.4	10.78	0.20	-5.16	3.80
α_Chlordane	409.8	843	478.4	78.4	0	0.2	399.9	0	0	2	0	0	0	2.24	27.6	9.70	0.49	-6.86	6.16
β_BHC	290.8	670	404.0	78.0	0	0	326.0	0	0	6	0	0	0	0.00	20.6	10.86	0.13	-6.08	3.78
δ_BHC	290.8	665	401.0	78.9	0	0	322.1	0	0	6	0	0	0	2.79	20.5	10.81	0.10	-4.46	4.14
γ_BHC	290.8	659	395.8	82.5	0	0	313.3	0	0	6	0	0	0	3.73	20.2	10.63	0.31	-4.60	3.72
γ_Chlordane	409.8	847	481.4	73.8	0	0.2	407.4	0	0	2	0	0	0	3.24	27.7	9.74	0.53	**-6.69**	**6.30**
Benzophenone	182.2	676.5	418.6	0	40.6	378.0	0	0	0	0	0	0	2	3.91	23.2	9.94	0.54	-3.12	3.18
Bisphenol A	228.3	805.6	476.3	115.2	109.0	252.1	0	0	0	0	2	2	1.5	2.93	25.9	8.98	-0.27	-3.28	3.32
Hydrocortisone	362.5	1086.4	578.0	359.5	190.7	27.8	0	0	0	0	5	3	8.15	6.91	34.4	10.14	0.08	-3.05	1.61
Indolebutyric acid	203.2	728.1	448.1	96.3	136.0	215.9	0	1	0	0	2	2	2	4.94	22.5	8.26	-0.02	-2.91	2.30
Isobutylparaben	194.2	731.6	450.7	198.7	101.9	150.2	0	0	0	1	1	1	2.75	1.44	22.0	9.51	0.34	**-2.93**	**2.34**
Propylparaben	180.2	684.3	432.4	173.3	103.7	155.5	0	0	0	1	1	1	2.75	1.36	20.1	9.52	0.34	-2.56	3.04
Triclocarban	315.6	888.2	538.8	0	61.9	274.9	201.9	0	1	0	0	2	2	5.05	30.7	8.79	0.41	**-4.62**	**3.73**

APPENDIX G
BIOFILTRATION SITE MAPS

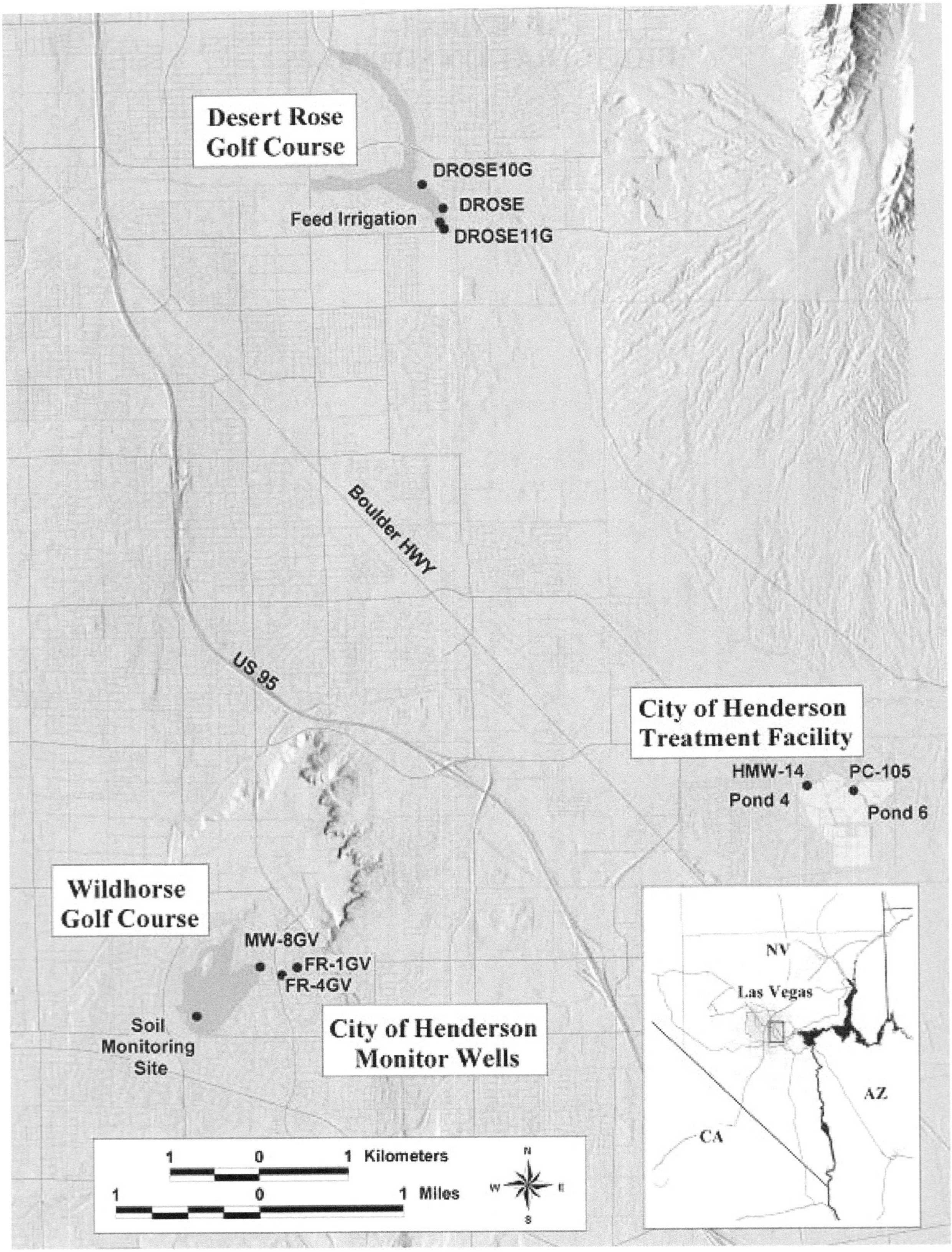

Figure G.1 Las Vegas valley field sites overview map (Source: Snyder et al. 2004. Copyright 2004 National Ground Water Association. Reprinted with permission)

Figure G.2 Desert Rose golf course aerial diagram

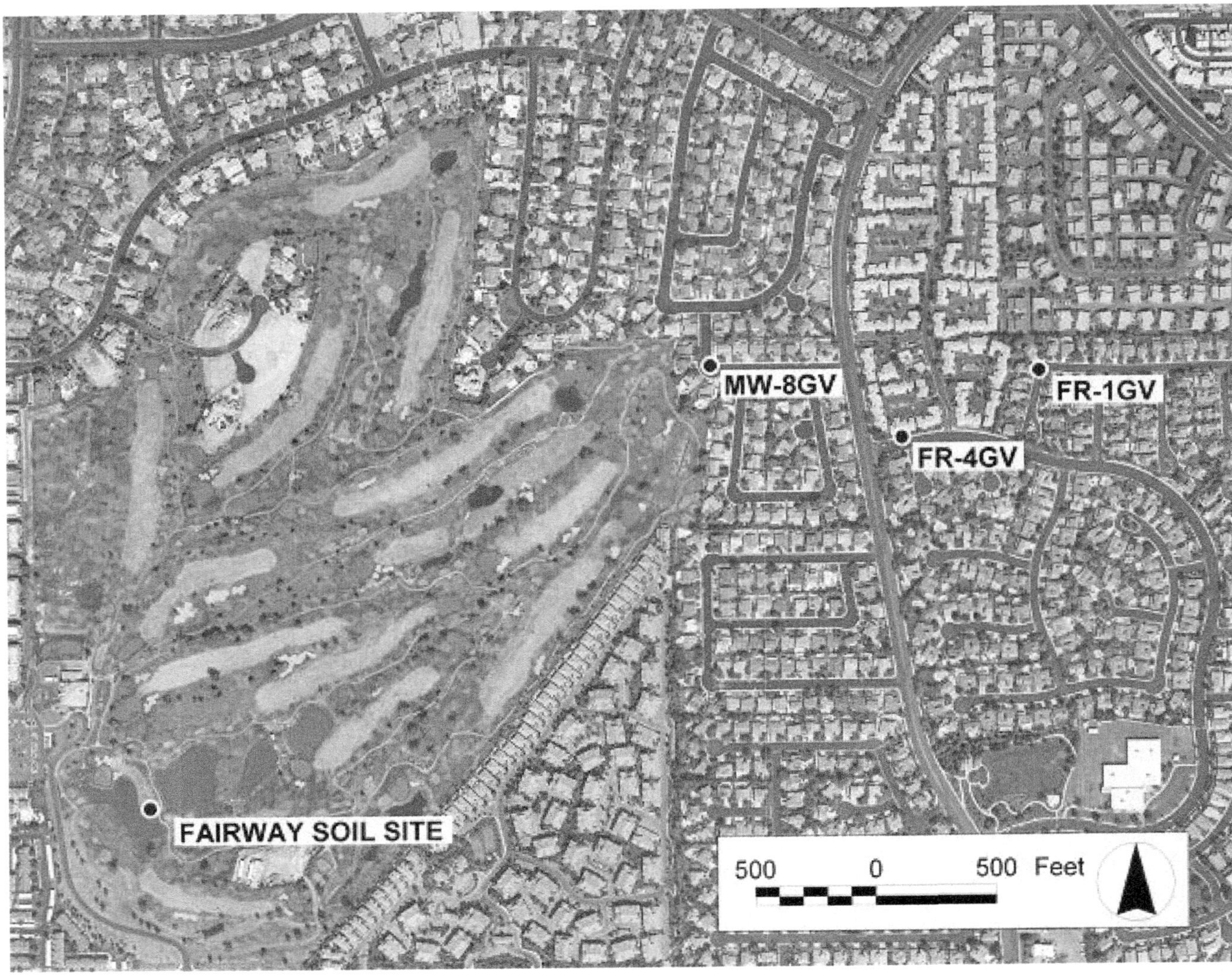

Figure G.3 Wild Horse gold course aerial diagram

Figure G.4 Henderson site aerial diagram

ABBREVIATIONS

AOP	Advanced oxidation process
APCI	Atmospheric pressure chemical ionization
APPI	Atmospheric pressure photoionization
ASU	Arizona State University
B&V	Black & Veatch
BAC	Biological active carbon
BAF	Biological active filtration
BAS	Biological active sand
BDOC	Biodegradable dissolved organic carbon
BOD	Biochemical oxygen demand
BTPP	Bench-top pilot plant
BVT	Bed volume treated
CI	Chemical ionization
COD	Chemical oxygen demand
CRW	Colorado River water
DBP	Disinfection by-product
DCM	Dichloromethane
DDT	dichlorodiphenyltrichloroethane
DI water	Deionized water
DOC	Dissolved organic carbon
DOM	Dissolved organic matter
DPD	N,N – diethyl-*p*-phenylenediamine
E2	Estradiol
EA	Electron affinity
EBCT	Empty bed contact time
ECD	Electron capture detector
EDC	Endocrine disrupting chemical
EDR	Electrodialysis reversal
EDSTAC	Endocrine Disruptor Screening and Testing Advisory Committee
EE2	17a-Ethynyl estradiol
EED	Electrical energy dose
E_{EO}	Electrical energy per order
EI	Electron ionization
ESI	Electrospray ionization
ET	Evapotranspiration
FID	Flame-ionization detector
FISA	Hydrophilic surface area
FOSA	Hydrophobic surface area

FPD	Flame-photometric detector
GAC	Granular activated carbon
GC	Gas chromatography
Hc	Henry's Law constant
HDPE	High density polyethylene
HLB cartridge	Hydrophilic-lipophilic balance cartridge
HOMO	Highest occupied molecular orbital
HRT	Hydraulic retention time
IP	Ionization potential
LC	Liquid chromatography
LFER	Linear free energy relationships
LFER	Linear free energy relationships
LLE	Liquid/liquid extraction
LUMO	Lowest unoccupied molecular orbital
LPHO lamp	Low-pressure high-output lamp
LT2ESWTR	Long Term 2 Enhanced Surface Water Treatment Rule
LWC	Louisville Water Company
MBR	Membrane bioreactor
MC	Monte Carlo
MGD	Million gallons per day
MIEX	Magnetic ion exchange resin
mM	Millimolar
MRL	Method reporting limit
MRM	Multiple reaction monitoring
MS	Mass spectrometer
MTBE	Methyl *tert*-butyl ether
MW	Molecular weight
MWCO	Molecular weight cut-off
MWD	Metropolitan Water District (at Southern California)
MWH	Montgomery Watson Harza
NF	Nanofiltration
nM	Nanomolar
NOM	Natural organic matter
NPD	Nitrogen-phosphorus detector
OE	Ozone exposure
ORW	Ohio River water
PAC	Powder activated carbon
PISA	pi surface area

PPCP	Pharmaceutical and personal care product
PSU	Pennsylvania State University
PRW	Passaic River water
QPlogPC16	Hexadecane/gas partition coefficient
QPlogPoct	Octanol/gas partition coefficient
QPlogPo/w	Octanol/water partition coefficient
QPlogPw	Water/gas partition coefficient
QPpolrz	Polarizability
QSAR	Quantitative structural-activity relationship
QSPR	Quantitative structural-property relationship
RBF	Riverbank filtration
RO	Reverse osmosis
RSD	Relative standard deviation
RSSCT	Rapid small-scale column test
SASA	Molecular surface area
SAT	Soil-aquifer treatment
SNWA	Southern Nevada Water Authority
SPE	Solid-phase extraction
SPMD	Semipermeable membrane device
SPME	Solid-phase microextraction
SRC	Syracuse Research Corporation
SRW	Suwannee River water
SUVA	Specific UV absorbance
TCE	Trichloroethylene
TFC membrane	Thin film composite membrane
TOC	Total organic carbon
TOF	Time-of-flight
UF	Ultrafiltration
US EPA	United States Environmental Protect Agency
UV	Ultraviolet
WPSA	Weakly polar surface area
WTP	Water treatment plant
WWTP	Wastewater treatment plant

IWA Publishing's authorised EU representative for General Product Safety Regulations is Diane D'Arras, 15 rue Duret, 75116 Paris, France, e-mail: safety@iwap.co.uk.

Printed and bound by CPI Group (UK) Ltd, Croydon, CR0 4YY
17/09/2026
02218802-0007